American Standard to U.S. Metric Mass (Weight)

	Milligrams	Grams	Kilograms	Ounces	Pounds
1 ounce =	2835	28.35	.028		
1 pound =	"Lots"	454	.454		
1 milligram =	.1	.001	.000001		
1 gram =	1000	1	.001	.032	.002
1 kilogram =	1,000,000	1000	1	.000032	2.2

Liquid Measure Volume Equivalents

60 drops	= 1 teaspoon
1 teaspoon	= ⅓ tablespoon
1 tablespoon	= 3 teaspoons
2 tablespoons	= 1 fluid ounce
4 tablespoons	= ¼ cup or 2 ounces
5⅓ tablespoons	= ⅓ cup or 2⅔ ounces
8 tablespoons	= ½ cup or 4 ounces
16 tablespoons	= 1 cup or 8 ounces
¼ cup	= 4 tablespoons
⅜ cup	= ¼ cup plus 2 tablespoons
⅝ cup	= ½ cup plus 2 tablespoons
⅞ cup	= ¾ cup plus 2 tablespoons
1 cup	= ½ pint or 8 fluid ounces
2 cups	= 1 pint or 16 fluid ounces
1 pint, liquid	= 16 fluid ounces
1 quart, liquid	= 2 pints or 4 cups
1 gallon, liquid	= 4 quarts

Stocking Up

New Revised and Expanded Edition

Stocking Up

How to Preserve the Foods You Grow, Naturally

By the staff of Organic Gardening and Farming®

Edited by

Carol Hupping Stoner

 **Rodale Press** Emmaus PA

Printed on recycled paper

Photographs by the Rodale Press Photo Lab

Library of Congress Cataloging in Publication Data
Main entry under title:

Stocking up.

 Includes index.
 1. Food—Preservation. I. Stoner, Carol.
II. Organic gardening and farming.
TX601.S7 1977   641.4   77-3942
ISBN 0-87857-167-1 hardcover
ISBN 0-87857-221-X deluxe

             18   20   19   hardcover
             18   20   19   deluxe

CONTENTS

VEGETABLES AND FRUITS

and storing dried foods—rehydrating—what freeze-drying
means and why it can't be done at home—fruit leather—sun-
cooked preserves—dried vegetable flours—dried tomato paste—
harvesting and drying herbs

DAIRY PRODUCTS

NUTS, SEEDS, AND GRAINS

Introduction to the New Edition

The writing and researching we here at Rodale Press have been doing in home food storage these past 30-odd years didn't stop when the first edition of STOCKING UP was published. In some ways it just marked the beginning of our involvement.

Since then, people in Rodale's editorial offices, its experimental kitchen, its research center, and on its farms have been busier than ever developing more information to aid homesteaders and gardeners in putting away their harvests. We've been testing for safety the new brands of canning lids that suddenly appeared on the market after the lid shortage of two summers ago. In response to the boom in food drying we have been experimenting with commercial dehydrators and have designed a few simple models that can easily be put together at home. When low-methoxyl pectin became readily available again we developed more recipes for using this substitute for regular pectin, and we are continuously testing pickle and jelly recipes sent to us by readers.

This edition of STOCKING UP reflects all our new work in ways to store food at home. Revised and expanded, this book is more extensive and in many ways better than the first. We think—and hope you agree—that it will prove to be a very useful aid in the harvest kitchen.

Like the STOCKING UP that came out in 1973, this edition represents the efforts of many people. I'd like to thank my husband Mike, for his help in researching and writing several chapters in both the original and in this new edition. ORGANIC GARDENING AND FARMING food editor Nancy Albright, deserves credit for the many recipes she developed and her valuable suggestions for improving the book throughout. My gratitude also to Rodale home economist Anita Hirsch, Research and Development editorial coordinator Diana Branch, and editors Ray Wolf and Carole Turko for their contributions. And a special note of thanks to Julie Ruhe, who helped me get the manuscript to the printer in one piece—and on time.

CHS

Introduction

Since ORGANIC GARDENING AND FARMING's earliest days we have witnessed a growing need for a book that thoroughly covers the subject of natural food preservation. That need is at its peak right now, because not since the "Victory Gardens" of World War II have so many people been raising and preserving so much of their own food as today. And people are not just raising tomatoes and green beans and strawberries, but a whole variety of vegetables and fruits, nuts, grains, and livestock as well. The number of organic gardeners and homesteaders has grown in recent years because people are discovering that just about the only way they can control the quality of the food they eat is to grow it themselves.

Each year with seasonal regularity, ORGANIC GARDENING AND FARMING's Reader Service Department receives hundreds of letters asking questions about food storage. Spring brings letters requesting building plans for outdoor storage and ideas for converting basements into root cellars. In early summer readers want information on freezing early vegetable varieties and on making jams and jellies with berries and honey. The letter flow builds through the peach and tomato seasons, and by fall, it has developed into a virtual flood: How can I best keep my squash? Should I can or freeze my corn? Do you have a good recipe for apple butter? How do I make sauerkraut using little or no salt? In the winter, readers want to know how to smoke meats and how to make natural cheese from their goat's or cow's milk.

To find the information our readers wanted we first turned to modern books on home food preservation. But we found some problems with the information these books had to offer. Those written in the last 30 years or so contain advanced storage techniques that best preserve the taste, appearance, and nutritive value of the fresh foods, but they are far from complete guides to home food storage. Rather, the modern literature available falls into one of two categories: Either it is books on fruit and vegetable storage for the backyard gardener who wants to can or freeze his or her garden surplus, or it is booklets and manuals on large-scale crop storage for farmers who have several hundred acres in field crops to hold over the winter months until they are put on the market or fed to livestock. There is pitifully little information for the person who has more than a backyard garden, but less than a big farm operation.

What's more, we didn't feel comfortable advising ORGANIC GARDENING AND FARMING readers to consult any of the books we did find, because none of them expressed our concern for organically grown and naturally preserved and prepared foods. ORGANIC GARDENING AND FARMING readers are very particular about their food. Lots of care, thought, and hard work go into its raising. Organic growers don't take any shortcuts because they know that chemicals that do a quick and easy job of fertilizing soil and speeding up animals' rates of growth may increase the quantity of food produced, but only at a sacrifice in quality, to say nothing of the possible dangers of the chemical residues found in these foods. Organic gardeners and farmers take the time and effort to raise their crops and animals in as natural an environment as possible, because they know that the great taste and nutritional value of a tomato or an apple or a chicken that was raised by organic methods is head and shoulders above that of their chemically grown counterparts. These people know that they have high-quality food and they want to keep it that way. They certainly don't want to preserve it with overprocessed or highly refined ingredients or use a method of storage that would unnecessarily destroy any of their food's natural qualities.

In an attempt to find more complete information on natural ways to preserve homegrown food, we looked to books written years ago, when raising and preserving one's own food without the aid of chemical fertilizers, sprays, hormones, medications, and additives were a real part of life for millions of American families. We found that these books are fairly complete guides to home food storage. They provide information on just about everything that could be raised in rural America. Unfortunately, these books are long out of print, and although some copies might exist in small public libraries, secondhand bookstores, and in the homes of rare book collectors, they are hard to find.

In addition, a lot of the information in these old books is dated. Old books don't include modern methods of food preservation which have added so much to the ease and safety of home food storage. Hampered by a lack of modern equipment, like steam-pressure canners and home freezers, and a limited knowledge of nutrition, farmers in the past had to rely on methods of food storage that did not always do the best job of retaining the essential nutrients of the foods they were keeping.

We soon realized that if we wanted a book that was a complete guide to food storage—one that had solid information on preserving everything that could be raised on a homestead and one that would express our philosophy about organic foods—we would have to write it ourselves.

And this is just what we did. We have combined what we feel to be the best of the new and the traditional methods of food preservation into a book designed to be a complete guide for organic gardeners, homesteaders, and family farmers who know the satisfaction of raising and preserving their own food as naturally as possible, without the use of any chemicals or overprocessed ingredients.

We are grateful to many people for supplying us with information for this book. We would like to thank the Library of Congress for helping us locate both new and long out-of-print, rare books; the U.S. Department of Agriculture and local extension stations for providing us with booklets and pamphlets; and Erewhon Trading Company, Walnut Acres, and other food processors and distributors for the information they have given us on the storage of natural and organically grown foods.

Our most important source of information, however, was our ORGANIC GARDENING AND FARMING readers, because they are the people who are out there raising and harvesting and storing their own food. Our special thanks go to all the organic growers who contributed ideas for food preservation through the articles they have written for our magazine and through the interviews they granted us especially for this book.

Carol Hupping Stoner

Equipment for Your Harvest Kitchen

If you don't know it already you'll soon learn that there is special equipment you're going to need to process your food for storage. You probably already have much of the simple equipment you will need for storing food in one way or another. If not, canning jars, vegetable scales, measuring cups, steaming baskets, freezer containers, and other common pieces of kitchenware are usually available in supermarkets and hardware and department stores.

There are other things, though, that you may have trouble finding in stores, things that are designed to make preparing special foods (like vegetable juice, dried fruits, and homemade ice cream and butter) or large quantities of food (like quarts of stoned cherries or cut corn) easier. There isn't room here to list all these devices and where you can get them, but there is space to list some of the mail-order catalogs that stock the harder-to-find kitchen equipment.

Please keep in mind that the list below does not include all catalogs that sell such goods; it's only a good representation of the ones that are available. While some of these catalogs are free, most are sold for a nominal price. Prices here are up-to-date at the time of publication, but you may want to check before you send for yours to make sure there has been no price change since then.

Barth's of Long Island
Valley Stream, New York 11582

This vitamin and mineral supplements and natural foods and cosmetics catalog also lists a few kitchenwares, like a grain mill, yogurt-maker, juicer, and seed sprouter. **No charge.**

Cook's Catalog, *edited by James Beard, Milton Glaser, and Burton Wolf.*
New York: Harper and Row, 1975.

> Not a mail-order catalog, but a 565-page book devoted to all kinds of kitchen equipment. Those who love to cook will like browsing through this book, and if you're interested in knowing what knives are best and why microwave ovens are an impractical luxury, you'll enjoy the authors' candid comments. **Price: $15.95.**

Countryside General Store
Highway 19 East SU
Waterloo, Wisconsin 53594

> The 80-page catalog's emphasis is on books and tools for livestock, but there are plenty of other products for the kitchen and homestead as well. Food equipment includes a butter paddle, can sealer, fruit crusher, and bottle capper. **Price: $1.00.**

Cumberland General Store
Route 3, Box 470
Crossville, Tennessee 38555

> Stoneware crocks, butcher block tables, oak kegs, knives, and nut shellers are just some of the kitchen equipment sold through this 250-plus-page catalog. Plenty of other tools, too, for farm and home use. **Price: $3.00.**

Garden Way Country Kitchen Catalog
48 Maple St., P.O. Box 944
Burlington, Vermont 05401

> A small catalog that features nothing but hard-to-find food preserving equipment and kitchen utensils. It sells products like cheese rennet tablets, a corn sheller, fruit press, electric food dryer, and a peanut butter machine. **No charge.**

Glen-Bel's Country Store
Route 5
Crossville, Tennessee 38555

> A 300-page catalog that includes everything from flower pots to cement mixers and stage coaches. There's a fairly good assortment of kitchen equipment as well, like kraut cutters, sausage stuffers, and ice cream-makers. **Price: $3.00.**

The "Good Neighbor" Heritage Catalog
Lehman Hardware and Appliances, Inc.
Box 41R
Kidron, Ohio 44636

> This 40-page catalog specializes in wood, gas, and coal stoves, heating equipment, and lamps, but sells some tools for the kitchen, like dough mixers, noodle makers, apple parers, and butter churns. **Price: $1.00.**

Hoosier Health House
Route 1, Box 369
Alexandria, Indiana 46001

> Stone grain grinders, yogurt-makers, juicers, bread mixers, steam-juicers, and several other kitchen items in this small catalog. **No charge.**

Mother's General Store Catalog
Box 506
Flat Rock, North Carolina 28731

> In addition to camping equipment, books, and other homestead-oriented items, Mother stocks a number of food processing devices like grain mills and food grinders, scales, and cast-iron cookware. **Price: 35¢.**

Sears, Roebuck and Co.
925 S. Homan Ave.
Chicago, Illinois 60607

> The Sears catalog sells canning and freezing equipment as well as grain mills, cider presses, and steam-juicers in both its regular catalog and the smaller Suburban, Farm, and Ranch Catalog. **Both are free.**

The Vermont Country Store
Weston, Vermont 05161

> There are a few practical pieces of kitchen equipment in here like a hand-cranked ice cream-maker and a bread mixer, but most are gourmet and gift items. **Price: 25¢.**

Pictured here is some of the equipment you'll need for preserving your harvest. Most of the items here are easy to find in stores.

Vegetables
and Fruits

Choosing Vegetable and Fruit Varieties

As the grower of your own food, you have many advantages over the supermarket shopper. Not only can you choose the fruits and vegetables that you want, but you can also choose the particular *varieties* of fruits and vegetables that best suit you.

If you page through any seed catalog you'll discover that each food is usually available in a number of varieties. Some of these varieties may be particularly good for freezing; others maintain their quality best when canned. Other varieties have been designed for drying, and some hold their flavor and texture well in underground storage. If you're planning to preserve a good part of your harvest, you'd do well to decide how you will store your garden surplus before you order your seeds, and then choose those fruit and vegetable varieties developed specifically for your method of storage.

Some of the more popular fruits and vegetables, like tomatoes, sweet potatoes, lima beans, beets, cabbage, carrots, potatoes, cauliflower, apples, blueberries, and strawberries are available in varieties particularly high in vitamin content. Although no kind of preservation will increase the vitamin content of food, fruits and vegetables that go into storage with an especially high vitamin A or C content will come out of storage with a higher vitamin content than conventional varieties, providing they are stored properly.

The charts that follow list those varieties of vegetables and fruits that are generally recognized as being best for freezing, canning, pickling, drying, and keeping in some kind of cold storage, be it in a root cellar, basement, or outdoor area. High vitamin varieties are also listed.

After each variety you'll find letters that represent seed companies in this country that sell that variety. By no means have we listed every seed company that sells these varieties. We have noted some of the larger companies located in different parts of the United States. We realize that many smaller seedsmen sell the same varieties and may also offer other varieties equally as good for these particular methods of storage. We also assume that many growers have had success preserving varieties different from those listed here. Although our listing is up-to-date at the time of publication, new varieties could possibly be added to the list now because improved varieties and hybrids are always being developed. We suggest that you contact your county agent and check the most current issues of seed catalogs for new varieties that are available to you.

The letters on the charts are abbreviations for the following seed companies:

Bg   Burgess Seed & Plant Company
        P. O. Box 218, Galesburg, MI 49053

Bp   W. Atlee Burpee
        Phila., PA 19132
        Clinton, IA 52732
        Riverside, CA 92502

E   Earl May Seed & Nursery Company
        Shenandoah, IA 51601

F   Farmer Seed & Nursery
        Faribault, MN 55021

G   Gurney Seed & Nursery Company
        Yankton, SD 57078

GB   Gill Brothers Seed Company
        Box 16128, Portland, OR 97216

H   Joseph Harris Company, Inc.
        Moreton Farm, Rochester, NY 14642

HF   Henry Field Seed & Nursery Company
        Shenandoah, IA 51601

J   J.W. Jung Company
        Randolph, WI 53956

M   J.E. Miller
        Canandaigua, NY 14424

N   The Natural Development Company
        Bainbridge, PA 17502

P   George Park Seed Company, Inc.
     Greenwood, SC 29646

R   Rayner Brothers
     Salisbury, MD 21801

S   R.H. Shumway
     P.O. Box 777, Rockford, IL 61101

SB   Stark Brothers Nurseries & Orchard Company
     Rt. 2, Louisiana, MO 63353

SQ   Schells Quality Seeds
     Harrisburg, PA 17101

VEGETABLE VARIETIES

Asparagus

Mary Washington	good freezer and canner	**Bp, Bg, F, G, H, P, S, SB**
California 500	good freezer	**HF**
Roberts	good freezer	**E, G**

Beans, bush, green

Top Crop	good freezer and canner	**N, Bg, Bp, E, F, G, J, HF, P, S**
Royalty Purple	good freezer	**Bg, F, P**
Green Isle	good freezer and canner	**F**
Tenderette	good freezer and canner	**G, HF, JH, S, N, P**
Leka Lake	good freezer and canner	**Bp, N**
Improved Tendergreen	good freezer and canner	**E, F, G, HF, S, P**
Burpee's Stringless	good freezer	**Bp, G, HF, P, S**
Blue Lake	good freezer and canner	**E, G, HF, J, H**
Garden Green	good freezer and canner	**G**
(Early) Contender	good freezer and canner	**E, G**

Spartan Arrow	good freezer	**F, HF, P**
Greencrop	good freezer	**F, HF**
Wade	good freezer	**F**
Bluecrop	good freezer and canner	**H**
Tendercrop	good freezer	**H, N, Bg**
Burpee's Tender Pod	good freezer and canner	**Bp**
Canyon	good freezer and canner	**Bp**
Greensleeves	good freezer	**Bp**
Roma	good freezer and canner	**Bp, P**
White Half Runner	good freezer and canner	**Bp, N, P**
Tender Green	good freezer and canner	**G, HF**
Avalanche	good freezer and canner	**P**

Beans, bush, yellow

Golden Wax	good freezer	**G, HF, P**
Pencil Pod Wax	good freezer	**E, G, HF, J, N, Bp, F**
Brittle Wax	good freezer and canner	**Bp, FS, J, N, G**
Kinghorn Wax	good freezer and canner	**Bp, G, H, P**
Cherokee Wax	good freezer	**E, G, H, B, F**
Butter Wax	good canner	**E, G, H**
Goldencrop	good freezer and canner	**Bp, J**

Beans, pole

Kentucky Wonder (Old Homestead)	good freezer	**Bp, E, F, G, H, HF, J, P, S, B, N**
Romano Italian	good freezer	**H, Bg, Bp, N**
Blue Lake, White Seeded	good freezer and canner	**E, HF, J, Bp, N, P**
Purple Pod	good canner	**HF, G**
New Pole	good freezer and canner	**F**

Beans, for drying

White Mexican	S
Red Kidney	E, F, G, HF, S, P, Bp
Improved White Navy	E, G, S, N
Great Northern (white)	E, F, HF, S, G
White Marrowfat	S, Bp
Dwarf Horticultural or Wren's Egg	E, S, N, Bp, B
McClasan Pole	S
Black Eye	E, G, HF, S
White Wonder	HF
Michlite	F
Sanilac	J
Pinto	HF, Bg
Redkloud	H
Garbanzo (chick-peas)	Bg
Lentil	Bg

Beans, lima

Fordhook 242 (Fordhook Potato)	good freezer and canner	Bg, Bp, E, G, H, F, HF, J, S, N, P
Henderson's	good freezer and canner	Bp, E, HF, S, Bg, N, P
Improved Giant Bush	good freezer	Bp, E, S, N
Baby Bush	good freezer and canner	Bp, HF, J, Bg, G, N
Romano Italian	good freezer and canner	Bp, S
Burpee's Best	good freezer, high vitamin C content	Bp
Clark's Green Seeded	good freezer and canner	G, HF, P
Thaxter	good freezer	H
Thorogreen (baby lima)	good freezer and canner	F
Fordhook Improved	good freezer and canner	F
Kingston	good freezer	Bp

Beets

Detroit Dark Red	good keeper	**Bg, Bp, E, F, G, H, HF, J, S, P, N**
Red Ball	good freezer and canner	**Bp**
Early Wonder or Model	good canner	**E, F, G, HF, J, S, Bg, P, N**
Dark Red Canner	good pickler	**HF**
Red Ruby Queen	good canner and pickler	**E, G, H, HF, P, N, F**
(Hybrid) Pacemaker	good canner	**G, H**
Sweetheart	good freezer and canner, high vitamin content	**F**
Baby Canning	good pickler	**E**
Tendersweet	good pickler	**B**
Golden Beet	good pickler	**HF, P**
Formanova	good canner and pickler	**P**
Burpee's Golden Beet	good pickler	**N, Bp**
Lutz Greenleaf Winter Keeper	good keeper	**N, Bp**
Firechief Beet	good canner	**G**
"Mono" King Explorer	good pickler	**G**
Garnet	good canner	**J**
Hybrid Redcross Beets	good keeper	**F**
Long Season	good keeper	**H**

Broccoli

Green Sprouting or Calabrese	good freezer	**Bg, F, G, HF, Bp**
Zenith	good freezer	**FM**
Greenbud	good freezer	**Bp**
New Spartan	good freezer	**E**
Waltham	good freezer	**H**
Premium Crop Hybrid	good freezer	**Bp**
De Cicco	good freezer	**Bp, N, P**

Brussels Sprouts

Jade Cross Hybrid	good freezer	**Bp, G, H, P, Bg, N**

Cabbage

Lightning Express	high vitamin content	**Bg**
Hybrid Stonehead	good kraut	**Bg, F**
Jumbo or Large Late Drumhead	good keeper (in underground storage)	**Bg**
Danish Ballhead or Roundhead	good keeper	**Bg, Bp, E, G, HF, H, S, H, J**
Copenhagen Market	good keeper	**Bg, Bp, E, HF, S, N**
Early Flat Dutch	good keeper	**Bp, F, HF, S**
Surehead	good keeper	**Bp, HF**
Premium Late Flat Dutch	good keeper	**F, HF, S, G, N**
Autumn King	good keeper	**S**
Wisconsin Red Hollander	good keeper	**F, S**
Wisconsin All Season	good keeper, kraut	**F, S, G**
Mammoth Red Rock	good keeper, kraut	**E, F, G, N**
Golden Acre	good kraut	**E, F, G, H**
Marion Market	good kraut	**E, F**
Penn State Ballhead	good keeper, kraut	**F**
Michihli	good keeper	**Bg, H, Bp, N, P**
Savoy Perfection	good keeper	**F**
Sanibel	good keeper, kraut	**F**
Gurney's Giant	good kraut	**G**
Emerald Cross	good keeper	**P**
Red Rock	good keeper	**HF**
Stonehead Hybrid	good kraut	**Bg**

Carrots

Red Cored Chantenay	good freezer	**Bp, F, G, HF, J**

Danvers Half Long	good keeper (in underground storage)	Bp, E, G, F, HF, H, S, Bg, P
Goldinheart	good freezer	N, Bp
Oxeheart	good keeper	G
Nantes Half Long	good freezer and canner	Bp, H, Bg, HF, P, S, N
Gold Pak	good keeper, freezer, and canner	F, G, HF, J, S, Bp
Tendersweet	good freezer	E, G, HF
Royal Chantenay	good freezer	HF, H, P, G, Bp
Imperator	high vitamin C and carotene content	Bp, F, S, N, G
Morse Bunching	high vitamin C and carotene content	Bp, F, S
Coreless	good freezer	E, G, J
Hipak	good freezer	H

Cauliflower

Early Snowball	good freezer and high vitamin A content	Bp, E, F, J, S
Super Snowball	good freezer	Bp, F, J, N, S, P, HF
Purple Head	good freezer	F, H, S, Bp
Snowball Imperial	good freezer	H
Snow Crown	good freezer	H, F, Bp
Burpeeana	good freezer	Bp, N

Celery

Fordhook	good keeper (in underground storage)	Bp
Giant Pascal	good keeper	Bg, Bp, S
Golden Self-Blanching	good keeper	Bg, F, HF, S, E

Corn

Country Gentleman Hybrid	good canner	Bp, E, G, HF, S

Stowell's Evergreen	good freezer and canner	**Bp, E, F, G, S**
Golden Bantam	good freezer	**Bp, E, G, HF, S**
Golden Cross Bantam	good freezer and canner	**Bp, F, G, HF, S, F, N**
Illini or Xtra Sweet	good freezer	**Bp, F, G, HF, N**
Iochief	good canner	**Bp, E, F, G, HF, N, P, H, S**
Silver Queen	good freezer	**H, P**
Six Shooter Sugar Corn	good freezer	**S**
White Evergreen	good canner	**S, Bp**
Golden Delicious	good freezer	**HF**
Most Tender	good freezer and canner	**HF**
Golden Midget	good freezer	**Bg**
Bantam Evergreen	good freezer	**Bg**
Ihinichief	good freezer	**Bg, F**
Faribo Golden Sugar	good freezer and canner	**F**
Early Golden Maincrop	good freezer and canner	**HF**
Hybrid Truckers	good freezer	**G**
Marcross Hybrid	good freezer and canner	**G**
Kanner King	good freezer and canner	**F**
Golden Beauty	good freezer and canner	**F**
Carmel Cross Hybrid	good freezer	**G**
Jubilee	good freezer and canner	**F**
Butter Nugget	good freezer	**F**
Early Sunglow Hybrid	good freezer	**G, N, P**
Wonderful	good freezer	**H**
New Cheddar Cross Hybrid	good freezer and canner	**J**
Harris Gold Cup	good freezer	**H**
Early Xtra Sweet	good freezer and canner	**J, P**
Corn N-K-199	good freezer and canner	**J**
Burbank Hybrid	good freezer	**Bp**
Barbecue	good freezer	**Bp**

Ioana	good freezer and canner	**Bp**
Stylepak	good freezer	**Bp**
Y-81	good freezer and canner	**HF**
Candystick	good freezer	**P**

Kale

Dwarf Blue Curled (Scotch)	high vitamin content	**Bp, H, J, P, N**

Onions

Ringmaster	good keeper (in underground storage)	**Bg, G**
Hybrid Yellow Sweet Spanish	good keeper	**Bg, Bp, E, F, G, P, HF, H, S**
Crystal White Wax	good pickler	**Bg, Bp, S**
Ebenezer	good keeper	**Bp, Bp, G, H, S, P**
White Sweet Spanish	good keeper	**G, S, N, HF, Bg**
Red Wethersfield Hamburger	good keeper	**S**
Southport Red, White, and Yellow Globes	good keeper	**E, F, G, H, J, S, N, P**
White Sweet Slicer	good keeper	**G**
Downing Yellow Globe	good keeper	**G, H, F**
White Sweet Keeper	good keeper	**E**
Buccaneer	good keeper	**H**
White Portugal or Silver Skin	good pickler	**J, Bp**
Burpee Yellow Globe Hybrid	good keeper	**Bp**
Snow White Hybrid	good keeper	**P**

Parsnips

All America(n)	good keeper (in underground storage)	**G, H**

Peas

Laxton's Progress	good freezer and canner	**Bg, F, G, HF, S, P**
Alaska	good canner and high vitamin C content	**Bg, G, S, P, Bp**
Green Arrow	good freezer	**Bp, HF, P, H**
Little Marvel	good freezer and canner	**Bg, Bp, E, F, G, N, H, HF, J, S, P**
Giant Stride	good freezer and canner	**Bg**
Freezonian	good freezer and canner	**Bg, Bp, F, HF, P, H, S, N**
Wando	good freezer and canner	**Bg, Bp, F, HF, H, J, N**
Blue Bantam	good freezer	**Bp, J, N**
Tall Telephone	good freezer	**F**
Lincoln	good freezer	**F, H**
Burpeanna Early	good freezer	**Bp, N**
Sweet Green	good freezer	**E, F**
Early All-Sweet	good canner	**G**
Frosty	good freezer	**E, G, H, P**
Thomas Laxton	good freezer	**Bp, G, S**
Victory Freezer	good freezer and canner	**G**
Perfected Freezer	good freezer	**H**
Miragreen	good freezer	**J**
Progress No. 9 (Early Giant)	good freezer	**E, Bp**
Beagle	good freezer and canner	**P**
Midseason Freezer	good freezer	**P**
Early-Frosty	good freezer	**Bp**
Fordhook Wonder	good freezer and canner	**Bp**
Alderman	good freezer	**H**

Peppers, sweet

Tasty	good freezer	**F**
Sweet Chocolate	good freezer	**F**

Worldbeater	good pickler and canner	**Bp**
Sunnybrook or Sweet Salad Tomato Pepper	good canner	**S**
Mammoth Ruby King	good pickler	**S**
Cherry Sweet	good pickler	**HF, P, Bg**
Early Thick Meat	good freezer	**E**
New Ace Hybrid	good freezer	**Bp**

Peppers, hot

Hungarian Wax or Yellow	good canner	**Bg, Bp, G, HF, P, S, H, N**
Pimiento	good canner	**Bg, E, HF, S, P**
Red Chili	good pickler and drying	**Bg, E**
Long Hot Cayenne	good pickler and drying	**Bp, J, S, G, N, Bg**
Anaheim Chilio	good pickler and drying	**S, Bp, N**
Jalapeno	good pickler	**HF, G, P**
Small Red Chili Pepper	good pickler	**S**

Potatoes

Norgold Russet	good keeper (in underground storage)	**F, G, HF**
Kennebec	good keeper, high vitamin C content	**E, F, G, HF**
Anoka	good freezer	**F, G**
Red Pontiac	good keeper	**E, F, G, HF**
White Cobbler	good keeper, high vitamin C content	**E, G, HF**
Superior	good keeper	**S**

Pumpkins

Small Sugar	good keeper (in underground storage)	**Bg, Bp, HF**

Jack O'Lantern	good canner	**Bg, Bp, E, G, HF,** **J, S, N, P**
Winter Luxury or Queen	good keeper	**S, G**
Yellow Connecticut Field or Big Tom	good canner	**E, F, G, HF, H, J, S,** **Bp, N**
Early Sweet Sugar	good canner	**G**

Radish

Round White	good keeper (in under- ground storage)	**Bg**
White Chinese or Celestial Winter	good keeper	**Bg, Bp**
Comet	good keeper	**Bp, N**
Cherry Belle	good keeper	**Bp, E, F, HF, H, J,** **S, Bg, P, N, G**
White Giant or Hailstone	good keeper	**HF, S, J**
Stop Lite	good keeper	**F**
China Rose (Winter)	good keeper	**E, HF, J, S, G**
Round Black Spanish	good keeper	**F, H, S, Bp**
Long Black Spanish	good keeper	**S, G**
Funny Face	good keeper	**S, HF**
Howden's Field	good keeper	**H**
Sugar Pie	good keeper	**F**
Luxury Pie	good keeper	**J**
Small Sugar	good keeper	**P**

Rutabaga

American Purple Top	good keeper (in under- ground storage)	**E, G, H, F, P, HF**
Laurentian Neckless	good keeper	**F, J**
Red Chief	good keeper	**F**
Macomber	good keeper (in under- ground storage)	**H**
Burpee's Purple-Top Yellow	good keeper (in under- ground storage)	**Bp**

Myers Beauty	good keeper	N

Soybeans

Bansei	good freezer	**Bg, Bp, GB, N**
Kanrich	high vitamin content	**Bp, N**

Spinach

Bloomsdale (Long-Standing)	good freezer	**Bp, E, F, HF, J, S, H, N**
Hybrid No. 7	good freezer and canner	**Bp, F**
Virginia Blight Resistant	good freezer and canner	**Bp, N**
Giant Thick Leaf	good freezer and canner	**HF**
New Hybrid	good freezer and canner	**G**
America	good freezer	**E, Bp**
Northland	good freezer and canner	**G**
King of Denmark	good freezer	**E, S**
Early Hybrid	good freezer and canner	**Bg**

Squash

Table Queen, Acorn, or Des Moines	good keeper (in underground storage)	**Bg, Bp, F, G, HF, S, H, J, E**
Hubbard (True or Warted)	good keeper	**Bg, Bp, F, HF, H, J, E**
(Golden or Red)	good keeper, freezer, and canner	**Bg, Bp, F**
Sweet Meat Squash	good keeper	**Bg**
Banana Squash	good keeper	**Bg, G, HF, S, J**
Royal Acorn	good keeper	**Bp**
Buttercup	good keeper and freezer	**Bg, Bp, F, G, S, H, J, E**

Butternut	good keeper	**Bg, Bp, F, G, H, J, E, S**
Waltham Butternut	good keeper	**Bp, F, HF, E**
Prolific Straightneck	good freezer	**H**
Sweet Nut	high vitamin content	**F**
Gold Nugget	good keeper	**F, G, H, J, E**
Golden Delicious	good freezer and canner	**G, H**
Hybrid Gold	good keeper	**G**

Sweet Potatoes

Gold Rush	good keeper (in underground storage)	**Bg**
All Gold	high vitamin A content	**F, E, P**
Centennial	good keeper	**E, P**
Vineless Puerto Rico	good keeper	**P, F**

Swiss Chard

| Lucullus | good freezer | **Bg, Bp, F, G, S, P** |

Tomatoes

Colossal	good canner	**Bg**
Pinkshipper	good canner	**Bg**
Rutgers	good canner and juice	**Bg, Bp, F, HF, S, P, H, E, N**
Marglobe	good canner	**Bg, Bp, HF, S, N**
Red (Yellow) Sugar	good canner, juice, and preserves	**Bg**
Beefsteak	good canner	**G, S, P, Bg, N**
Abraham Lincoln	juice	**S**
Ponderosa	good canner	**F, G, P, E**
Roma (Red)	good canner, juice, sauce	**Bp, S, H, E**
Italian Canner	good canner, juice, sauce	**Bp, S**

Queen's	juice	S
Red Pear	good canner, preserves	HF, S
Yellow Pear	good canner, preserves	E
Garden State	good canner	S
Jubilee	juice	S, N
Tiny Tim Midget	preserves	F, J
Droplet	preserves	F
Bellarina	sauces, purees	G
Crimson Giant	good canner	G
Yellow Husk	preserves	G
Crack-Proof	good canner	Bg
Climbing Tree Tomato	good canner	HF
White Beauty	good canner, juice	Bg
Golden Sunray	juice	HF, G, H
Little Pear	preserves	G
Caro Red	high vitamin C content	Bg, J
Doublerich	high vitamin C content	Bg
Pink Gourmet	good canner, catsup	E

Turnips

Purple Top White Globe	good keeper (in underground storage)	Bg, Bp, F, G, HF, H, J, E

FRUIT VARIETIES

Apples

Anoka	good keeper (in underground storage)	G, HF, E, Bp
Northern Spy	good keeper, high vitamin C content	Bg, M, SB

Fireside or Minnesota Delicious	good keeper	**Bg, F**
Yellow (Red) Delicious	good keeper, sauce	**Bg, HF, SB, M, G, E, Bp, S**
McIntosh	good keeper, cider	**F, G, HF, SB, J, Bg, M, B**
Jonadel	good keeper, juice	**F, G, HF, S, E, Bp**
Transparent	sauce	**E**
Red Winesap (Crimson)	good keeper, juice, high vitamin C content, sauce	**HF, R, SB, E, S**
Chieftan	good keeper	**HF, E**
Stayman's Winesap	good keeper and freezer, sauce	**HF, M, Bp**
Jonathan	good keeper and freezer, sauce	**G, HF, Bg, M, Bp**
Grimes Gold	good keeper, sauce	**HF**
Haralson	good keeper	**F, G**
Baldwin	good keeper, high vitamin C content	**M, Bp**
Arkansas Black Twig	good keeper	**SB**
Prairie Spy	good keeper	**F, G**
Stark Red Winesap	good canner	**SB**
Macoun	good keeper	**M**
Redwell	good keeper	**F**
R.I. Greening	good keeper	**M**
Red Rome	good keeper	**M**
Red Baron	good keeper	**G, F**
Red Duchess	sauce	**G, F, S**
Spartan	good keeper	**J**
Connell Red	good keeper	**F, J**
Rome Beauty	good keeper	**SB**
Red Dolgo Crab	pickler, jelly, sauce	**Bg, F, G, S, E**
Whitney Crab	pickler, jelly, sauce	**F, G, HF, J, E**
Blushing Golden	good keeper	**SB**

Splendor	good canner	**SB**
Jona Delicious	good keeper	**SB**
Jon-a-Rich	good keeper	**SB**

Apricots

Manchu	good canner, preserves	**Bg, G, S**
Hardy Iowa	good canner, preserves	**HF**
Hardy Superb	good canner, preserves	**HF**
Moongold	good canner, preserves	**F, G, E**
Scout	good canner, preserves	**G**
Chinese Golden	good canner	**J, Bp**
Wilson Delicious	good canner and freezer	**SB**
Hungarian Rose	dried	**SB**
Stark Early Orange	good canner, freezer, dried	**SB**

Blackberries

Darrow	good canner, preserves	**SB, E, HF, M, Bp**
Desoto	good canner	**Bg**
Baily	preserves	**HF**
Thornfree	good canner	**G, Bp**
Ebony King	good freezer and canner	**G**

Blueberries

Mammoth Cultivated	preserves	**S**
Jersey	good freezer	**Bp, J**
Coville	good freezer, preserves	**HF, E, SB, Bp**
Saskatoon	preserves	**G**
Blue Ray	high vitamin C content	**R, J, E, HF, M, Bp, F**

| Rubel | high vitamin C content | **R, Bg** |

Boysenberries

| New Thornless | good canner, preserves | **Bg, G, E, S** |

Cherries, sour

Early Richmond	good canner, preserves	**Bg, HF, S, SB**
North Star	good canner	**J, F, HF, M, SB, Bg**
Large Montmorency	good canner, preserves	**Bg, G, HF, S, M**
Meteor	good canner, preserves	**Bg, F, HF, SB**

Cherries, sweet

Royal Ann or Napoleon	good freezer and canner	**Bg, S, SB**
Kansas Sweet	good freezer	**Bg, HF**
Yellow Glass	good freezer and canner	**G, Bp**
Bing	good freezer and canner	**SB, Bp**
Vista	good canner	**SB**
Stark Gold	good canner and freezer	**SB**
Van	good canner and freezer	**SB**
Stark Lambert	good canner and freezer	**SB**
Suda Hardy	good canner and freezer	**SB**
Emperor Francis	good canner and freezer	**SB, M**
Black Tartarian	good canner	**G, SB, Bp**

Cherries, bush

| Black Beauty | good canner | **G, HF** |
| Hansen | good canner, preserves | **G, HF, M** |

Giant Red-Fleshed	good canner	**HF**
Nanking Cross	good canner, preserves	**G**
Drilea	preserves	**G**
Brooks	preserves	**G**

Currants

Wilder Currant	jelly	**G, HF**
Red Lake	jelly	**F, HF, E, G, J, M**

Gooseberries

Welcome	jelly, preserves	**HF, Bp**

Grapes

Concord	juice, jelly	**Bg, F, G, HF, S, SB, J, M, Bp**
Seibel	juice, jelly	**HF**
Buffalo	juice, jelly	**HF, SB**
Red Caco	jelly	**G, HF, J, F**
Beta Grape	juice, jelly	**F**
Steuben	juice, jelly	**SB, M, Bp, G**
Van Buren	juice, jelly	**M**
Burgess Red	juice, jelly	**Bg**
Suffolk Red Seedless	jelly	**SB**
Delaware	jelly	**SB**
Vinered	jelly, juice	**SB**
Stark Blue Boy	jelly, juice	**SB**

Melons

Crenshaw	good freezer	**Bg, HF, Bp**
Honey Dew	good keeper (in underground storage)	**Bg, Bp, S**
Hale's Best Muskmelon	good freezer	**Bp, F, HF, J, E**
Garrisonian Watermelon	good freezer	**HF**
Golden Beauty Casaba	good keeper	**Bp**
Citron	good pickler	**J**

Winter Melon	good keeper	**F, G**
Crimson Sweet	good keeper	**E, HF**

Peaches

Rochester	good keeper	**Bg**
Late Glo	good canner, freezer	**SB**
Eden	good canner	**Bg, S**
Golden Jubilee	good canner, pickler	**Bg, HF, S, M, Bp**
Wisconsin or Balmer	good canner, pickler	**HF**
Hale (Haven)	good freezer and canner	**Bg, HF, SB, E, M, Bp, G**
Ranger	good freezer and canner	**SB**
Fuzzless Gold	good keeper	**HF**
Polly	good canner	**HF, E**
Elberta (Queen)	good freezer and canner	**Bg, G, SB, E, M, HF, Bp**
Stark Hal-Berta Giant	good canner	**SB**
Babygold	good canner	**SB**
Cresthaven	good canner and freezer	**SB**
Stark EarlyGlo	good canner and freezer, preserves	**SB**

Pears

Seckel	preserves	**Bg, HF, SB, E, Bp**
Duchess	preserves	**S, SB, HF, M**
Colette	preserves	**HF**
Bartlett	good keeper (in underground storage), canner	**Bg, G, HF, SB, J, E, Bp, M**
Magness	good keeper and canner	**HF, SB**
Golden Spice	good pickler	**G**
Moonglow	good canner	**SB, M**
Tyson	good canner	**SB**

Persimmons

Ozark	preserves	HF, G

Plums

Blue Damson	good canner, preserves	Bg, HF, E, Bp, SB
Green Gage	good canner	S, SB
Superior	jelly, preserves	F, HF, E
Stanley Prune	good canner	Bg, F, HF, SB, E, G, Bp, M, S
Underwood Plum	good canner	F
Pipestone Plum	good canner	F
Fellemberg	good canner, preserves	M
Ozark Premier	good canner	SB
Santa Rosa	good canner, jelly	SB
Redheart	good canner, preserves, jelly	SB
Burbank Elephant Heart	good canner, freezer	SB

Prunes

Kaga	good canner, preserves	G
Idaho Prune	good canner	G
South Dakota	preserves	G
Sapalta	preserves	F
Mt. Royal Blue	preserves	Bg, F, J
Delicious	good canner, jelly	SB

Raspberries, red

Indian Summer Everbearing	good canner	G, HF, S
Latham	good freezer and canner	Bg, F, S, SB, J, SB, Bp, G
New Fall Red Everbearing	good freezer and canner	F, S, E, Bg, HF, J, G

| September Red Everbearing | jelly and preserves | **Bg, G, SB, J, Bp** |
| Boyne | good freezer and canner | **J** |

Raspberries, black

Black Hawk	good freezer	**Bg, G, HF, SB, J, E**
Bristol	good freezer	**SB, M, Bp**
John Robertson	good freezer and canner	**G**

Raspberries, purple

| Amethyst | good freezer | **Bg, HF, S, E** |
| Sodus | good freezer and canner, preserves | **G, HF, J, F** |

Rhubarb

Tenderstalk	good freezer	**HF**
Flare	good freezer	**G**
Victoria	good freezer	**SB, E, Bp, F**
Canada Red	good freezer and canner	**Bg, S, HF**

Strawberries

(Senator) Dunlap	good canner, preserves	**F, G, HF, S, SB, J**
Ozark Beauty Everbearing	good freezer and canner, preserves	**Bg, F, G, HF, S, SB, J, E, M, Bp**
Cyclone	good freezer	**G, HF, E**
Vesper	preserves	**G, F**
Surecrop	good freezer and canner	**Bg, HF, SB, E, M, Bp**
Streamliner Everbearing	good freezer, preserves	**F, G, HF, E, S, M**
Sunrise	preserves	**SB, Bp**
Premier	good canner, preserves, high vitamin C content	**F, R**

Midway	good freezer, preserves	**Bg, HF, J, SB, Bp**
Fairfax	high vitamin C content	**R, Bp**
Catskill	high vitamin C content	**R, M**
Midland	high vitamin C content	**R**
Tennessee Beauty	high vitamin C content	**R**
Robinson	high vitamin C content	**R**
Gem	high vitamin C content	**R, SQ**
Blakemore	good freezer	**HF**
Ogallala Everbearing	good freezer, preserves, high vitamin C content	**Bg, F, G, HF, E, Bp**
Trumpeteer	good canner, freezer	**F, G, E**
Jumbo	good freezer, high vitamin C content	**F**
Sparkle	good freezer and canner, high vitamin C content	**F, G, R, J, E, Bp**
Wisconsin 537	good freezer	**F, J**
Paymaster	good freezer	**F**
Pocahontas	good freezer, preserves	**SB**
Stark Red Giant	good freezer, canner, preserves	**SB**
Sequoia	good freezer, preserves	**HF**
Badgerglo	good freezer	**J**
Darrow	good freezer	**Bp, F**
Earlydawn	good freezer	**Bp**
Red Chief	good freezer	**Bp**
Superfection	good freezer and canner, preserves	**Bp, G**

Vegetables
for Vitamins

One of the important yardsticks for measuring the nutritional worth of any food is the contribution in terms of *vitamins* that it makes to our diet and our health. Vitamins are organic food substances—that is, substances existing only in living things, plant or animal. Although they exist in foods in minute quantities, they are absolutely necessary for proper growth and the maintenance of health. Plants manufacture their own vitamins. Animals manufacture some vitamins and obtain some from plants or from other animals that eat plants.

Vitamins are not foods in the sense that carbohydrates, fats, and proteins are foods. They are not needed in bulk to build muscle or tissue. However, they are essential, like hormones, in regulating body processes. As in the case of trace minerals (iodine, for instance), the presence or absence of vitamins in very small amounts can mean the difference between good and bad health. Many diseases and serious conditions in both human beings and animals are directly caused by a specific or combined vitamin deficiency.

The green leaves of plants are the laboratories in which plant vitamins are manufactured. So the green leaves and stalks of plants are full of vitamins. Foods that are seeds (beans, peas, kernels of wheat and corn, etc.) also contain vitamins which the plant has provided to nourish the next generation of plants. The lean meat of animals contains vitamins; the organs (heart, liver, etc.) contain even more, which the animal's digestive system has stored there. Milk and the yolk of eggs contain vitamins which the mother animal provides for her young. Fish store vitamins chiefly in their livers.

The vitamins we know most about are called by a letter and also a chemical name. These are vitamin A (carotene); vitamin B_1

(thiamin), B_2 (riboflavin), B_6 (pyridoxine), the other members of the vitamin B group (biotin, choline, folic acid, inositol, niacin, pantothenic acid, para-aminobenzoic, B_{12}); vitamin C (ascorbic acid); the several D vitamins, D_2 (calciferol) and D_3 (7-dehydrocholesterol); vitamin E (tocopherol); vitamins F; K; L_1; L_2; M; and P.

Researchers have established approximate estimates of the daily allowances of most of the vitamins needed to maintain health. These amounts are usually spoken of in terms of milligrams. (A milligram is 1/1000 of a gram. A gram is 1/32 of an ounce.) You may also find daily vitamin allowances expressed in terms of International Units.

Vegetables are an especially valuable source of many of the vitamins. Yellow and green leafy vegetables, along with tomatoes, contain appreciable amounts of carotene, the plant substance which is changed into vitamin A in the body. Ascorbic acid (vitamin C) is plentiful in tomatoes, peppers, and many of the raw leafy vegetables. While potatoes have only a fair amount of ascorbic acid, the quantities in which they are eaten by many people make them a material source of it. Several of the B vitamins, too, are present in a variety of vegetables. The green leafy ones, legumes, peas, and potatoes provide some of the needed thiamin, riboflavin, and niacin in a well-balanced diet.

Let's not forget that the *way* vegetables are grown has a definite role in the nutritive values—including vitamins—they will contain. As far back as 1939, the United States Department of Agriculture yearbook FOOD AND LIFE stated:

Underneath all agricultural practices there is a guiding principle . . . to carry out this cycle of destruction and construction economically—to see that plants, animals and man utilize raw materials efficiently to build up the products of life and that these products are broken down efficiently into raw materials that can be used again . . . By the proper use of fertilizers and other cultural practices it might be possible to insure the production of plant and animal products of better-than-average nutritive value for human beings.

In selecting vegetables, planning and varying those you serve at meals, it's important to remember that fresh vegetables offer higher overall nutrient values than either canned or frozen, and in season are usually less expensive. Within economic reason, it is preferable that

Vegetables Containing
the Largest Amounts of Vitamin A
(Recommended Daily Allowance is 5,000 Units)

Vegetables	International Units of Vitamin A
Asparagus, fresh	1,000 in 12 stalks
Beans, snap	630 to 2,000 in 1 cup, cooked
Beet greens	6,700 in ½ cup, cooked
Broccoli	3,500 in 1 cup, cooked
Carrots, fresh	12,000 in 1 cup, cooked
Celery cabbage	9,000 in 1 cup
Collards	6,870 in 1 cup, cooked
Dandelion greens	13,650 in 1 cup, cooked
Endive (escarole)	10,000 to 15,000 in 1 head
Kale	7,540 in ½ cup, cooked
Lettuce, green	4,000 to 5,000 in 6 large leaves
Parsley	5,000 to 30,000 in 100 sprigs
Peas, split	1,680 in 1 pound
Peppers, green	3,000 in 2 peppers
Peppers, red	2,000 in 2 peppers
Pumpkin	1,200 to 3,400 in 1 cup, cooked
Spinach, fresh	9,420 in ½ cup, cooked
Spinach, canned	5,500 in ½ cup
Squash, winter	4,950 in ½ cup, cooked
Sweet potatoes	7,700 in 1 medium potato, baked
Tomatoes, fresh	1,100 in 1 medium tomato
Turnip greens	9,540 in ½ cup, cooked
Watercress	4,000 in 1 bunch

as much of any family's vegetable diet as possible be of the fresh variety. Of course, the sooner they are eaten after picking or purchase, the more vitamin and other food values are retained. Careful storing and refrigeration of those that must be held are essential for vitamin retention.

Vegetables Containing
the Largest Amounts of B-Complex Vitamins

Vegetable	Milligrams of			
	Thiamin (B₁) (RDA is 5,000 units)	Riboflavin (B₂) (RDA is 1.2-1.8 mg)	Pyridoxine (B₆) (RDA is 2.0 mg)	Choline (RDA not established)
ARTICHOKES, Jerusalem				
ASPARAGUS, fresh	.16 in 12 stalks			130 in 12 stalks
BEANS, dried lima	.60 in ½ cup, cooked	.24 to .75 in ½ cup	55 in ½ cup	
BEANS, dried soy		.31 in ½ cup		340 in ½ cup
BEANS, green				340 in 1 cup
BEETS			.11 in ½ cup	8 in ½ cup
BEET TOPS		.17 to .30 in ½ cup		
BROCCOLI				
CABBAGE			.29 in ½ cup	250 in 1 cup
CARROTS				95 in 1 cup
CAULIFLOWER	.09 in ¼ small head			
COLLARDS	.19 in ½ cup, cooked	.22 in ½ cup, steamed		
CORN, yellow	1.9 in 1 lb.	0.5 in 1 lb.		
DANDELION, greens	.19 in 1 cup, steamed			
ENDIVE (escarole)		.20 in 1 head		
KALE				
MUSHROOMS				
MUSTARD GREENS		.20 to .37 in 1 cup, steamed		252 in 1 cup
PEAS, green, fresh	.36 in 1 cup, cooked	.18 to .21 in 1 cup, steamed	1.9 in 1 cup	260 in 1 cup
POTATOES, Irish			.16 in 1 medium potato	105 in 1 med. potato
POTATOES, sweet			.32 in 1 medium potato	35 in 1 med. potato
SPINACH		.24 to .30 in ½ cup, steamed		240 in ½ cup
TOMATOES, fresh				
TURNIPS			.10 in ½ cup	94 in ½ cup
TURNIP, greens	.10 in ½ cup, steamed	.35 to .56 in ½ cup		245 in ½ cup

Milligrams of			
Folic Acid (Micrograms) (RDA not established)	Inositol (RDA not established)	Niacin (RDA is 12-18 mg)	Pantothenic Acid (RDA not established)
			.40 in 4 artichokes
120 in 12 stalks			
330 in ½ cup	170 in ½ cup	9.6 in 1 lb.	.83 in ½ cup
71 in 1 cup			
42 in ½ cup, diced	21 in ½ cup		
25 in ½ cup			
90 to 110 in 1 cup			
	95 in 1 cup		1.4 in 1 cup
97 in 1 cup	48 in 1 cup		.2 in ½ cup
44 in ¼ small head	95 in ¼ head		.92 in ¼ head
		7.8 in 1 lb.	2.3 in 1 lb.
62 to 75 in 1 head			
100 in 1 cup			.30 in 1 cup
98 in ½ cup, cooked	17 in 7 mushrooms	6.0 in 7 mushrooms	1.7 in 7 mushrooms
22 in 1 cup	80 in ½ cup		.60 in 1 cup
140 in 1 small potato	29 in 1 med. potato		65 in 1 med. potato
	66 in 1 med. potato		.95 in 1 med. potato
225 to 280 in ½ cup	27 in ½ cup		
12 to 14 in 1 sm. tomato	46 in 1 sm. tomato		
	46 in ½ cup		

Vegetables Containing
the Largest Amounts of Vitamin C

(Recommended Daily Allowance is 70.75 mg.)

Vegetables	Milligrams of Vitamin C
Asparagus	20 in 8 stalks
Beans, green lima	42 in ½ cup
Beet greens, cooked	50 in ½ cup
Broccoli, flower	65 in ¾ cup
Broccoli, leaf	90 in ¾ cup
Brussels sprouts	130 in ¾ cup
Cabbage, raw	50 in 1 cup
Chard, Swiss, cooked	37 in ½ cup
Collards, cooked	70 in ½ cup
Dandelion greens, cooked	100 in 1 cup
Kale, cooked	96 in ¾ cup
Kohlrabi	50 in ½ cup
Leeks	25 in ½ cup
Mustard greens, cooked	125 in ½ cup
Parsley	70 in ½ cup
Parsnips	40 in ½ cup
Peas, fresh cooked	20 in 1 cup
Peppers, green	125 in 1 medium pepper
Peppers, pimiento	100 in 1 medium pepper
Potatoes, sweet	25 in 1 medium potato
Potatoes, white, baked	20 in 1 medium potato
Potatoes, white, raw	33 in 1 medium potato
Radishes	25 in 15 large radishes
Rutabagas	26 in ¾ cup
Spinach, cooked	30 in ½ cup
Tomatoes, fresh	25 in 1 medium tomato
Turnips, cooked	22 in ½ cup
Turnips, raw	30 in 1 medium turnip
Turnip tops, cooked	130 in ½ cup
Watercress	54 in 1 average bunch

Vegetables Containing
the Largest Amounts of Vitamin E
(Recommended Daily Allowance has not been established)

Vegetables	Milligrams of Vitamin E
Beans, dry navy	3.60 in ½ cup steamed
Carrots	.45 in 1 cup
Celery	.48 in 1 cup
Lettuce	.50 in 6 large leaves
Onions	.26 in 2 medium raw onions
Peas, green	2.10 in 1 cup
Potatoes, white	.06 in 1 medium potato
Potatoes, sweet	4.0 in 1 medium potato
Tomatoes	.36 in 1 small tomato
Turnip greens	2.30 in ½ cup steamed

Harvesting Vegetables and Fruits

If you grow your own food you've got it made over those who must rely on the grocery store or the supermarket for their daily sustenance, because you can pick and process the food that grows from your soil when its quality is at its very best. This means that you can harvest fruits and vegetables when they have reached just the right stage of maturity for eating, canning, freezing, drying, or underground storage, and you don't have to lose any time in getting the food from the ground into safe keeping, either.

The desired stage of maturity can vary from food to food and depends a great deal on what you intend to do with the produce once you've harvested it. In most cases, vegetables have their finest flavor when they are still young and tender: peas and corn while they taste sweet and not starchy; snap beans while the pods are tender and fleshy before the beans inside the pods get plump; summer squash while their skins are still soft. Carrots and beets have a sweeter flavor, and leafy vegetables will be crisp, but not tough and fibrous, when they are young. This is the stage at which you'll want to preserve their goodness.

Fruits, on the other hand, are usually at their best when ripe, for this is when their sugar and vitamin contents are at their peak. If you're going to can, freeze, dry, or store them, you'll want them firm and mature. But if you plan to use your fruits for jellies and preserves, you will not want them all fully ripe because their pectin content—which helps them to gel—decreases as the fruit reaches maturity. In order to make better jellies some of the guavas, apples, plums, or currants you are using should be less than fully ripe.

It is nearly impossible to control just when your peaches, pears, apples, and berries will be mature. Once planted, fruit trees and berry plants will bear their fruit year after year when the time is right.

You're at their mercy and must be prepared to harvest just when the pickings are ready if you want to get the fruit at its best.

Vegetables are a different story. Because most are annuals and bear several weeks after they are planted, you can plan your garden to allow for succession plantings that extend the harvesting season for you and furnish you with a continued supply of fresh food at the right stage of maturity. This means eating fresh vegetables for the whole time your garden is producing and having vegetables just right for preserving several times—not all at once—for off-season months.

By planting three smaller crops of tomatoes instead of one large crop, you won't be deluged with more tomatoes than you can possibly eat and process at one time. Space your three pea plantings ten days apart in early spring and you'll have three harvests of peas and still plenty of time to plant a later crop of something else in the same plots after all the peas are picked. Vegetables like salad greens that do not keep well should be planted twice. Plant early lettuce about a month before the last frost and follow it with cauliflower. After the onions are out of the ground, put some fall lettuce in their place for September salads. If corn is one of your favorites and you've been waiting out the long winter for the first ears to come in, by all means, eat all the early-maturing corn you want, but make sure that enough late corn has been planted for freezing later on.

Vegetables that keep well stored fresh at low temperatures, like cabbage, squash, and the root crops, should be harvested as late in the season as possible so you won't have to worry about keeping vegetables cool during a warm September or early October. Some vegetables like carrots, parsnips, and Jerusalem artichokes can be left right in the ground over the winter. It is wise to plant some late crops of these vegetables. Snap beans, planted in early May, can be followed by cabbage in mid-July. Beets planted in the beginning of April may be followed by carrots in July that can be stored right in the ground over the winter and into the early spring.

The charts that follow will give you a good idea as to the right time to harvest and how best to harvest for good eating and good keeping.

HARVESTING VEGETABLES

Vegetable	How and When to Harvest
Asparagus	Usually not until third year after planting when spears are 6 to 10 inches above ground

Beans, lima while head is still tight. Harvest only 6 to 8 weeks to allow for sufficient top growth.
When the seeds are green and tender, just before they reach full size and plumpness. If you intend to dry your beans, harvest them when they are past the mature stage, when they are dry.

Beans, snap Before pods are full size and while the seeds are about one-quarter developed, or about 2 to 3 weeks after first bloom. Snap pole beans just below the stem end and you will be able to pick another bean from the same spot later in the season. Bush beans yield only one harvest, so it doesn't matter where you snap the pod from the bush. If you want to dry your beans, delay the harvest until the beans are dry on their stems.

Beets When 1¼ to 2 inches in diameter.

Broccoli Before dark green blossom clusters begin to open. Side heads will develop after central head is removed, until frost.

Brussels sprouts When full-sized and firm, before sprouts get yellow and tough. Lowest sprouts generally mature first. Sprouts may be picked for many months, even if temperatures go below freezing.

Cabbage When heads are solid and before they split. Splitting can be prevented by cutting or breaking off roots on one side with a spade after a rain.

Carrots Anytime when 1 to 1½ inches in diameter.

Cauliflower Before heads are ricey, discolored, or blemished. Tie outer leaves above the head when curds are 2 to 3 inches in diameter; heads will be ready in 4 to 12 days after tying.

Celery Self-blanching celery must be blanched 2 to 3 weeks before harvest in warm weather, and 3 weeks to a month in cool weather. If to be used immediately, cut plant root right below soil surface. For winter storage, plants are lifted with roots.

Chinese cabbage After heads form, cut as needed. For storage, pull up plants with roots attached.

Corn	When kernels are fully filled out and in the milk stage (break a kernel open and check to see if corn milk flows when the kernel is pressed). Use before kernels get doughy. Silks should be dry and brown and tips of ears filled tight.
Cucumbers	When fruits are slender and dark green before color becomes lighter. Harvest daily at season's peak. If large cucumbers are allowed to develop and ripen, production will be reduced. For pickles, harvest when fruits have reached the desired size. Pick with a short piece of stem on each fruit.
Eggplant	When fruits are half-grown, before color becomes dull.
Endive (escarole)	To remove bitterness, blanch by tying outer leaves together when plants are 12 to 15 inches in diameter. Make sure plants are completely dry when this is done to prevent rot. Leave this way for 3 weeks before harvest. For storage, pull up plant with roots intact before a hard freeze.
Garlic	Pull when tops are dry and bent to the ground.
Jerusalem artichoke	Tubers can be dug anytime after the first frost and anytime throughout the winter.
Kohlrabi	When balls are 2 to 3 inches in diameter. If bulbs are close together, cut them off below bulb in order not to cut off tangled roots of an adjacent bulb.
Lettuce	Pick early in the day to preserve crispness caused by the cool night temperatures. Wash thoroughly but briefly as soon as harvested, then towel or spin dry to prevent vitamin loss. Loose leaf types, if cut off at ground level without disturbing roots, will send up new leaves for a second crop.
Okra	Pick a few days after flowers fall, while they are still young and not woody. Pods will be 1 to 4 inches long, depending upon variety. Freeze, can, or dry at once because they quickly become woody once mature.
Onions	For storage, pull when tops fall over, shrivel at neck of the bulb, and turn brown. Allow to

mature fully but harvest before a heavy frost. Allow them to cure for a few days in the sun. Bring in when outer skin is dry.

Parsnips   Delay harvest until a heavy frost. Roots may be safely left in ground over the winter and used the following spring before growth starts. (They are not poisonous if left in ground over winter.)

Peanuts   In the South, dig vines before frost. Peanut shells should be veiny and show color, and the foliage slightly yellow. In the northern areas, peanuts are left in ground until mid-October.

Peas   When pods are firm and well filled, but before the seeds reach their fullest size. For drying, allow them to dry on the bush.

Peppers   When fruits are solid and have almost reached full size. For red peppers, allow fruits to become uniformly red.

Potatoes   When tubers are large enough. Tubers continue to grow until vine dies. Skin on unripe tubers is thin and easily rubs off. For storage, potatoes should be mature and vines dead.

Pumpkins and Squash   Summer squash is harvested in early immature stage when skin is soft and before seeds ripen, before they are 8 inches long. Patty pans may be picked anytime from 1 to 4 inches in diameter. Skin should be soft enough to break easily with press of finger. If picked in this early stage, the vines will continue to bear. Winter squash and pumpkins should be well matured on the vine. Skin should be hard and not easily punctured with the thumbnail. Cut fruit off vine with a portion of the stem attached. Harvest before heavy frost.

Radishes   Summer radishes should be pulled as soon as they reach a good size. Leaving them in the ground after maturity causes them to become bitter. Winter radishes may be left in the ground until after frost.

Rutabagas   After exposure to frost but before heavy freeze.

Salsify   Leave until after frost, as freezing of the roots improves flavor. Also may be dug out in spring.

Soybeans	Green beans should be picked when the pods are almost mature, but before they start to yellow. Harvest period lasts for only about a week. Dry soybeans are allowed to dry on the vines and are picked just as they are dry, while stems are still green; otherwise the shells will shatter and drop their beans.
Spinach	May be cut off entirely when fully mature (when 6 or more leaves are 7 inches long) or outer leaves can be cut from plant as they mature, leaving inner leaves on to ripen.
Tomatoes	When fruits are a uniform red, but before they become soft.
Turnips	When 2 to 3 inches in diameter. Larger roots are coarse and bitter.

HARVESTING FRUITS

Fruit	How and When to Harvest
Apples	Summer apples are picked when ripe, and should be used or preserved at once. They usually do not store well for more than a few days. Pick fall and winter apples at peak ripeness for best storage. Pick with the stems; if stems are removed, a break in the skin is left which will let bacteria enter and cause rot. When picking apples, be careful not to break off the fruiting spur, which will bear fruit year after year if undamaged.
Apricots	Apricots should be left on the tree until fully ripe, because once they are picked they do not increase their supply of sugar. For drying, they should be ripe and firm.
Blackberries and Boysenberries	Berries are ripe when they fall readily from bush into the hand, a day or two after they blacken. Berries should be picked in the cool of the morning, kept out of the sun, and refrigerated or processed as soon as possible.
Blueberries	Blueberries should be left on the bush several days to a week after they turn blue. When fully ripe they are slightly soft, come easily from the bush, and are sweet in flavor.

Cherries   In order not to damage the fruiting twigs, cherries should be picked without the stems. This will leave a break in the fruit; therefore the picked cherries must be processed at once or else spoilage will begin. If sour cherries are protected from birds with netting, fruit should be allowed to ripen on tree for 2 to 3 weeks. The longer it hangs, the sweeter it becomes.

Currants   The longer currants hang, the sweeter they become, so leave on bush for 4 to 6 weeks, unless you plan to make jelly from them. Then pick some of them when still a little green because they lose their pectin content as they ripen.

Dates   In dry weather dates should be left on the trees until they are thoroughly ripe. If the weather becomes wet, they must be picked before any rain touches them. The ripening process is then finished indoors.

Gooseberries   Because gooseberries of one variety all mature at the same time, they may be harvested in one day. The picker usually wears heavy leather gloves and strips the branches of their fruit by running his or her hand along the whole branch, catching the berries in an open container, such as a bushel basket. The small pieces of leaf and stem may be separated by rolling the fruit down a gentle incline made by tipping a piece of wood or cardboard. The leaves and stems will be left on the incline and the moderately clean fruit rolls to the bottom.

Grapes   Grapes should be picked when fully ripe. The fruit will be aromatic and sweet, and the stem of the bunch will begin to show brown areas. Grapes that are to be stored should be picked in the coolest part of the day. Clip the bunches from the stems with sharp shears and handle them by the stems rather than by the fruit.

Guavas   Guavas ripen in a period of about 6 weeks. If they are to be used for jelly or juice, some of them may be picked before they are quite ripe. They contain such a large amount of pectin that a pound of fruit will make more than 3 pounds of jelly.

Oranges   Color is not always a sign of maturity in oranges. Their skin contains a mixture of pigments: green, orange, and yellow. In the fall the green predominates until the weather turns cool enough to check their growth, when the green fades out and the other pigments predominate. But in the spring when growth begins, green pigments will again appear in perfectly ripe fruit as it hangs on the tree. Navel oranges, Valencia, and other varieties with tight skins are picked by pulling away from the stem. Those with loose skins, like mandarins, Temples, and tangerines are picked with clippers which clip the fruit with about a half inch of the stem remaining. Once clipped from the tree, the remaining stub is clipped off.

Peaches   Peaches are ripe when the fruit is yellow. Fully tree-ripened fruit will have more sugar and less acid than fruit which is picked when half ripe. Don't pull the fruit directly from the tree because it will cause a bruise which will make the fruit spoil quickly. Rather, remove the fruit from the tree by tipping and twisting it sideways.

Pears   Pears are harvested when they have reached their full size and the skins change to a lighter green. Seeds will be starting to turn brown at this stage, and the stems separate easily from the tree when lifted. The quality of the pears will be much better if they are picked before stony granules are formed through the flesh, during the last few weeks of ripening.

Persimmons   When fully ripe, persimmons are very soft and are very sweet. American persimmons may be harvested just before they are ripe, or they may be left hanging on the tree into the winter months. Fruit left hanging through January, even though it is frozen on the tree, retains its flavor when picked and thawed.

Plums   For canning and jelly-making plums may be harvested as soon as they have developed their bloom. At this point they are slightly soft, but still retain some of their tartness and firmness. Those prune varieties that are to be dried will hang on the tree long after they are ripe. They will develop more sugar as they hang.

Pomegranates   Fruit is picked in the fall after it has changed color. It will ripen in cold storage. It may also be permitted to ripen on the tree, so long as it does not split.

Quinces   Quinces may hang on the bush until after the first fall frost. If they are to be stored, they may be picked a few weeks earlier.

Raspberries   Because they become soft when fully ripe, raspberries must be picked every day, or at least every other day during harvesting time. Rain at harvesttime causes berries to become moldy. They should therefore be picked immediately after a rain and processed at once before they mold. If moldy berries are left on the plant the mold will spread to green berries and destroy them.

Rosehips   The fruit should be picked when it is fully mature in late fall. At this time the rosehips will be deep in color, have a mellow, nutlike taste, and the vitamin C content will be at its peak.

Strawberries   The berries should be picked early in the morning when the fruit is still cool. Gently twist the fruit off its stem; do not pull it off the stem. Fruits washed without their stems will lose more vitamins than fruits which are de-stemmed after washing.

Watermelons   Watermelons must ripen on the vine because they do not develop more sugar or better color after they have been taken from the vine while still green. Most melons are fully ripe when the tendril accompanying the fruits dies, but this is not always the case with all varieties. A ripe melon has a hollow sound and a green one has a metallic ring when knocked with the knuckles.

Handling Food After Harvest

Making sure that you harvest your food at the right time is only half the key to great tasting fruits and vegetables kept through the winter months. Handling the food after the harvest—during the time between picking and processing—is just as important. Although the actual growth of fruits and vegetables stops when they are plucked from the ground and cut off from their food supply, respiration and activity of enzymes continue. The physical and chemical qualities of the plants deteriorate rapidly. Not only will there be a deterioration of appearance and flavor as the freshness of food fades, there will also be a loss of nutrients, particularly of vitamin C.

Fruits and vegetables should be prepared and canned, put into the freezer, dried, or placed in cold storage as soon as is humanly possible after harvest.

If you cannot avoid a delay in preparing and storing, cool your food as soon as it is picked. Do not keep it at room temperature, or, even worse, expose it to the sun. The quickest way to cool it is to immerse the food in ice water. After draining, keep the food at low temperatures, preferably between 32° and 40°F. Covering the produce with cracked ice is another means of cooling and thereby slowing down the loss of quality. These aids, of course, do not replace the need for prompt processing.

Fruits and vegetables are at their best when first picked. Don't expect any kind of storage to make a great food out of an inferior one. If you've taken the effort to grow good food, make the extra effort to harvest it at the right time and get it into proper storage as soon as possible.

Freezing Vegetables and Fruits

For many, freezing is the best way of preserving the prides of their organic gardens. Because you have to purchase a freezer and keep it running year round, freezing may cost you more money than other ways of preserving, but the money is well spent when you consider food flavor, color, texture, and nutrients, and the time you save by freezing rather than canning or drying your vegetables and fruits.

A general rule to remember is that those vegetables most suited for freezing are those which are usually cooked before serving. These include asparagus, lima beans, beets, beet greens, cauliflower, broccoli, Brussels sprouts, peas, carrots, kohlrabi, rhubarb, squash, sweet corn, spinach, and other vegetable greens. Vegetables that are usually eaten raw, such as celery, cabbage, cucumbers, lettuce, onions, radishes, and tomatoes, are least suited for freezing. Almost all fruits, especially berries, freeze very well.

If you are concerned about preserving the vitamin C content in foods, do not can, but freeze those foods that supply us with most of our vitamin C. Fruits and vegetables rich in vitamin C include broccoli, spinach, cauliflower, Brussels sprouts, kohlrabi, turnip greens, strawberries, grapefruit, lemons, and oranges.

The freezing process itself does not destroy any nutrients in food. However, there can be some nutrient loss during blanching of vegetables and the cooling process that takes place right before food is frozen, when food combines with the oxygen of the air and goes through a process called oxidation, or from the "drip" which results from excessive thawing. Nutrient losses can be kept at a minimum if you are quick and efficient when blanching and freezing. Food should be prepared as soon as it is harvested, or kept at 40°F. or lower no longer than 24 hours before it is prepared and frozen.

Studies published by the North Dakota Cooperative Extension Service show that if foods are prepared and frozen properly, they retain their food value the same as fresh foods:

Carbohydrates show no change with the exception of the sugar (sucrose) being reduced to simple sugars (glucose and fructose) during long storage. This is of no importance. Minerals might be lost in solution during the blanching and cooking of vegetables, but this loss is usually no greater than when cooking fresh vegetables. Vitamin A is lost only when vegetables are not blanched.

B vitamins, such as thiamin and riboflavin, are lost by overstorage and through solution in cooling and blanching because they are water soluble. However, a greater loss is suffered if the vegetables are not blanched. Some thiamin is lost by the heat in cooking.

Vitamin C is easily lost through solution as well as oxidation. Vegetables lose some of their vitamin C through blanching, but without this process, the loss would be much greater. The same amount of vitamin C is lost in cooking fresh vegetables.

Containers to Use for Freezing

You can use any containers that will exclude air and prevent contamination and loss of moisture. Plastic containers are ideal. If tin cans with friction lids or glass jars are utilized, don't fill them to within more than an inch of the top. That space will allow room for the normal expansion of the food without accidents.

Fruits which are to be packaged dry and all vegetables can be placed in heavy plastic bags, heat-sealed with a hot iron or closed with a wire band at a point as close to the contents as possible so that there is a minimum of air in the package. (Do not, however, use boil-in-the-pouch-type plastic bags, as the label on the box these bags come in states.) It is a good idea to overwrap thin plastic packages with another plastic bag or stockinette to prevent the plastic material from tearing once inside the freezer. Oxygen entering frozen food through tears in packaging can ruin the best quality foods. To test for rips and holes, fill the bag with water. If you discover a leak, don't use the bag for freezer storage. Even the smallest hole can allow oxygen to enter and moisture to escape. Of course, plastic containers can be used over and over again. So can heavy-duty plastic bags, so long as they are in good shape.

Any container that is strong, moisture-proof, and can be sealed to exclude air makes a good container for freezing fruits and vegetables. Plastic containers, glass canning jars, and heavy-duty plastic bags are all popular freezer containers.

Wax containers made especially for freezer use can also be used.

All freezer containers should be marked before they are put into the freezer. The type of food and the date it was frozen should be marked so that you can pick the kind of fruit or vegetable that you want at a glance and can use up those that are oldest first. If you are freezing different varieties of the same food, mark the variety on the label as well, so that you'll be able to determine for the next season which varieties freeze best.

Thawing Vegetables and Fruits

Freezing is not a method of sterilization as is heat processing in canning. While many microorganisms are destroyed by freezing temperatures, there are some, most notably molds, that continue to live even though their growth is retarded and their activity rate is slowed down. When foods are removed from freezer storage, their temperatures rise and the dormant microorganisms begin to multiply. Even during thawing, the process of spoilage sets in, and the higher the thawing temperature, the faster the growth of spoilage microorganisms. Thus, the microbial population will increase at a slower rate if food is thawed at a low temperature, such as in a refrigerator, than at room temperature.

Frozen foods that need to be thawed, such as those that will be eaten raw or mixed with other foods for casseroles, should be thawed in the refrigerator and not on the kitchen counter or in hot water, whenever possible. Because decomposition of thawed foods is more rapid than fresh foods, they should be used as quickly as possible after thawing. Of course, if frozen food is to be cooked, there is no reason to thaw it first. Just remove it from the freezer and immediately place it in boiling water or in a preheated oven and cook it frozen, before microorganisms become active.

Refreezing

Foods still partially frozen or even those that are completely thawed but appear to be edible may be refrozen. Refrozen foods may lose some of their quality, but they will be safe to eat. Don't, however, refreeze any food that shows any signs of off-odor, color, or other indication of bacterial change. Mark foods which have been refrozen and use them as soon as is convenient.

How and Why to Blanch Vegetables

If you have ever tried to freeze vegetables without first blanching them, you may have discovered the horrible cardboard flavor they acquire after a few months in the freezer. They bear no relation to the succulence of the products you hopefully packed last summer. This is the work of enzymes.

Vegetables, as they come from the garden, have enzymes working in them. These break down vitamin C in a short time and convert starch into sugar. They are all slowed down (not stopped) by cold temperatures, but they are destroyed by heat—by blanching.

The blanching idea isn't new. Methods for scalding or steaming fresh produce in preparation for freezing were introduced over forty years ago. Since then, however, food specialists have been discovering more about the unrealized and subtle effects of using this pre-cold-storage process. They have found, for example, that blanching makes certain enzymes inactive which would otherwise cause unnatural colors and disagreeable flavors and odors to develop while the foods remain frozen. Then too, they've found that blanched vegetables are somewhat softened, so that they can be packed more easily and solidly into freezer containers.

The first and foremost reason for blanching, of course, is to help frozen produce keep better. What freezing and the right preparation for it do is literally to hold on to as much natural "freshness" as possible.

There is another benefit in the blanch-freeze method. Experiments have shown that ascorbic acid—or, as it's more commonly known, vitamin C—is retained in much greater amounts in many of those vegetables blanched before freezing. Some held two, three, and even four times more of this elusive element through periods ranging up to nine months. Aside from maintaining better quality, this nutritive advantage over vegetables frozen unblanched (or treated in other ways, such as with sulphur dioxide gas) is significant.

Concentrated research on food freezing, blanching, and quality and nutrient retention was carried out at the University of Illinois College of Agriculture for several years. The University's experiment station quarterly, ILLINOIS RESEARCH (Summer 1959), reported extensive experiments on the blanching questions.

Five vegetables—broccoli, peas, snap beans, spinach, and corn—were picked fresh, at optimum maturity for freezing, and processed promptly. Several lots of each were used. Part of these were given preliminary preparation, blanching, cooling, packaging, freezing, and freezer storage, according to standard directions. The rest were packaged and frozen without blanching.

The green vegetables were compared for ascorbic acid retention after freezer storage periods of one, three, six, and nine months. In one instance (broccoli after one month) the amount was equal. In all the others, analyses showed *more vitamin C was retained in the blanched vegetables than in those frozen unblanched,* at every period of testing.

After three months, for example, blanched broccoli had held 64 percent of its raw ascorbic acid content, compared to 57 percent for the unblanched samples. At six months, the difference had widened to 60 to 40 percent; and by nine months, 54 to 36 percent.

Peas revealed even greater losses where blanching had been omitted. After a month, blanched specimens tested 70 percent vitamin C retention against 63 percent for the unblanched. By the third month, it was 75 to 55; six months, 71 to 36; and nine months after freezing, 70 percent to 37 percent.

With snap beans, a larger disparity was found right at the start. Blanched beans had held 85 percent of their initial ascorbic acid after a single month in the freezer, while those unblanched had dropped back to 58 percent. In three months it was 83 to 44; at six, 64 to 15; and at nine months the blanched beans still had 43 percent—and the unblanched just 3 percent.

Spinach was tested after two weeks' freezer storage and showed a 52 to 28 percent advantage for blanching. Corn, which is low in vitamin C to begin with, was not tested for retention, but was included in the cooking and palatability comparisons since it is one of the more popular vegetables for home freezing.

In addition to checking for ascorbic acid variation in the blanched and unblanched produce, the Illinois research sought to compare what is called *palatability*, too. This includes such factors as appearance, color, texture, flavor (or off-flavor), and general acceptability. "During the entire nine months," states the report, "scores of blanched samples were 4.0 to 4.4, corresponding to ratings of good to high good. After cooking, the blanched vegetables were bright green and tender, and had a good flavor. No off-flavors were noted."

On the other hand, unblanched samples that had been stored for only one month received general acceptability scores of 2.4 to 2.9, which corresponds to ratings between poor and fair. Strong off-flavors developed and unblanched products lost color during the first month in the freezer.

Over the longer storage periods, the vegetables that were not blanched deteriorated still further, some of them so much that they were considered inedible. Most became faded, dull, or gray; all became tough or fibrous; and some, broccoli especially, developed an objectionable haylike flavor.

Tests with the frozen corn disclosed palatability ratings between good and very good for the blanched samples, while none of the unblanched was considered even fair. Although the appearance and color of the unblanched corn had held up, its flavor had become disagreeable and there was deterioration in texture.

Steps to Follow for Freezing Vegetables

To keep the largest possible amounts of vitamin C in the vegetables you freeze and to keep them tasty and appetizing, it pays to blanch carefully before freezing. Whether you want to store your own surplus for a healthful supply next winter or if you'd like to be sure of putting up enough naturally grown produce bought throughout this season, here are some tips on freezing:

1.   Line up everything needed for blanching and freezing *first*. Nothing counts more than speed in holding on to freshness, taste, and nutritive value. Plan a family operation deep-freeze; have all hands on deck to help quickly, and arrange equipment and containers in advance for a smooth production.

2.   Pick young, tender vegetables for freezer storage; freezing doesn't improve poor-quality produce. As a rule, it is better to choose slightly immature produce over any that is fully ripe; avoid bruised, damaged, or overripe vegetables. Harvest in early morning. Try to include some of the tastiest early-season crops; don't wait only for later ones.

3.   Blanch with care and without delay. Vegetables should be thoroughly cleaned, edible parts cut into pieces if desired, then heated to stop or slow down enzyme action. For scalding, use at least a gallon of water to each pound of vegetable, preheated to boiling point in a covered kettle or utensil (preferably stainless steel, glass, or earthenware). Steaming is better for some vegetables because it helps retain more nutritive value. Use a wire-mesh holder or cheesecloth bag over 1 inch of boiling water in an 8-quart pot. The same arrangement is handy for plunging vegetables into boiling water, 1 pound at a time. Start timing as soon as basket or bag is immersed or set in place for steaming. If you live in a high altitude area, add ½ minute to blanching time for each 2,000 feet above sea level.

4.   Cool quickly to stop cooking at the right point. Vegetables that are overblanched show a loss of color, texture, flavor, and nutritive value. Plunge blanched vegetables into cold water (below 60°F.); ice water or cold running water will do best.

Organic gardener Ruth Tirrell cools her blanched vegetables by what she calls the waterless freezing method. Instead of placing vegetables under running water or soaking them, which can wash away water-soluble vitamins and minerals, she pours her vegetables into a flat pan which is resting on another pan filled with ice cubes and cold water. Once the vegetables are in this top pan, she covers them with an ice bag or plastic bag which is filled with ice cubes. Miss Tirrell cautions that this method of cooling blanched vegetables takes a lot of ice, so have a good quantity ready before you begin.

5.   Package at once in suitable containers. Glass jars require a 1- to 1½-inch headspace; paper and plastic containers call for leaving a ½-inch headspace, except for vegetables like asparagus and broccoli that pack loosely and need no extra room. Work out air pockets by gently running a knife around the interior sides of the containers. Then seal tightly.

6. Label all frozen food packages; indicate vegetable, date of freezing, and variety. Serve in logical order—remember food value and appeal are gradually lowered by long storage. Maximum freezer periods for most vegetables are 8 to 12 months. Except for spinach and corn on the cob, cook without thawing. Avoid overcooking.

As mentioned earlier in this chapter, almost every vegetable that can be cooked before eating may be frozen successfully. The list that follows tells how to prepare for freezer storage just about every vegetable found in an American garden. Although all the vegetables listed

Blanch herbs before packing them for the freezer. Tie sprigs of fresh-picked herbs into small bunches and blanch them, one bunch at a time, in a steaming basket for 1 minute. Then plunge each bunch into ice water to cool it quickly. Shake off excess water, pack loosely in a plastic bag, and seal with wire twists.

can be frozen, there will most likely be some here that you will prefer to store in some other manner. Potatoes, cabbage, root crops, and squash store well in a root cellar or an outdoor storage area. Beets are preferred canned by most people. Herbs are easy to dry and convenient to use in dehydrated form. If your freezer space is limited, you'd be wise to consider these alternate ways of storage.

If you like to freeze mixed vegetables, we suggest that you cut and blanch each vegetable separately. Then mix and freeze them. Since blanching times vary according to the type of vegetable, blanching separately for the required time assures you that each vegetable is sufficiently precooked.

Check recipes later in this chapter and also at the end of the book for prepared foods and combination dishes for the freezer.

PREPARING VEGETABLES FOR FREEZING

Artichokes Select small artichokes or artichoke hearts. Cut off the top of the bud and trim down to a cone. Wash and blanch for 8 minutes in boiling water or 8 to 10 minutes in steam. Cool, pack, and freeze.

Asparagus Use young, green stalks. Rinse and sort for size. Cut in convenient, equal lengths to fit container. Blanch in steam or boiling water from 2 to 4 minutes, depending upon size of pieces. Cool, pack, and freeze.

Beans, lima Pick when pods are slightly rounded and bright green. Wash and blanch in steam or boiling water for 4 minutes, drain and shell. Rinse shelled beans in cold water. No additional blanching is necessary. Pack and freeze.

Beans, shelled Pick pods when they are well filled, bright green, and tender. Wash and blanch for 2 minutes in boiling water or 2 or 3 minutes in steam. Cool, pack, and freeze.

Beans, snap Pick when pods are of desired length, but before seeds mature. Wash in cold water and drain. Snip ends and cut, if desired. Blanch in steam or boiling water, 3 to 4 minutes depending upon size and maturity. Cool, pack, and freeze.

Beets Beets are usually preferred canned, but they may be frozen. Harvest while tender and mild flavored. Wash and leave ½ inch of the tops on. Cook whole until tender, skin and cut, if desired. No further blanching is necessary. Cool, pack, and freeze.

Broccoli Select well-formed heads. Buds that show yellow flowers are too mature and should not be frozen or canned. Rinse, peel, and trim. Split broccoli lengthwise into pieces not more than 1½ inches across. Soak in salt water for 30 minutes to drive out any worms. Blanch 5 minutes in steam or 3 minutes in boiling water. Cool, pack, and freeze.

Brussels sprouts Pick only green buds. Like broccoli, heads that are turning yellow are too mature to process. Rinse and trim. Remove outer leaves. Blanch in steam or boiling water 3 to 5 minutes, depending upon size. Cool, pack, and freeze.

Cabbage Trim off outer leaves. You can shred for tight packing or cut into wedges. Blanch the shredded cabbage in boiling water for 1½ minutes, or in steam for 3 minutes. Wedges should be blanched in boiling water for 3 minutes or in steam for 4 minutes. Cool, pack, and freeze.

Carrots Root cellar storage is preferred to freezing, but they may be frozen. Harvest while still tender and mild flavored. Trim, wash, and peel. Small carrots may be frozen whole. Cut others into ¼-inch cubes or slices. Blanch in boiling water: 2 minutes for small pieces, 3 minutes for larger pieces, and 5 minutes for whole carrots; or blanch in steam: 4 minutes for small pieces and 5 minutes for larger ones. Cool, pack, and freeze.

Cauliflower Select well-formed heads free of blemishes. Wash and break into flowerets. Peel and split stems. Soak in salt water for 30 minutes to drive out any worms. Blanch 4 minutes in steam or boiling water. Cool, pack, and freeze.

Celery If your space is limited, this is one food to leave out of the freezer because it does not freeze very well. But it may be frozen and then used in cooked casseroles, stews, and soups. Select crisp stalks. Clean well and cut across the rib into 1-inch pieces. Blanch in boiling water 3 minutes or in steam for 4 minutes. Cool, pack, and freeze.

Corn (whole kernel corn)

Pick ears as soon as the corn ripens. The natural sugars in corn turn to starch quickly after ripening, so good timing is important. Husk, silk, and wash the ears. Blanching and then cooling is a lot easier when the kernels are on the cob, so put about 3 ears in either steam or boiling water for 6 to 8 minutes. Then cool and remove the kernels from the cob with a sharp knife or a corn cutter. (If using a knife you may want to secure the cob upright as you cut by pushing one end of the cob through a slender nail which has been driven through a 3- or 4-inch square of 1-inch-thick odorless wood.)

To make cutting corn off the cob easier, you can place the end of the cob in a nail embedded in a piece of wood. The nail will hold the cob firm and prevent slips.

If you prefer to blanch the kernels after they have been cut from the cob, put about a pound of corn at a time in a large pot of boiling water or in a steamer for 6 to 8 minutes. Cool, pack, and freeze.

Corn (corn on the cob) Harvest corn as above. (The corn varieties with smaller cobs are especially good to freeze whole.) Then husk, silk, and wash. Blanch about 3 ears at a time in either steam or boiling water for 6 to 8 minutes. Cool and pack separately, or pack enough for one meal together. Cobs can be wrapped in freezer paper, double layers of aluminum foil, or in plastic freezer bags.

Eggplant Select firm, heavy fruit of uniform dark purple color. Harvest while seeds are tender. Wash, peel, and cut into ⅓- to ½-inch slices or cubes. Dip in solution of 1 tablespoon lemon juice to 1 quart water. Blanch 4 minutes in steam or boiling water. Dip again in lemon juice solution after heating and cooling. Pack and freeze.

Greens See spinach.

Herbs Harvest on a sunny morning right before plants blossom. Remove damaged portions and rinse under cold water. Blanch for 1 minute in steam. Cool, pack, and freeze.

Kohlrabi Harvest while tender and of mild flavor. Avoid any that are overmature. Wash and trim off trunk. Slice or dice in ½-inch pieces or smaller. Blanch in steam or boiling water for 1 to 2 minutes, depending upon size of cubes or slices. Cool, pack, and freeze.

Mushrooms Select firm, tender mushrooms, small to medium sized. Wash and cut off lower part of stems. Cut large mushrooms into pieces. Add ⅓ teaspoon lemon juice to 1 gallon of water. Blanch in boiling water: 2 minutes for small, whole mushrooms; 4 minutes for large, whole mushrooms; and 2 minutes for slices. Or blanch in steam: 3 minutes for small mush-

Okra

rooms and slices, and 5 minutes for large, whole mushrooms. Cool, pack, and freeze.

Select young, tender pods. Wash and cut off stems so as not to rupture seed cells. Blanch 2 to 3 minutes in boiling water, or 5 minutes in steam. Cool; freeze whole or slice crosswise.

Parsnips

Choose smooth roots. Woody roots should not be used for freezing; they will be tough and tasteless. Remove tops, wash, and peel. Cut into slices or chunks. Blanch in boiling water or steam for 3 minutes. Cool, pack, and freeze.

Peas

Pick when seeds become plump and pods are rounded. Freeze the same day they are harvested, as sugar is lost rapidly at room temperature. Discard immature and tough peas. Shell peas. Do not wash. Blanch 1½ minutes in steam or boiling water. Cool, pack, and freeze.

Peppers, hot

Wash and stem. Leave whole and pack fresh; there is no need to blanch them first.

Peppers, sweet and Pimientos

Select when fully ripe, either green or red varieties. Skin should be glossy and thick. Wash and halve. Remove seeds and pulp. Slice or dice. Peppers do not require blanching, but you may blanch for 2 minutes in steam or boiling water. This makes packing easier and will help to keep especially large peppers from getting tough during freezing. Cool, pack, and freeze.

Pumpkin, Summer and Winter squash

Harvest when fully colored and when shell becomes hard on pumpkins and winter squash. Summer squash should be harvested before rind becomes hard. Wash, pare, and cut into small pieces. Cook winter squash and pumpkins completely before packing. Do not add seasoning. Blanch summer squash in steam or boiling water 4 minutes, and blanch zucchini 2 to 3 minutes, depending upon size. Slice summer squash ½ inch thick. Cool, pack, and freeze.

Rhubarb

See fruits.

Soybeans Pick when pods are well rounded, but still green. Yellow pods are too mature for processing. Two to 3 days too long in the garden will result in overmaturity. Wash and blanch 5 minutes in steam or boiling water before shelling. Cool and shell with pea sheller or by hand. Rinse shelled beans in cold water. No additional blanching is necessary. Pack and freeze.

Spinach and other greens (beet, dandelion, mustard, turnip, carrot, collards, kale, etc.) Harvest while still small and tender. Cut before seed stalks appear. Harvest entire spinach plant. Use only tender center leaves from old kale and mustard plants. Select carrot, turnip, or beet leaves from young plants. Rinse well. Trim off leaves from center stalk. Trim off large midribs and leaf stems. Discard insect-eaten or injured leaves. Blanch 2 minutes in boiling water or 3 minutes in steam. Stir while blanching to prevent leaves from matting together. Cool, pack, and freeze.

Sweet potatoes Use smooth, firm sweet potatoes. Wash and cook in water or bake at 350°F. until soft. Cool, and remove skins if you wish. Pack whole, sliced, or mashed. To each 3 cups pulp mix 2 tablespoons lemon or orange juice, or dip whole or sliced potatoes in ½ cup lemon juice to 1 quart water to retain bright color. Cool, pack, and freeze.

Tomatoes Tomatoes can better than they freeze. If you do freeze tomatoes, use them within a few months for best taste and nutritive value. Tomato skins can get tough once frozen, so you may want to peel them first. Do so by first plunging whole tomatoes in boiling water until skins crack, about 1 minute or so. Then cool quickly in cold water and remove skins with a sharp knife. Pack tomatoes whole in freezer containers without blanching, or stew the tomatoes before freezing. Stew tomatoes by cutting them into quarters and simmering them slowly in a heavy pot (without adding water) until soft, about 20 minutes. Stir continuously to avoid scorching. Seasonings may be added before tomatoes are chilled, packed, and frozen, but don't add bread crumbs to thicken the stewed tomatoes until they are heated before serving.

Turnips   Use only young, tender roots. Cut off tops, wash, and peel. Slice into lengthwise strips or chop into small cubes. Blanch in boiling water or steam 2½ minutes. Chill, pack, and freeze.

Approximate Yield
of Frozen Vegetables from Fresh

Vegetable	Fresh, as Purchased or Picked	Frozen
Asparagus	1 crate (12 2-pound bunches) 1 to 1½ pounds	15 to 22 pints 1 pint
Beans, lima (in pods)	1 bushel (32 pounds) 2 to 2½ pounds	12 to 16 pints 1 pint
Beans, snap, green, and wax	1 bushel (30 pounds) ⅔ to 1 pound	30 to 45 pints 1 pint
Beet greens	15 pounds 1 to 1½ pounds	10 to 15 pints 1 pint
Beets (without tops)	1 bushel (52 pounds) 1¼ to 1½ pounds	35 to 42 pints 1 pint
Broccoli	1 crate (25 pounds) 1 pound	24 pints 1 pint
Brussels sprouts	4 quart boxes 1 pound	6 pints 1 pint
Carrots (without tops)	1 bushel (50 pounds) 1¼ to 1½ pounds	32 to 40 pints 1 pint
Cauliflower	2 medium heads 1⅓ pounds	3 pints 1 pint
Chard	1 bushel (12 pounds) 1 to 1½ pounds	8 to 12 pints 1 pint

Vegetable	Fresh, as Purchased or Picked	Frozen
Collards	1 bushel (12 pounds) 1 to 1½ pounds	8 to 12 pints 1 pint
Corn, sweet (in husks)	1 bushel (35 pounds) 2 to 2½ pounds	14 to 17 pints 1 pint
Eggplant	1 pound	1 pint
Kale	1 bushel (18 pounds) 1 to 1½ pounds	12 to 18 pints 1 pint
Mustard greens	1 bushel (12 pounds) 1 to 1½ pounds	8 to 12 pints 1 pint
Peas	1 bushel (30 pounds) 2 to 2½ pounds	12 to 15 pints 1 pint
Peppers, green	⅔ pound (3 peppers)	1 pint
Pumpkin	3 pounds	2 pints
Spinach	1 bushel (18 pounds) 1 to 1½ pounds	12 to 18 pints 1 pint
Squash, summer	1 bushel (40 pounds) 1 to 1¼ pounds	32 to 40 pints 1 pint
Squash, winter	3 pounds	2 pints
Sweet potatoes	⅔ pound	1 pint

U.S. Department of Agriculture

Freezing Fruits

Fruits lend themselves to freezing better, in most cases, than do vegetables, because fruits do not need to be blanched before they are frozen. It is true that certain changes in the texture of the frozen fruits, a cellular breakdown or softening, are similar to the changes in cooked fruits. Some fruits, such as papayas, pears, mangoes, bananas, watermelons, and avocados, are more subject to the loss of texture than others and cannot be frozen very satisfactorily. Whenever it is possible to freeze them, however, fruits can be made to retain more nutritive value and flavor than by any other method of preservation, and the process usually takes less labor than canning or drying. Fruits can be frozen safely for up to one year.

Freezing Fruits with Honey

Fruits are usually frozen in one of two ways—dry or floated in a sweet syrup. Most information on freezing fruits recommends freezing them with dry sugar or mixing them in a syrup of water and sugar. Fruits held at freezing temperatures will keep without the aid of a sweetener, but they may lose some of their flavor, texture, and color when packed alone. If you want to add something to preserve the taste and appearance of your frozen fruit, don't use sugar, use honey. While sugar adds little or no nutritional value to foods in its refined or unrefined state, honey, if uncooked and unfiltered, contains important vitamins, minerals, and enzymes that do add food value to your frozen fruit.

In general, you can substitute honey for sugar in recipes for freezing fruit. Just cut down the amount of sweetener you use by one half. This means that ¼ to ½ cup of honey is mixed with 1 pint of dry fruit when the recipe calls for ½ to 1 cup of sugar for each pint of dry fruit.

If you want to freeze your fruits in a sweet syrup, make one with honey instead of sugar. A thin to medium honey syrup is best to use with most fruits. A thin syrup can be made by blending 1 cup of honey with 3 cups of very hot water. A medium syrup can be made by blending 2 cups of honey with 2 cups of very hot water. Chill all syrups before using them. Use enough syrup to completely cover the fruit. If the fruit is packed tightly enough, ½ cup chilled honey syrup should be sufficient for pint containers and 1 cup should be enough for quart containers.

Because honey has a flavor of its own while sugar does not, it is important to use mild-flavored, light honeys for freezing (and canning, for that matter) unless you enjoy the taste of light, fruit-flavored honey instead of sweet frozen fruit. Early summer and spring honeys are generally milder than those collected in the fall. Clover, locust, and alfalfa honeys are certainly more suitable for freezing with fruit than dark honeys like buckwheat.

Preventing Fruits from Darkening

Changes in color, flavor, aroma, and vitamin C content of fruits during freezing and thawing are caused mainly by oxidation. No browning will occur in fruit tissues until practically all the ascorbic acid has been oxidized.

There are a few steps that you can take to preserve the qualities of fresh fruit. Freeze only mature fruit. Immature fruit is usually higher in tannins and other constituents involved in darkening; some

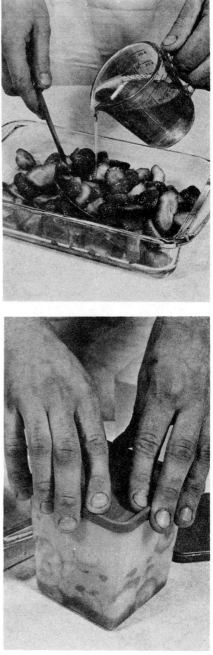

To freeze strawberries in a honey syrup, slice them into a dish, and pour a light honey syrup over them. One-half cup of syrup should be sufficient for each quart of berries. Gently toss the berries so that all are coated with syrup. To protect the berries from being crushed in the freezer, pack them in rigid or semi-rigid containers.

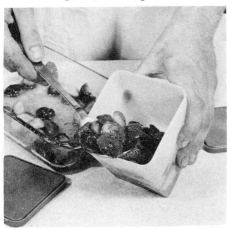

contain compounds which become bitter during freezer storage and thawing. Also, handle fruits and fruit products quickly during preparation for freezing, packing, partial thawing, and serving to minimize exposure to air. Cut directly into syrup any fruit that is likely to discolor, and place crumpled wax paper or foil on top to keep fruit under the syrup.

Pears, peaches, apricots, sweet cherries, and figs darken easily during freezing. To help retain their color, these fruits may be packed in a *pectin pack*. To make a pectin pack, boil for 1 minute 1 box of commercial pectin and 1 cup of water, stirring constantly. Add ¼ cup honey, stirring until dissolved. Then add enough cold water to make 2 cups of syrup. Place fruit and syrup in a freezer container. All the fruit must be covered by the pectin syrup. To make certain that all the fruit is submerged, stuff a piece of crumpled wax paper or foil under the container lid. This will keep the fruit below the syrup.

The addition of ascorbic acid will also help to prevent discoloration. A tablespoon or two of *rose hip concentrate,* in liquid or powder form, may be added to pure honey or a honey syrup before it is poured over the packed fruit. A tablespoon or two of *lemon juice* can be used in place of the rose hip concentrate or pectin. You can also crush 3 or 4 500-milligram *ascorbic acid* (or vitamin C) *tablets* in a quart of water and dip each piece of fruit into the solution before packing in honey.

Try these natural sources of ascorbic acid with a small quantity of fruit first. Freeze the fruit and, after a few days, thaw it and test it to determine the amount of ascorbic acid you should use to avoid browning. Ascorbic acid has a bitter taste, so use it sparingly. You may find that the amount of honey will have to be adjusted.

Thawing Frozen Fruits

In frozen fruit, the softening of the plant tissue permits a more rapid rate of spoilage after thawing than in the corresponding fresh fruit. Fruit should be thawed at a low temperature in its original container to best retain its nutritive value, appearance, and flavor. Fruit should be eaten before it is entirely thawed or immediately after it has thawed.

PREPARING FRUITS FOR FREEZING

Apples   To freeze in slices, peel, core, and slice apples. Pack them dry or mix with 2 to 4 tablespoons of honey, mixed with 2 tablespoons of lemon juice or 1 teaspoon rose hip concentrate (to prevent browning).

For sauce, core, but leave skins on (if organically grown) and either grind whole in blender, or cook until soft in open kettle and put through food grinder. Add honey and lemon juice to taste, if desired.

Apricots Skins tend to toughen during freezing. Unless you plan to use the apricots for pies, peel them before freezing. Dip a few of them at a time into boiling water for 15 seconds or until skins loosen. Chill them quickly in ice water and peel. Cut in half and remove pits. Add a few pits to each container for flavor. Trickle honey thinned with warm water over fruit. Add 1 teaspoon rose hip concentrate if desired. Apricots may also be packed in a thin or medium honey syrup or in a pectin pack.

Avocados Choose those that are ripe and perfect. Peel, cut in halves, remove pits. Scoop out the pulp and mash it. Pack and freeze.

Blackberries If organically grown, pick out leaves and debris, but do not wash. Pack dry or trickle small amount of honey over the berries in the container. Seal and shake until well mixed.

Blueberries If organically grown, do not wash. Pick out stems and leaves. If wild, blanch in steam or boiling water for 1 minute to prevent toughening of skins. Pack dry or with small amount of honey trickled over fruit.

Cantaloupe Cut flesh in slices, cubes, or balls. Add honey and lemon juice if desired. The texture of cantaloupe can best be captured if the fruit is served before entirely thawed.

Carambola Wash and slice. The tough rind is not softened by freezing, so it is best used as a garnish. Pack in thin honey syrup.

Cherries, bush Sort, wash, pit, or pack whole. Add honey to taste.

Cherries, sour Wash and chill in ice water before pitting to minimize loss of juice. Stem and pit. Add a small amount of pure honey. Mix and pack.

Cherries, sweet   Wash and chill in ice water before pitting to minimize loss of juice. Add lemon juice or rose hip concentrate to hold color. Light varieties need more lemon juice or rose hip concentrate to retain color. Light varieties may also be packed in a pectin pack to retain their color.

Coconut   Drain out milk. Cut away the hull, but not the smooth inner skin which contains many minerals. Leave in large pieces or grate in blender or meat chopper. Pack dry. It will keep well for one year.

Cranberries   Choose plump, glossy berries. Sort, wash, and drain. Pack dry or make a purée by cooking berries in 1 cup water to each pint of berries until skins burst. Put in blender or through food mill and add honey syrup. Pack and freeze.

For raw relish, grind 1 orange with 1 pound of berries. If they are organically grown, grate with skin on. If they have been sprayed, remove the skins and discard. One cup crushed pineapple may also be added.

Currants   Choose the larger varieties for freezing. Stem and wash. Pack dry or add honey to taste.

Dates   Choose ripe, firm fruit. Wash and remove pits. Pack whole or purée dates in blender or food mill.

Figs   Wash, sort, cut off stems, peel, and leave whole or slice. Cover with a thin syrup. For crushed figs, wash and coarsely grind in blender. Add honey if desired. Figs may also be packed in a pectin pack.

Gooseberries   Sort, remove stems and blossom ends. Pack dry or in a thin or medium honey syrup.

Grapefruit   Peel and remove sections from heavy membrane. Smaller membranes may be left on; they contain important vitamins. Pack dry or add honey to taste.

Grapes   Wash and stem. Leave seedless grapes whole. Cut in half and remove seeds from others. Pack dry or in thin syrup.

Guavas   For purée, remove seedy portion and strain to remove seeds. Sweeten with honey, if desired.

For slices, pare, halve, and slice. Cover with a thin honey syrup.

Lychees   Wash. Leave about ¼ inch stem on fruit. Pack dry.

Mulberries   Wash, if necessary, and stem. Pack dry or in honey syrup.

Oranges   Peel and remove sections from heavy membrane. Pack dry or add honey to taste.

Peaches   Use only fruit that is ripe enough so that skins may be pulled off without blanching. To avoid discoloration, prepare only fruit enough for one container at a time. Wash, skin, pit, and freeze in halves or slices. Add honey mixed with a small amount of lemon juice or rose hip concentrate if desired. Peaches may also be packed in pectin pack.

Pears   Pears retain better appearance and texture when they are canned. If you wish to freeze them, choose ripe, but firm (not hard) fruit. Wash, peel, and remove cores. Prepare only enough fruit at one time to fill a container to avoid unnecessary discoloration. Slice pears directly into a honey syrup mixed with a small amount of lemon juice or rose hip concentrate, or pack in a pectin pack.

Persimmons   Sort, wash, slice, and freeze or press through a food mill for purée. Add 2 tablespoons of lemon juice per pint. Sweeten to taste with honey.

Pineapple   Use only fruit ripened on the plant. Pare, trim, core, and slice or cut in wedges. Pack in own juice or in a thin honey syrup.

Plums and Prunes   If freestone, wash and pit, halve, or quarter. If clingstone, crush slightly, heat just to boiling, cool, and purée in food mill or blender. Add 2 tablespoons lemon juice or 1 teaspoon rose hip concentrate per pint. Sweeten with honey or add honey syrup.

Raspberries   Clean and remove stems. Pack dry or fill containers and trickle in 2 tablespoons of honey per container. Seal and shake to mix.

Rhubarb   Choose crisp, tender red stalks. Early spring rhubarb freezes best. Remove leaves and dis-

	card any wooden ends. Wash and cut into 1-inch pieces. Blanch for 1½ minutes in steam or boiling water and pack dry; or pack fresh and cover with syrup or favorite sauce.
Soursap	Peel and cut lengthwise through the center. Remove and discard seeds. Force through food mill. Sweeten to taste with honey.
Strawberries	Wash and slice, cut in half, or freeze whole. Sweeten to taste with honey or use a thin honey syrup. One-half cup syrup is enough for 1 quart of berries. Strawberries may also be packed dry so long as you are freezing fully ripe berries.

Approximate Yield
of Frozen Fruits from Fresh

Fruit	Fresh, as Purchased or Picked	Frozen
Apples	1 bushel (48 pounds) 1 box (44 pounds) 1¼ to 1½ pounds	32 to 40 pints 29 to 35 pints 1 pint
Apricots	1 bushel (48 pounds) 1 crate (22 pounds) ⅔ to ⅘ pound	60 to 72 pints 28 to 33 pints 1 pint
Berries[1]	1 crate (24 quarts) 1⅓ to 1½ pints	32 to 36 pints 1 pint
Cantaloupes	1 dozen (28 pounds) 1 to 1¼ pounds	22 pints 1 pint
Cherries, sweet or sour	1 bushel (56 pounds) 1¼ to 1½ pounds	36 to 44 pints 1 pint
Cranberries	1 box (25 pounds) 1 peck (8 pounds) ½ pound	50 pints 16 pints 1 pint
Currants	2 quarts (3 pounds) ¾ pound	4 pints 1 pint
Peaches	1 bushel (48 pounds) 1 lug box (20 pounds) 1 to 1½ pounds	32 to 48 pints 13 to 20 pints 1 pint

Fruit	Fresh, as Purchased or Picked	Frozen
Pears	1 bushel (50 pounds)	40 to 50 pints
	1 western box (46 pounds)	37 to 46 pints
	1 to 1¼ pounds	1 pint
Pineapple	5 pounds	4 pints
Plums and Prunes	1 bushel (56 pounds)	38 to 56 pints
	1 crate (20 pounds)	13 to 20 pints
	1 to 1½ pounds	1 pint
Raspberries	1 crate (24 pounds)	24 pints
	1 pint	1 pint
Rhubarb	15 pounds	15 to 22 pints
	⅔ to 1 pound	1 pint
Strawberries	1 crate (24 quarts)	38 pints
	⅔ quart	1 pint

[1] Includes blackberries, blueberries, boysenberries, dewberries, elderberries, gooseberries, huckleberries, loganberries, and youngberries.
U.S. Department of Agriculture

What to Do When the Freezer Stops Running

If, for any unfortunate reason (power failure, freezer breakdown, accidentally pulling the plug), your freezer stops functioning, don't run to it and open the door to check the contents. This is probably the worst thing you can do. First, try to approximate how long the freezer will be out of service. This may mean calling the electric company or the repairman. Do it as soon after you discover the problem as possible. If your freezer will be off for 48 hours or less and it is packed full, you have nothing to worry about. The food will stay frozen, so long as you don't open the door and let warm air enter. A freezer packed less than half full should cause more concern; it may not keep food frozen for more than 24 hours.

If you expect the food to start thawing before the freezer resumes operation, place dry ice (with gloves on—dry ice "burns") inside the freezer, out of direct contact with food. Place pieces of cardboard or wood on top of the food packages and put the dry ice on top of these. Close the freezer door and open only to add more ice. If the dry ice is placed in the freezer soon after it has stopped running, 25 pounds should keep the food frozen 2 to 3 days in a 10-cubic-foot freezer half full and 3 to 4 days in a fully loaded freezer.

If it is necessary to remove some food to make room for the ice, store it in the refrigerator and cook the food as it thaws. Some foods moved to the refrigerator may be frozen again if your freezer starts operating sooner than you expected. Foods that have thawed may be refrozen as long as they show no signs of spoilage. Mark refrozen foods appropriately and use them before you pull other foods from the freezer. Food that has been removed from the freezer and then cooked may be frozen again, so long as it is frozen soon after cooking.

If no dry ice is available, you can always move your frozen food to a commercial locker plant or to a friend's freezer. Pack the food in ice chests or wrap it in thick layers of newspaper to prevent thawing during transportation.

Canning Vegetables and Fruits

Before freezers were around canning was the most popular method of preserving. In many households, especially in rural areas, canning is still the primary method of storing garden produce. More than 40 percent of all the families in the United States can food at home. An obvious advantage of canning is that there is practically no storage problem. You may can until your basement bulges, whereas your freezer space is definitely limited. And you need only invest in a pressure canner and canning jars (unless you plan to use tin cans and a sealer), both of which can be used over and over again, through many harvests.

Sterilizing Food

In simple terms, canning food means sterilizing it and keeping it sterile by sealing it in containers made from either glass or tin. Although commercial canning operations use tin cans almost exclusively, glass canning jars are preferred for home canning because they require no equipment for sealing. Sterilization is achieved by heating the food and the canning container sufficiently to kill all pathogenic and spoilage organisms that may be present in raw food. The duration of the processing time and the temperature at which the food and its container are held during processing depend upon the food being canned.

High-acid foods like pickled vegetables, fruits, and tomatoes are vulnerable only to heat-sensitive organisms. Boiling temperatures are sufficient to sterilize these foods and make them safe for consumption. Low-acid foods, like vegetables, meats, and dairy products, however, are not only susceptible to heat-sensitive organisms, but also to bacteria that can withstand temperatures above the boiling point of water. *Clostridium botulinum* forms a dangerous toxin that causes botulism, and this may be present in low-acid foods even after long boiling. Such foods must be processed at 240°F. to kill any possible traces of *Clostridium botulinum*. Temperatures above the boiling point of water (212°F.) cannot be reached under ordinary conditions. A pressure canner must be used to process low-acid foods so that the necessary high temperature can be obtained.

Sealing the container immediately after heat processing makes it impossible for destructive organisms to invade the food and reinfect it. In addition, sealing creates a vacuum inside the container. This vacuum protects the color and flavor of the product, helps to retain the vitamin content, prevents rancidity due to oxidation, and assists in retarding corrosion of tin cans and corrosion of the closures on glass jars.

You can see why the success of preserving the taste, appearance, flavor, nutritional value, and safety of foods by canning depends upon the complete sterilization of foods and their containers and perfect seals. Extreme care must be taken to follow the canning instructions presented later in this chapter.

The Tomato Controversy   If you've been keeping up with the USDA or your State Extension Service you know that in the last few years there has been some controversy about whether or not all tomatoes are acidic enough to be canned by the traditional boiling-water bath method. Some people in the know claim that, depending upon your soil, climate, and the tomato varieties you grow, you may have tomatoes that are slightly less acidic than are ideal for this type of processing.

If you suspect that your tomatoes are lower in acid than most (and unfortunately there is no real way for a home canner to determine this him- or herself—check with your County Extension Service and see what the seed company says about the particular variety you've grown) or if you just want to play it safe, we suggest that you add one of the following to tomatoes, tomato sauce, or tomato juice to be canned in a boiling-water bath. (These suggestions were supplied by the Kentucky State Extension Service.)

For quarts:	For pints:
4 teaspoons lemon juice	2 teaspoons lemon juice
or	or
2 tablespoons vinegar	1 tablespoon vinegar
or	or
½ teaspoon citric acid	¼ teaspoon citric acid

Choosing and Preparing Your Vegetables and Fruits

There are a few more things to remember when canning. As when freezing, use only fresh food in tip-top condition. Sort foods for size and maturity so that they will heat up evenly and pack well. Don't bring hot food into contact with copper, iron, or chipped enamelware. If possible, use soft water for syrups. Some foods darken or develop a gray tinge during the canning process. This is a chemical reaction between the food and the minerals in hard water or in metal utensils. Although such a discoloration does not mean that the food is unfit to eat, it may make canned food unattractive.

For the best results, food should be canned quickly, preferably on the day it is harvested. There should be no time lag between steps, so have all equipment clean and at hand before you begin. Food spoilage called "flat sour" can result if vegetables, particularly starchy ones, like corn, have stood too long between steps. The canned food may look all right and smell fine, but it has an unpleasant, sour taste. It is not fit to eat, even though it is not poisonous.

Checking Seals

After processing and cooling canned foods, check their seals. Press down on the center of the lids on glass jars. If the lids do not "give" when you press on them, the jar is sealed properly. Check the tin cans by examining all seams and seals. Properly sealed cans should have flat, not bulging, ends, and the seams should be smooth with no buckling. If you suspect a container of having a faulty seal, don't take any chances. Either discard the food or open the container and process the food over again for the required time.

Labeling Containers

After processing and cooling the jars or cans, the type of food and the canning date should be marked on the top of the containers. Use the foods in the order in which they were canned. If you have canned different varieties of the same food, the variety of the food (for example: Peaches—Elberta, or Peaches—Hale) should be marked on the label so that you can compare different varieties and determine which ones best retain their taste, flavor, and appearance after canning and storage.

Storing Canned Foods

Once canned, fruits and vegetables should be stored in a cool, dry place for best keeping. The higher the temperature of the storage area, the more chance of vitamin loss in the canned product. The U.S.

Department of Agriculture tells us in the HANDBOOK OF AGRICULTURE (1959) that canned fruits and vegetables will lose insignificant amounts of vitamin C when stored at 65°F. Losses are about 2 to 7 percent after 4 months and increase slowly to about 10 percent after 1 year's time. When canned fruits and vegetables are stored at 80°F., however, 15 percent of the vitamin C value can be lost after 4 months, 20 percent after 8 months, and up to 25 percent after 12 months.

Don't Discard the Liquid

The liquid in canned fruits and vegetables is an important source of food value. If you discard it, you're throwing out a good part of the vitamins and minerals found in the can or jar. Normally, fruit and vegetable solids make up about two-thirds of the total contents of the container; the rest is water. Soon after canning, the water-soluble vitamins and minerals distribute themselves evenly throughout the solids and liquid. It follows, then, that about one-third of the water-soluble nutrients are in the liquid portion.

Spoiled Food

If, when opening canned food for use, you suspect that it has spoiled, do not test it by tasting it. Some spoilage bacteria, like those which cause botulism, are so toxic that a taste may be fatal. Boil the suspected food rapidly for a few minutes. If you notice an unusual and unappetizing odor developing, you can be certain that the food is not safe to eat. Burn the food or bury it deep enough so that no animal can uncover it and eat it.

Canners

Boiling-Water Canner   High-acid foods, which include all fruits, tomatoes, and pickled vegetables, should be processed in a boiling-water canner. Any large vessel will do for a boiling-water canner so long as it meets these requirements: It should be deep enough to have at least 1 inch of water over the top of the jars and an inch or two extra space for boiling. It should have a snug-fitting cover. And there should be a rack to keep the jars from touching the bottom of the pot. If your steam-pressure canner is deep enough, you can use it for boiling-water processing. Set the cover in place without fastening it. Be sure to have the petcock open wide or remove the weighted gauge so that the steam escapes and no pressure is built up.

Steam-Pressure Canner   All vegetables, except tomatoes, are low-acid foods and must be processed in a steam-pressure canner because they require temperatures higher than that of boiling water for sterilization. Here you can't improvise; you must have a canner especially made for steam-pressure processing, and it must be in good working condition. The safety valve and petcock opening on the canner should be checked each time the steam-pressure canner is used to make sure that they are not stopped up with food or dirt. They can be cleaned by drawing a string or piece of cloth through them. Check the pressure gauge each year for accuracy. Instructions for checking should come with the canner. A weighted gauge needs only to be thoroughly cleaned. Make sure the canner is clean before using it.

Containers

Glass Jars To play it safe, use only jars made especially for canning. Some people, however, do use other jars—like those in which mayonnaise, peanut butter, and jelly are sold—for canning. If you want to use such jars, use them *only for boiling-water processing* and only if the jars pass the following test:

Place jars for testing in a pot of water, making sure that water fills the jars and covers them. Heat the water to exactly 190°F. (check with a dairy or candy thermometer). Then remove the jars from the pot *without emptying the water out of them* and place them in another pot of exactly 70°F. water, making sure the water is covering the jars. Leave the jars in this water for 5 minutes. Then remove them and check for cracks around the bottom and especially along the seams. If the jars are still in perfect shape, they're safe to use. Jars need only be tested once and, if they pass, can be used over and over.

No matter what jars you use, be sure they are in good condition, clean, and sterile before packing food in them.

Closures There are three types of closures for glass jars. Be sure to follow the sealing directions that come with each type of closure. Some general suggestions are given in the following paragraphs.

Mason Top. If porcelain lining is cracked, broken, or loose, or if there is even a slight dent at the seal edge, discard the cover. Opening these jars by thrusting a knife blade into the rubber and prying ruins many good covers. Each time you use a jar, have a new rubber ring of the right size.

Three-piece Cap. This fits a deep-threaded jar with or without a shoulder. The metal band holds the cap in place during processing and cooling. Remove it when the contents of the jar are cold, usually after 24 hours. Use a new rubber ring each time.

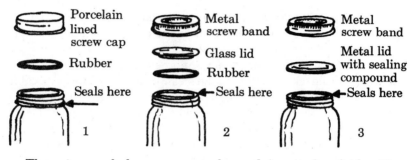

Three types of closures commonly used in canning foods: (1) mason top; (2) three-piece cap; and (3) two-piece cap.

Two-piece Cap. Use the metal lid only once. The metal band is needed only during processing and cooling. Do not screw it farther after taking the jar from the canner. Remove the band after the contents of the jar are cold, usually after 24 hours.

Lids. When the first edition of STOCKING UP came out in 1973, there were only three companies producing lids for the popular two-piece cap. Now, in 1977, after the canning lid shortage of a few years ago, there are at least 15 companies producing lids. Most of these new lids are fine and work well, but after testing in our Fitness House experimental kitchen, we did discover some types of lids that did not hold the seal as well as they should.

We suggest that when buying lids you check the inner sealing ring to see that it is uniform in thickness and covers the entire area of the lid that will come into contact with the jar. One thing to look for in a good lid is a series of concentric circles, at different levels. This form of construction allows the vacuum to pull down the middle part of the lid, forcing the outer ring into good contact with the jar. Lids that do not have these rings tend to have the entire lid bow under pressure, still giving a seal, but one that is much more fragile, as less of the lid is in contact with the jar. Another point to look for in a good lid is the quality of the coating on the inside of the lid. Lids that do not have a thick, uniform coating may corrode much more quickly than quality lids. One last feature we found desirable is a small dimple in the middle of the lid. These dimples give the lid a louder clicking sound when a proper seal is made, and make it easier for you to see that the lid is pulled down, the sign of a good seal.

Paradoxically, when buying canning lids you may well not get what you pay for. Large established companies are able to sell quality lids at a low price due to the size of their production. Some of the smaller companies therefore sell less quality, for more money.

If you are in doubt about the quality of the lid, look to see who makes it. If the company lists their name and mailing address on the package you can write them and ask for test results. Ask your local extension service's home economist if there are any tests results available on the lids. No government agency is responsible for the quality of lids. Several government agencies test the sealant used in lids for safety, but no one agency is responsible for seeing that lids work well.

If you buy a new brand of lid and you aren't 100 percent sure of its quality, make this simple test to see how (or if) it works:

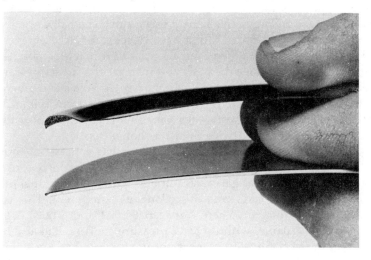

When shopping for lids pick the kind with a small dimple in the middle that tells you when a proper seal is made.

Follow the directions for using the lid, and can a few test quarts of tap water. Keep the jars in the boiling-water bath 10 minutes. Open one jar after it cools to see if the indentation of the jar mouth is centered in the lid's ring of sealing compound. Let the other jars set 2 or 3 weeks to see if they become unsealed. When you open a properly sealed jar there should be a slight whooshing sound as air rushes into the vacuum created in the jar. If you do not hear the vacuum breaking, or the lid is bulged upward instead of downward, you do not have a good seal. Remember to follow the lid manufacturer's instructions exactly when using a new style of lid.

Community Canning Centers

During the World War II years there were literally millions of "Victory Gardens." From these backyard vegetable plots more than

3,800 community canning centers developed. The theory behind the centers was similar to an old-fashioned barn raising, a quilting bee, or a cornhusking party: when people work together using the proper equipment, a lot of work gets done quickly and fairly easily. With backyard gardens becoming popular again as a means of producing food, community canning centers began to be revived a few years ago.

Canning centers are modern units designed to preserve foods using the same methods you use at home, only they enable you to do large amounts of food at a time, greatly reducing the total time and labor involved in large-scale food preservation. The Ball Corporation is one of the two companies that have developed such centers. The Ball people have redesigned their original unit that was popular in the 1940s to both reduce the costs and increase the capacity. Current Ball units (which number over 100 in the United States and several in 10 foreign countries) have the capacity to process from 200 to 2,400 quarts a day. Although estimates vary, the consensus holds that a canning center is about twice as fast as home canning.

Ball's basic community canning center has 4 pressure cookers with a capacity of 16 quarts or 24 pints each, an atmosphere cooker, a blancher/sterilizer, a 10-gallon steam-jacketed kettle, a pulper/juicer, cooling tank, table carts, and miscellaneous equipment. This is the basic unit, costing anywhere between $4,500 and $6,000. Larger models are available, with up to 12 pressure cookers. Ball also has a large variety of options, from meat-cutting saws and sausage stuffers to temperature-monitoring devices to record exact temperatures at which food is processed. (Without such a temperature-recording device, low-acid foods cannot be processed for resale.)

It is usually necessary to make an appointment to use the facilities. The disadvantage of lugging your produce over to the center may be more than offset by the opportunity to use first-rate equipment under ideal conditions, and, often, to have a good sociable time, and possibly get some good advice while processing your food.

To cover the yearly operating costs, 500 families would need to spend about $13 per year at the center. Most canning centers charge 10¢ per quart and 7¢ per pint for all items processed. Finding a means of finance is one of the main topics of discussion among people trying to establish a center. In most cases, it is sponsored by a cooperative, a nonprofit organization, a school, or other neighborhood group. The sponsoring group then has to find a source of financial aid. Usually funds to help purchase the unit and get the center established come from local, state, or federal grants. Church, fraternal, and other local organizations all provide money, as do private industries. In most cases a canning center will serve from 50 to 500 families.

As food costs continue to rise, more and more government officials are looking at food preservation centers as a self-help program to help people secure good food at low prices. In some pilot centers, food stamps are used to buy canning supplies and bulk purchases of food, and canning center use is free. Estimates indicate this more than doubles a person's food dollar.

The other manufacturer of canning units is the Dixie Canner Equipment Company. They make what is actually a small-scale commercial cannery. These units cost more than the Ball units, anywhere from $25,000 to $45,000 and up. This unit uses tin cans instead of glass containers.

The Dixie setup is capable of a much higher volume than the Ball system and has monitoring equipment to make resale of processed foods possible. The main use of the Dixie unit is in schools, with the cost being paid by both the county and the school district, and the vocational agriculture teachers instructing and running the centers. The centers are then opened to all community residents for use. The states of Virginia and Georgia both have extensive school programs with the canning centers, their use dating back to the early 1930s.

The other main use of the Dixie unit is for privately owned custom canneries, or cooperatively owned canneries. These operations are usually owned either by growers, consumers, or community groups. Their goal is to have more local produce sold and processed in their area instead of selling to brokers, having the food processed in large factories, and resold in the area at a high price. Most attempts at a regional food chain using small canneries are in their infancy, and it will take a few more years before the economics of such a venture are known.

Although canning centers are formed for different reasons, the basic goal is always the same. By setting up a small food-processing center, you are able to can a great quantity of food that you either grow yourself or buy directly from a grower, without a large personal investment in canning equipment.

For additional information on canning centers contact:

Food Preservation Program
　Ball Corporation　　　　or
　345 South High Street
　Muncie, Indiana 47302

Dixie Canner Equipment Company
　P.O. Box 348
　Athens, Georgia 30601

Raw Packs and Hot Packs for Glass Jars

Most fruits and vegetables may be packed either by the raw-pack or hot-pack method. More food can be packed into the jars when the hot-pack is used because the food has already shrunk slightly during heating. Hot-packing is best used for foods that tend to discolor during canning. However, foods may lose some of their food value when they are hot-packed because of the additional heating. Both methods are described for individual fruits and vegetables in the charts that follow this section.

Canning Fruits

How to Prevent Fruits from Discoloring

Fruits have a tendency to darken in the canning process. To keep apricots, apples, nectarines, peaches, and pears from darkening, they may be placed in a solution of 2 tablespoons lemon juice or vinegar and 2 tablespoons salt to 1 gallon of water. The fruits should be dropped in this solution as soon as they are washed and sliced. Let them soak for 20 minutes, then rinse them in cold water before packing.

Making Honey Syrup

Most people who can fruits pack them in a sweet syrup. However, fruit may be canned without sweetening. It may be packed in juice made from the fruit itself (by blenderizing or extracting) or even purchased apple, orange, etc., juice, or in water. Sugar helps canned fruit hold its shape, color, and flavor, but it is not needed to prevent spoilage. Unsweetened fruit should be processed just as sweetened fruit.

For your health's sake, if you wish to use a sweet syrup with your canned fruit, make it with honey, not with sugar or corn syrup. A very satisfactory syrup can be made by blending 2 cups honey with 4 cups of very hot water. Just as when freezing fruit with honey, choose a light-flavored honey for making your syrup so that the honey taste will not overpower the flavor of the fruit. To prevent the fruit from darkening, pack fruit in the hot honey syrup as soon as it is peeled and sliced.

Processing Fruits, Tomatoes, and Pickled Vegetables in Boiling-Water Bath

1. Fill boiling-water canner over half full, deep enough to cover containers. Turn on the heat.
2. If you are using jars, wash and rinse them and put them into

hot water until needed. Pour boiling water over the lids and set them aside. If you are using rubber rings, do not pour boiling water over them.

[If you are using tin cans and lids, have new ones ready, and have sealer at hand.]

3. Prepare syrup, if it is to be used.

4. Prepare fruit (or high-acid vegetables) for canning.

5. If you are using jars, pack fruit, either raw or hot, into jars and add hot syrup or water to fill jars no more than ½ inch from the top. Remove the air from the jars by running a spatula or knife along the side, pressing fruit as you do so. Wipe the jars and screw the tops on tightly.

[If you are using tin cans, fill the cans with raw or hot fruit and cover the fruit with syrup or water. To exhaust the cans of air, the food inside must be heated to at least 170°F. Place the open, filled cans in a large pot with water about 2 inches below the tops of the cans. Cover the kettle and bring the water back to boiling. Boil until the food reaches 170°F. (about 10 minutes). To be sure that the food is heated enough, test the temperature with a thermometer, placing the bulb in the center of one of the cans. Remove the cans from the water one at a time. Replace any liquid spilled from the cans by filling them with boiling syrup or water. Place a clean lid on each can and seal at once.]

6. Place the closed jars or sealed cans upright in the canner. Water should be 2 inches over the tops of the containers. Add boiling water, if needed, being careful not to pour water directly on the containers. Allow 2 inches extra space at top of canner for boiling.

7. Put the lid on the canner and bring the water to boiling.

8. Begin to count time as soon as the water starts to boil and process for time recommended for each specific food in the timetable charts. (The processing times given on the chart are for altitudes less than 1,000 feet above sea level. For high-altitude areas, increase the processing time 2 minutes for each 1,000 feet above sea level. For example, if you live 3,000 feet above sea level, process 6 minutes longer than recommended time.) Leave the lid on the canner, but check periodically to make sure the water is boiling gently and steadily. Add more boiling water as needed to keep containers covered.

9. As soon as processing time is up, remove the containers from the canner.

10. The common two-piece cap is self-sealing, but if you're using another cap that doesn't self-seal, complete the seals as soon as you take them out of the canner. You can tell if the two-piece lid is sealed by pressing down on the center of the lid. If it is down already or stays down when pressed, the seal is good. If it fails to stay down, reprocess the jars, place in refrigerator and use within the next few days, or freeze.

11. Set jars right side up to cool, making sure that they are far enough apart from one another so that air can circulate freely around them. For better circulation, place them on racks. Do not put them in a cold place or cover them while they are cooling. Keep them out of drafts.

[Tin cans should be cooled quickly in cold water, using as many changes of water as is necessary. Remove cans from the water while they are slightly warm so that they will dry in the air.]

Canning Vegetables

Preparing Your Vegetables

It is not necessary to precook or blanch vegetables intended for canning as you do for freezing and drying, since the enzymes that would otherwise break down the food are killed by the heat in the canning process.

Vegetables may be packed salted or unsalted. The small amount of salt used in canning does not prevent spoilage; it is only used as a seasoning. Can unsalted vegetables just as salted ones.

Processing Vegetables in the Steam-Pressure Canner

1. Put 2 or 3 inches of hot water in the bottom of the canner.
2. If you are using glass jars, wash and rinse them and put them into hot water until needed. Pour boiling water over the lids and set them aside.

   [If you are using tin cans and lids, have new ones ready and have a sealer at hand.]
3. Prepare vegetables for canning. (See timetable charts elsewhere in this chapter for directions for specific vegetables.)
4. If you are using jars, pack vegetables, either raw or hot, into jars, leaving a ½ to 1-inch headspace (see specific directions in the timetable charts that follow). Pour in enough boiling

Whereas you can fill tin cans right to the rim, fill glass canning jars with food and boiling water only to within ½ to 1 inch of the top. This headspace allows for expansion of food and bubbling of the liquid during processing.

water to cover vegetables. Wipe tops and threads of jars and screw tops on tightly.

[If you are using tin cans, fill the cans with raw or hot vegetables and cover them with water. To exhaust the air in the cans the food must be heated to at least 170°F. Place the open, filled cans in a large pot with boiling water about 2 inches below the tops of the cans. Cover the pot and bring the water back to boiling. Boil until the food reaches 170°F. (about 10 minutes). To be sure that the food is heated enough, test the temperature with a thermometer, placing the bulb in the center of one of the cans. Remove the cans from the water one at a time. Replace any liquid spilled from the cans by filling them with boiling water. Place a clean lid on each and seal at once.]

5. Set the closed jars or sealed cans on a rack in the canner so that steam can circulate around them freely. Cans may be staggered without a rack between layers, but jars should have a metal rack between layers.

6. Fasten the cover of the canner securely so that no steam escapes except at the open petcock or weighted gauge opening.

7. Allow steam to escape from the opening for 10 minutes so all the air is driven out of the canner. Then close the petcock or put on the weighted gauge and let the pressure rise to 10

(con't. on page 94)

Timetable for Processing Fruits, Tomatoes, and Pickled Vegetables in Boiling-Water Bath

Product	Glass Jars		Tin Cans	
	Pints	Quarts	#2	#2½
	Min.	Min.	Min.	Min.
Raw pack or hot pack foods following directions. Put filled jars into canner containing hot or boiling water. For raw pack have water in canner hot but not boiling; for hot packs have water boiling. Add boiling water to bring water 1 or 2 inches over tops of jars, but don't pour boiling water directly on glass jars. Put on cover of canner. Count processing time when water in canner comes to a rolling boil. (See step-by-step directions elsewhere in chapter.)				
Apples HOT PACK Pare, core, cut into pieces. To keep from darkening, place in water containing 2 tablespoons each of salt and vinegar per gallon. Drain, then boil 5 minutes in thin syrup, juice, or water. Pack apples in jars to ½ inch of top. Cover with hot syrup or water, leaving ½ inch at top.	15	20	10*	10*
Applesauce HOT PACK Make applesauce according to your recipe or one that appears later in this chapter; pack hot to ¼ inch of top.	10	10	10*	10*
Apricots See Peaches				
Beets, pickled (For other pickled products, see the Pickles and Relishes chapter.) HOT PACK Cut off beet tops, leaving 1 inch of stem and root. Wash beets, cover with boiling water, and cook until tender. Remove skins and slice. For pickling syrup use 2 cups vinegar to 1 cup honey. Heat to boiling. Pack beets in jars to ½ inch of top. Add ½ teaspoon salt to pints, 1 teaspoon to quarts. Cover with boiling syrup, leaving ½ inch at top.	30	30	—	—

Berries, except Strawberries	RAW PACK This is preferred for raspberries, blackberries, boysenberries, dewberries, and loganberries. Wash berries and drain. Fill jars to ½ inch of top, shaking berries down gently. Cover with juice or syrup (thin or medium is recommended) leaving ½ inch at top.	10	15	15†	20†
	HOT PACK This is preferred for blueberries, cranberries, currants, elderberries, gooseberries, and huckleberries. Wash berries and drain well. Add ¼ cup honey to each quart fruit. Cover pan and bring to boil. Pack berries to ½ inch of top.	10	15	15†	20†
Cherries	RAW PACK Wash; remove pits in sour or pie cherries. Sweet cherries need not be pitted, but do prick their skins with a pin or tip of a knife so that they don't burst during processing. Fill jars to ½ inch of top, shaking cherries down gently. Cover with boiling juice or syrup (thin or medium), leaving ½ inch at top.	20	25	20†	25†
	HOT PACK Wash; remove pits if desired (see raw pack). Add ¼ cup honey to each quart of fruit. Add a little water to unpitted cherries. Cover pan and bring to a boil. Pack hot to ½ inch of top.	10	15	15†	20†
Figs	HOT PACK Use ripe figs only. Wash, but do not remove skins or stems. Cover with boiling water and simmer for 5 minutes. Pack hot figs and add boiling syrup or juice to ½ inch of top. Add 2 teaspoons lemon juice to pints and 4 teaspoons to quarts.	85	90	85†	90†
Fruit purée	HOT PACK Use sound, ripe fruit. Wash; remove pits if desired. Cut large fruit in pieces. Simmer until soft, add a little water if needed. Put through strainer or food mill. Add honey to taste. Heat to simmering and pack to ¼ inch of top.	10	10	10†	10†

Timetable for Processing Fruits, Tomatoes, and Pickled Vegetables in Boiling-Water Bath

Product	Glass Jars		Tin Cans	
	Pints	Quarts	#2	#2½
	Min.	Min.	Min.	Min.
Raw pack or hot pack foods following directions. Put filled jars into canner containing hot or boiling water. For raw pack have water in canner hot but not boiling; for hot packs have water boiling. Add boiling water to bring water 1 or 2 inches over tops of jars, but don't pour boiling water directly on glass jars. Put on cover of canner. Count processing time when water in canner comes to a rolling boil. (See step-by-step directions elsewhere in chapter.)				
Grapes				
RAW PACK Wash and stem seedless grapes. Pack tightly, but be careful not to crush. Add boiling juice or medium syrup, leaving ½ inch at top.	15	20	20*	25*
HOT PACK Prepare as for raw pack. Bring to a boil in medium syrup or juice. Pack without crushing, add syrup or juice to ½ inch of top.	15	20	20*	25*
Grapefruit, Oranges, Nectarines, Tangerines				
RAW PACK Remove fruit segments, peeling away the white membrane that could develop a bitter taste in the canning process. Seed carefully. Pack fruit in jars and cover with boiling juice or a thin syrup. Leave ½ inch at top.	10	10	—	—
Juices				
See the Juicing Your Harvest chapter				
Mixed fruit				
HOT PACK Prepare pineapples, pears, and peaches by peeling and cutting into uniformly sized pieces. Add slightly underripe seedless grapes if you wish. Cook in fruit juice or a syrup for 3 to 5 minutes, until slightly limp. Pack hot into jars and cover with hot syrup to within ½ inch of top.	20	25	15*	20*

Food	Instructions				
Peaches or Apricots	**RAW PACK** — Wash peaches or apricots and remove skins. Remove pits. To keep from darkening place in solution (same as apples). Drain, pack fruit in jars to ½ inch of top. Cover with boiling juice or syrup (light or medium), leaving ½ inch at top.	25	30	30*	35*
	HOT PACK — Prepare fruit as for raw pack. Heat fruit through in hot juice or syrup. If fruit is very juicy you may heat it with ½ cup of honey to 1 quart of raw fruit adding no liquid. Pack fruit to ½ inch of top.	20	25	15*	20*
Pears	Peel, cut in halves, and core. Follow directions for peaches either raw pack or hot pack using same timetables.				
Pineapple	**RAW PACK** — Peel, core, and cut into uniformly sized chunks. Pack tightly and cover with boiling syrup or fruit juice to within ½ inch of top.	30	30	—	—
	HOT PACK — Prepare as for raw pack. Simmer fruit in a syrup or fruit juice for about 10 minutes. Pack hot and cover with syrup to within ½ inch of top.	20	20	—	—
Plums (and Italian prunes)	**RAW PACK** — Wash. To can whole, prick skins. Freestone varieties may be halved and pitted. Pack fruit in jars to ½ inch of top. Cover with boiling juice or syrup, leaving ½ inch at top.	20	25	15†	20†
	HOT PACK — Prepare as for raw pack. Heat to boiling in syrup or juice. If fruit is very juicy, you may heat it with honey, adding no liquid. Pack hot fruit to ½ inch of top. Cover with boiling juice or syrup, leaving ½ inch at top.	20	25	15†	20†
Rhubarb	**HOT PACK** — Wash and cut into ½-inch pieces. Add ¼ cup honey to each quart rhubarb and let stand to draw out juice. Bring to boiling. Pack hot to ½ inch of top.	10	10	10†	10†

Timetable for Processing Fruits, Tomatoes, and Pickled Vegetables in Boiling-Water Bath

Product	Glass Jars		Tin Cans	
	Pints Min.	Quarts Min.	#2 Min.	#2½ Min.
Product Raw pack or hot pack foods following directions. Put filled jars into canner containing hot or boiling water. For raw pack have water in canner hot but not boiling; for hot packs have water boiling. Add boiling water to bring water 1 or 2 inches over tops of jars, but don't pour boiling water directly on glass jars. Put on cover of canner. Count processing time when water in canner comes to a rolling boil. (See step-by-step directions elsewhere in chapter.)				
Sauerkraut (For other ways to keep sauerkraut, see the Pickles and Relishes chapter.) HOT PACK Heat well-fermented sauerkraut to simmering (185° to 210°F.). Pack hot kraut to ½ inch of top. Cover with hot juice, leaving ½ inch at top.	15	20	20*	25*
Strawberries Strawberries are more difficult to can than other berries, and we recommend freezing them or making jams and preserves instead, when possible. HOT PACK To can, wash and then hull berries. Using ¼ to ½ cup honey for each quart of berries, spread berries one layer deep in pans and drizzle honey over them. Cover and let stand at room temperature for 2 to 4 hours. Then place in saucepan and simmer in their own juice for 5 minutes, stirring to prevent sticking. Pack without crushing and cover with extra boiling thin syrup if berries didn't produce enough juice of their own. Leave ½ inch at top.	10	15	15†	20†

Tomatoes (For sauces and combination foods, see individual recipes at end of this chapter.)	RAW PACK Use only perfect, ripe tomatoes. Scald just long enough to loosen skins; plunge into cold water. Drain, peel, and core. Leave tomatoes whole or cut in halves or quarters. Pack tomatoes, pressing gently to fill spaces. Cover with water or juice to ½ inch of top. Add 4 teaspoons lemon juice, 2 tablespoons vinegar, or ½ teaspoon citric acid to quarts, and half that amount of any one to pints if desired (see page 68).	35	45	55*	55*
	HOT PACK Quarter peeled tomatoes. Bring to boil and pack to ½ inch of top, adding extra boiling water or juice if tomatoes have not made enough juice of their own to cover. Add salt as for raw packed tomatoes.	10	15	45*	45*

* Use plain tin for apples, apricots, light grapes, peaches, pears, sauerkraut, and tomatoes.
† Use R enamel cans for berries, cherries, dark grapes, plums, and rhubarb.

Timetable for Processing Low-Acid Vegetables

Product		Use 10 Pounds Pressure (unless you live in a high-altitude area)			
		Glass Jars		Tin Cans	
		Pints Min.	Quarts Min.	#2 Min.	#2½ Min.
Product	Work rapidly. Raw pack or hot pack foods following directions, adding if desired ½ teaspoon salt for pints and 1 teaspoon salt for quarts. Place jars on rack in steam-pressure canner containing 2 to 3 inches of boiling water. Fasten canner cover securely. Let steam escape 10 minutes or more before closing petcock. (See step-by-step directions and special instructions for high-altitude areas elsewhere in this chapter.)				
Artichokes	HOT PACK Trim and wash artichokes. Be sure that you trim enough so that the chokes can fit into a wide-mouthed jar. Cook for 5 minutes in a solution of ¾ cup vinegar in 1 gallon water. Discard the solution and pack chokes to ½ inch of top. Cover with a brine made from mixing ¾ cup lemon juice and 3 tablespoons salt per gallon of water, leaving the ½-inch headspace. (To make pulling the chokes out of the jar easier and prevent them from falling apart when you do, tie a string firmly around the petals.)	25	30	—	—
Asparagus	RAW PACK Wash asparagus; trim off scales and tough ends and wash again. Cut in 1-inch pieces. Pack asparagus as tightly as possible without crushing to ½ inch of top.	25	30	20*	20*
	HOT PACK Prepare as for raw pack; then cover with boiling water. Boil 2 or 3 minutes. Pack asparagus loosely to ½ inch of top. Cover with boiling water, leaving ½ inch at top.	25	30	20*	20*
Beans, dry, with tomato or molasses sauce	HOT PACK Sort and wash dry beans. Cover with boiling water; boil 2 minutes, remove from heat, and let soak 1 hour. Heat to boiling and drain, saving liquid for sauce. Fill jars ¾ full with hot beans. Add small piece of salt pork, ham, or bacon if you wish. Fill to ½ inch of top with hot tomato or molasses sauce.	65	75	65*	75*

Beans, fresh, lima	RAW PACK Shell and wash beans. Pack the small type loosely to 1 inch of top of jar for pints and 1½ inches for quarts; for large beans fill to ¾ inch of top for pints and 1¼ inches for quarts. Cover with boiling water.	50	50	40*	40*
	HOT PACK Shell the beans, then cover with boiling water and bring to a boil. Pack beans loosely in jar to 1 inch of top. Cover with boiling water, leaving 1 inch at top.	40	50	40*	40*
Beans, snap or green	RAW PACK Wash beans. Trim ends and cut into 1-inch pieces. Pack tightly in jars to ½ inch of top. Cover with boiling water, leaving ½ inch at top.	20	25	25*	30*
	HOT PACK Prepare as for raw pack beans. Then cover with boiling water and boil 5 minutes. Pack beans in jars loosely to ½ inch of top. Cover with boiling-hot cooking liquid and water, leaving ½ inch at top.	20	25	25*	30*
Beets (For pickled beets see Boiling-Water Bath Processing Timetable.)	HOT PACK Sort beets for size. Cut off tops, leaving a 1-inch stem and root, and wash. Boil until skins slip easily. Skin, trim, cut, and pack into jars to ½ inch of top. Cover with boiling water, leaving ½ inch at top.	30	35	30‡	30‡
Broccoli and Brussels sprouts	*Not recommended for canning because the processing intensifies the strong flavor and discolors the vegetable. Much better frozen or pickled.* HOT PACK Cut off woody, tough stems and old leaves and yellowing blossoms. Soak in cold, salted water (about 1 tablespoon salt to each quart water) for 10 to 15 minutes to drive out clinging bugs. Rinse well and pick over. Cut into 2-inch pieces. Cover cut vegetables with boiling water and boil 3 minutes. Drain, reserving liquid. Pack tightly and cover with boiling liquid, leaving 1 inch at top.	25	30	30†	30†

Timetable for Processing Low-Acid Vegetables

Product		Use 10 Pounds Pressure (unless you live in a high-altitude area)			
		Glass Jars		Tin Cans	
		Pints Min.	Quarts Min.	#2 Min.	#2½ Min.
Product	Work rapidly. Raw pack or hot pack foods following directions, adding if desired ½ teaspoon salt for pints and 1 teaspoon for quarts. Place jars on rack in steam-pressure canner containing 2 to 3 inches of boiling water. Fasten canner cover securely. Let steam escape 10 minutes or more before closing petcock. (See step-by-step directions and special instructions for high-altitude areas elsewhere in this chapter.)				
Cabbage	*Not recommended for canning, except for sauerkraut, which is described in the Boiling-Water Bath Processing Timetable. Fresh cabbage is much better kept in cold storage.* Clean, cut up into small wedges, and process as for broccoli and Brussels sprouts.	25	30	40†	40†
Carrots	RAW PACK Wash and scrape carrots. Slice, dice, or leave whole. Pack tightly in jars to 1 inch of top. Cover with boiling water.	25	30	20*	25*
	HOT PACK Prepare as for raw pack, then cover with boiling water and bring to boil. Pack carrots in jars to ½ inch of top. Cover with boiling-hot cooking liquid and water, leaving ½ inch at top.	25	30	20*	25*
Cauliflower	*Not recommended for canning. Much better frozen.* Prepare and process like broccoli and Brussels sprouts.				
Celery	HOT PACK Wash and trim off tough leaves and woody bottoms. Cut into 1-inch pieces. Cover with boiling water and boil 3 minutes. Drain, reserving liquid. Pack jars and cover with hot liquid, leaving 1 inch at top.	30	35	30*	30*

	Directions				
Corn—cream style	RAW PACK Husk corn and remove silk. Wash. Cut corn from cob at about center of kernel and scrape cobs. Pack corn loosely in pint jars to 1 inch of top. Cover with boiling water.	95	—	105†	—
	HOT PACK Prepare as for raw pack. Add 1 pint boiling water to each quart of corn. Heat to boiling. Pack hot corn to 1 inch of top.	85	—§	105†	—
Corn—whole kernel	RAW PACK Husk corn and remove silk. Wash. Cut from cob at about 2/3 the depth of kernel. Pack corn loosely to 1 inch of top with mixture of corn and liquid.	55	—§	60†	—
Eggplant	*Not a particularly good canned or frozen food by itself. Eggplant is better pickled or frozen by itself. Eggplant is better in a casserole.* HOT PACK Wash and peel, then slice or cube. To draw out the bitter juice, line a colander with eggplant, sprinkle with salt, put over that another layer of eggplant, then salt, and so on. Let stand 1 hour. Then press the eggplant against the sides of the colander before taking out and rinsing off well. Boil in fresh water for 5 minutes. Drain, reserving liquid. Pack into jars and cover with hot liquid, leaving 1 inch at top.	30	40	35*	40*
Mushrooms	HOT PACK Select tender, young mushrooms. Discard any that have opened. Wash thoroughly and trim off tough stalks. Cut in slices or leave small mushroom caps whole. Steam mushrooms for 4 minutes. Pack into hot jars, leaving 1 inch at top. You may add 1/2 teaspoon salt. Cover mushrooms with boiling water.	30	—	35	—

Timetable for Processing Low-Acid Vegetables

Product		Use 10 Pounds Pressure (unless you live in a high-altitude area)			
		Glass Jars		Tin Cans	
		Pints Min.	Quarts Min.	#2 Min.	#2½ Min.
Product	Work rapidly. Raw pack or hot pack foods following directions, adding if desired ½ teaspoon salt for pints and 1 teaspoon for quarts. Place jars on rack in steam-pressure canner containing 2 to 3 inches of boiling water. Fasten canner cover securely. Let steam escape 10 minutes or more before closing petcock. (See step-by-step directions and special instructions for high-altitude areas elsewhere in this chapter.)				
Okra	HOT PACK Choose young, tender pods only. Wash and trim stems. Leave whole or cut into 1-inch slices. Cover with boiling water and boil 1 minute. Drain, reserving liquid. Pack into jars and cover with hot liquid, leaving 1 inch at top.	20	40	25*	35*
Onions (small white)	HOT PACK Choose onions of uniform size, about 1 inch in diameter. Peel, trim off roots and stalks and wash if necessary. (If you push the hole in the onion end downward through the middle with a sharp object like a clean finishing nail, the centers will cook with less chance of shucking off outer layers.) Cover with boiling water and cook gently for 5 minutes. Pack hot onions loosely in jar and cover with the boiling liquid to within ½ inch of the top.	25	30	—	—
Peas, green	RAW PACK Shell and wash peas. Pack peas loosely in jars to 1 inch of top. Cover with boiling water, leaving 1 inch at top.	40	40	30*	35*
	HOT PACK Prepare as for raw pack. Cover with boiling water and bring to a boil. Pack peas loosely in jars to 1 inch of top. Cover with boiling water, leaving 1 inch at top.	40	40	30*	35*

		(at 5 pounds pressure)			
Peppers, bell	HOT PACK Remove stem, core, and remove seeds and inner white membrane. Remove skins by first plunging them in boiling water for a few minutes, then running them under cold water, and finally taking off the now split skins with a sharp knife or potato peeler. Slice peppers or flatten whole halves and pack carefully in layers. Cover with boiling water to within ½ inch of the top. You can add ½ tablespoon of lemon juice or 1 tablespoon of vinegar per pint if you wish. PROCESS AT 5 POUNDS PRESSURE, as a higher pressure injures both flavor and texture.	50	60	—	—
Peppers, hot	Hot peppers are usually pickled before they are canned. See the chapter on pickles and relishes.				
Potatoes, white	HOT PACK (cubed) Wash, pare, cut into ½-inch cubes. Dip cubes in brine (1 teaspoon salt to 1 quart water) to prevent darkening. Drain. Cook 2 minutes in boiling water. Pack hot and cover with boiling water to within 1 inch of top.	35	40		—
	HOT PACK (whole) Use potatoes 1 to 2½ inches in diameter. Wash, pare, and cook in boiling water 10 minutes. Pack hot and cover with hot cooking liquid or boiling water, to within 1 inch of top.	30	35		—
Pumpkin or Winter squash	HOT PACK (cubed) Wash pumpkin or winter squash, remove seeds, and pare. Cut into 1-inch cubes. Steam until tender (about 25 minutes). Put through food mill or strainer. Pack hot in jars to ½ inch at top.	55	90	50‡	75‡
	HOT PACK (strained) Wash pumpkin or winter squash, remove seeds, and pare. Cut into 1-inch cubes. Steam until tender (about 25 minutes). Put through food mill or strainer. Simmer until heated. Pack hot in jars to ½ inch of top.	65	80	75‡	90‡

Timetable for Processing Low-Acid Vegetables

Product		Glass Jars Use 10 Pounds Pressure (unless you live in a high-altitude area)		Tin Cans	
		Pints Min.	Quarts Min.	#2 Min.	#2½ Min.
	Work rapidly. Raw pack or hot pack foods following directions, adding if desired ½ teaspoon salt for pints and 1 teaspoon for quarts. Place jars on rack in steam-pressure canner containing 2 to 3 inches of boiling water. Fasten canner cover securely. Let steam escape 10 minutes or more before closing petcock. (See step-by-step directions and special instructions for high-altitude areas elsewhere in this chapter.)				
Salsify (Oyster plant)	HOT PACK To prevent this root vegetable from discoloring as the outer surface oxidizes, scrub the roots well and slice. Then immediately drop each slice in a half-gallon of water into which 1 tablespoon of vinegar and 1 tablespoon of salt have been added. When all have been cut and set in this solution, rinse them quickly but well, place in a pot, and cover immediately with boiling water. Boil for 2 minutes then drain, reserving liquid, and pack into jars. Cover with hot liquid, leaving 1 inch at top.	30	35	30†	30†
Soybeans	HOT PACK Shell beans, then cover with boiling water and bring to a boil. Drain, reserving liquid. Pack beans loosely in jar and cover with hot liquid, leaving 1 inch at top.	55	65	50*	60*
Spinach and other greens	HOT PACK Pick over and wash thoroughly. Cut out tough stems and midribs. Place about 2½ pounds of spinach in cheesecloth bag and steam about 10 minutes or until well wilted. Pack loosely to ½ inch of top. Cover with boiling water, leaving ½ inch at top.	70	90	65*	75*
Sweet potatoes	DRY PACK Wash and sort for size. Boil or steam 20 to 30 minutes to facilitate slipping of skins. Cut into uniform pieces. Pack hot sweet potatoes lightly, pressing gently to fill air spaces to within 1 inch of jar top. Do not add salt or liquid.	65	90	—	—

Vegetable	Directions				
	WET PACK Prepare as directed for dry pack. Pack hot sweet potatoes to within 1 inch of jar top and cover with either boiling water or medium syrup, as preferred, to within 1 inch of top.	55	90	—	—
Tomatoes	See Boiling-Water Bath Timetable				
Turnips, Parsnips, and Rutabagas	*These are not recommended for canning; much better kept in cold storage.* Wash and peel, cube, or slice. Cover with boiling water and boil 3 minutes. Drain, reserving liquid. Pack into jars and cover with hot liquid, leaving 1 inch at top.	25	30	30†	30†
Vegetable mix (Carrots, green beans, celery, and lima beans)	HOT PACK Use almost any mixture or as described. Prepare each accordingly. Mix vegetables together; boil 3 minutes. Pack hot and cover with boiling water to within 1 inch of top.	Process pints and quarts for the time needed for the vegetable requiring the longest processing.			
Zucchini and Yellow squash (Summer squash)	RAW PACK Wash and slice; do not peel unless the squash is large and the skin is tough. Cut into 1/2-inch slices and halve or quarter slices that are extra large in diameter. Pack tightly in jars and cover with boiling water, leaving 1 inch at top:	25	30	20*	20*
	HOT PACK Prepare as for raw pack. Cover with boiling water and bring to a boil. Drain, saving liquid. Pack loosely and cover with hot liquid, leaving 1/2 inch at top.	30	40	20*	20*

* Use plain tin.
† Use C enamel cans.
‡ Use R or sanitary enamel.
§ The U.S. Department of Agriculture recommends all corn be canned in pints rather than quarts since processing time required for quarts tends to darken it.

pounds. (If you live in a high-altitude area, you need to increase the pressure by ½ pound for each 1,000 feet above sea level. A weighted gauge may need to be corrected for high altitude by the manufacturer.)

8. Start counting time as soon as 10 pounds pressure is reached, and process for the required time (see timetable chart). Keep the pressure as uniform as possible by regulating heat under the canner.

9. At the end of processing time, gently remove the canner from the heat.

10. If you are using glass jars, let the canner stand until pressure returns to zero. Wait a minute or two, then slowly open petcock or remove weighted gauge.

Unfasten cover and tilt far side up so steam escapes away from you. As you take jars from the canner, complete the seal if the jars are not the self-sealing type. If they are the common two-piece self-sealing lids, press down on the center of the lid. If it is down already or stays down, the jar is sealed. If jars fail to seal, reprocess them, put into the refrigerator for use within the next few days, or freeze them. Set jars upright on a rack, placing them far enough apart so that the air can circulate around all of them.

[If you are using tin cans, release steam in the canner at the end of processing time by slowly opening the petcock or taking off the weighted gauge. When no more steam escapes, remove the cover. Cool tin cans in cold water, changing water often enough to cool them quickly. Take cans out of cooling water while still slightly warm so they can air-dry.]

Drying Vegetables and Fruits

One of the oldest methods of preserving food—drying—has been largely abandoned in favor of modern food processing. But in the last several years many people have rediscovered drying fruits, vegetables, and herbs as a perfect and natural storage method, one that is low cost and energy efficient.

Nothing is added, and only water is taken away. The drying process removes 80 to 90 percent of the moisture content so that spoilage bacteria can't develop during storage. Dried food conserves storage space. Four pounds of fresh food will yield approximately 1 pound of dried. You can dry food from your garden with what was once common knowledge. Drying is not difficult, but to have a good dried product you must follow directions carefully. The faster you work, the higher will be the vitamin content of the dried food, and the better the flavor and cooking quality.

In the drying process, most of the water content is, of course, removed. Dried fruit is sweeter than fresh, due to the high concentration of fruit sugar. On a pound-for-pound basis, the dried food then has a substantially increased concentration of many nutrients, especially minerals. However, there is a loss of vitamins. Some, like vitamins A, E, and many B vitamins, are broken down by oxygen or light. Vitamin C is destroyed by heat, and heat also hastens the action of oxygen on other vitamins.

Temperature and Ventilation

Temperature is the factor of greatest concern to people concerned about nutrient loss in the drying process. And there is much controversy about what is the ideal temperature. The dispute ranges from 95° to 145°F. At the higher temperatures, less oxygen passes over the food since drying time is much faster. But at the high temperatures, the rate of oxidation is faster, too. Lower temperatures seem to save more vitamins, but at very low temperatures (90°F. or so) mold or bacteria can spoil the food right in the dryer in humid climates. A popular low-medium temperature is 110°F.

There's one more factor to consider in choosing a drying temperature. The enzymes naturally present in the foods will continue to work slowly even after drying is finished. Eventually they too will change the color and flavor of the food. Temperatures around 140°F. will destroy a lot of these enzymes and keep palatability high for long-term storage. On short-term storage, however, there's no argument. Low drying temperatures will keep foods fine for 6 to 9 months.

For fast, efficient drying, good ventilation is essential. The aim is not to heat the food, but to remove moisture from it. The more warm, dry air moving over the maximum surface area the better. This is why the best drying trays are those that let air through; top and bottom, and also why dryers should be well ventilated.

Drying on Trays Outdoors

The cheapest and usually easiest way to dry is to let the sun do all the work. But drying outdoors works well only if you live in an area that enjoys long, hot, sunny days of low humidity. And only if you live in an area that has clean, unpolluted air. If you can't depend upon several good drying days in a row or you live in an industrial area or near a heavily traveled highway, perhaps you ought to move on to the next section and read about indoor drying.

If drying is to be done outdoors, plan to set the food out early in the morning so that it won't be too wet the first night. Place the trays in a comparatively dust-free location, on racks raised above the ground. Raising the racks permits air to circulate freely under as well as over the food.

This free air movement under the racks is very important if you're covering the racks with a solid material, like glass or polyethylene sheeting. Although both glass and plastic don't allow for much air circulation above the food, they are good in that they allow you to take advantage of the greenhouse effect. That is, capturing heat from the sun's radiation and holding it in. You'll find that drying under glass or polyethylene is good in areas of higher humidity and lower temperature where you need all the help you can get. Be careful, however, when using glass or polyethylene in hot, dry areas,

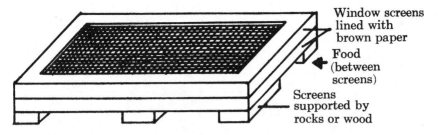

Window screens
lined with
brown paper

Food
(between
screens)

Screens
supported by
rocks or wood

Clean window screens also make suitable drying trays. Cover them with paper and then place food only one piece deep on the screens or trays for best results.

as the excessive heat buildup around your food could actually cook or burn it! To increase air circulation when using glass or polyethylene, raise the cover slightly with blocks and cover the opening gap with screening or netting to keep bugs out.

If the trays are not protected by polyethylene or glass, they must be covered at night to prevent dew from settling on the food. Put them in a sheltered place or cover them with cardboard (cartons are fine), heavy towels, an old shower curtain, or anything that will keep moisture out. Obviously, if you expect rain, make sure the cover is really waterproof. And cover tightly, as many insects are nocturnal and will try to get into the food after the sun goes down. Be sure to exclude insects and animals from the drying area. Turn the food often, and dry only on sunny days. When the food is dry enough to bring in, try to do it in the heat of the day, on a hot day. Any insects that might be on the food are driven off by the heat to a shadier, cooler area.

Drying Trays

Just about anything that has a good-sized flat surface can work as a drying tray, but the best trays are those that have ventilated bottoms made of cheesecloth or wooden slats. Although convenient, wire mesh or window screening should not be used alone. The metal can interact with the food (especially high-acid foods like fruits and tomatoes) and either destroy some of the vitamins or introduce a questionable metal onto the food itself. If you're using wire mesh or window screening, cover it with brown or freezer paper, or clean grocery bags opened inside out.

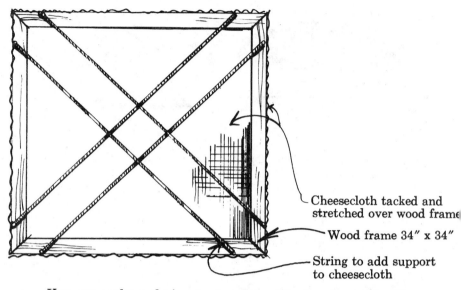

Cheesecloth tacked and stretched over wood frame

Wood frame 34″ x 34″

String to add support to cheesecloth

You can make a drying tray such as this from spare lumber, string, and cheesecloth or wire mesh. Place food directly on the cheesecloth, but if you are using wire mesh, put brown paper or freezer wrap over the wire before placing the food on it. Then put the entire tray on rocks or wooden blocks to allow air to circulate over and under the food.

If you plan to dry your food outdoors you may want to make your trays so that they'll fit into your oven or a food dryer, too, just in case you hit a rainy spell and decide to dry indoors instead or become impatient with the sun and want to speed things up with extra, controlled heat. Trays made especially for drying food may be purchased, but they are easy to make. Construct wooden frames and stretch fiberglass or nylon mesh or cheesecloth over them. Clean sheeting or other thin material can also be used. While sturdier than cheesecloth, it won't let as much air circulate around the food. If you're using cheesecloth, make the frames at least an inch or two under 36 inches, as most cheesecloth comes in bolts that are 3 feet wide. The mesh or cheesecloth should be reinforced underneath with string that is tacked diagonally between the corners of each frame.

Put food on the tray one piece deep, and place a piece of cheesecloth, other fine meshed material, wire mesh, or window screening over, but not touching the food, keeping it slightly above the food with blocks of wood or clean stones. When done out in the open, carefully lay strips of wood or stones on the cloth edges to prevent the material from blowing off. When you're finished drying a batch, remove the cloth or mesh and wash it well and let it dry before you set out more food. Scrub any sticky wooden surfaces as well.

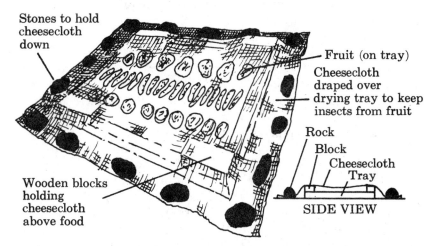

Stones to hold cheesecloth down

Fruit (on tray)

Cheesecloth draped over drying tray to keep insects from fruit

Rock

Block

Cheesecloth

Tray

SIDE VIEW

Wooden blocks holding cheesecloth above food

Such a drying setup permits maximum air circulation while protecting fruit from insects. It does not, however, protect food from moisture. Put trays outside on sunny days after the morning dew has settled, and cover them or bring them indoors at dusk to protect the food from the evening dew.

Drying Indoors

When the atmosphere is dry and sunny, drying may be accomplished outside, but when heavy dews and frequent rains are normal, drying must be done indoors in an oven, in an attic, or in a specially constructed dryer.

Drying with controlled heat in dryers or in a kitchen oven has several advantages. The drying goes on day and night, in sunny or cloudy weather. Controlled heat dryers shorten the drying time and extend the drying season to include late-maturing varieties. Vegetables dried with controlled heat cook up into more appetizing dishes than do sun-dried vegetables and have a higher vitamin A content and a better color and flavor.

Oven Drying

One oven can take about 6 pounds of prepared fruit or vegetable pieces at a time. Food should be exposed top and bottom. Place food directly on oven racks, one piece deep, or, if the slats are too far apart, cover them first with nylon mesh or cheesecloth and then place the food on top. Regular cookie sheets can be used, but because they are solid, they will not expose the food to drying heat on all sides. Special drying trays, either purchased or made of mesh or wooden slats for drying food in the sun (see previous section) may also be used in the oven. Separate trays in the oven by placing 3-inch blocks of wood at each corner when stacking them.

It's very difficult to give more than general guidelines for time and temperature. Set your oven no higher than 145°F. This may be difficult to do in some ovens since the lowest setting is often 200°F. If this is the case, set your oven to "warm" and use an oven thermometer to check the real temperature inside. If you can't get the temperature to stay below 145°F., you ought to consider other ways to dry food. Food drying in the oven should be checked often. Leave the oven door open slightly (as you do when broiling) to provide good ventilation which is so important for food drying. Stir the food often (every hour or so) and move the trays or oven racks from time to time. Don't place any food closer than about 3 inches from either the top or bottom of your oven. Sliced fruits and vegetables and small whole berries can take from 4 to 12 hours to dry in a warm oven.

An oven is a handy place to dry food, and if you can keep the temperature below 145°F., a perfectly acceptable place. However, it's not usually an economical means of food drying, since keeping the oven on with the door ajar for several hours at a time can be energy inefficient and pretty costly.

Home Dryers That You Can Buy

The popularity of drying food has grown in the last few years, and so has the number of commercial food dryers. Because there are so many home drying units to choose from, the Research and Development Group here at Rodale Press decided to do some comparative testing on several dryers on the market. Hopefully their observations and generalizations will help you select the most appropriate unit for your particular needs.

Drying techniques fall into three categories: electrical, stove top, and solar. Electrical units tend to be expensive but for many the shorter drying time and the high dependability seem worth the price. Stove top dryers are of two varieties: those requiring a primary heat source (electric coil, gas flame, or hot stove surface), and those requiring heat convected from the corner of a wood-burning stove. Most solar drying is of the build-it-yourself type. It's less costly but more risky.

Electrical Units

Size and Appearance First decide what the greatest volume is you plan to dry at any one time and convert that volume into drying space (10 square feet per ½ bushel). To determine the drying space of any dryer, follow this formula:

shelf length × width (inches) × number of
shelves, divided by 144 = sq. ft. of drying space

It's important to know where you want to put the dryer when balancing cost, size, and drying space factors. Some models are made with a wood veneer or colored porcelain finish to match any kitchen decor. These tend to be the more compact, deluxe units ranging in price from $130 to over $200. They're costly but attractive and handy to your work area.

You'll find a greater selection of dryers open to you if you don't confine yourself to the kitchen. With thermostatic temperature control and an electrically safe unit (watch for the UL stamp of approval), there's no need to keep the dryer under constant surveillance. If you've got space in the basement or pantry for a large dryer and aren't particular about appearances, you can find something nice for around $100.

Here are some of the commercially available electric food dryers Rodale's Research and Development Group tested.

Unit Construction Among the units R and D tested were dryers made of particle board, Masonite, aluminum, and sheet metal. Wood, acting as a good insulator, confines heat within the dryer while the exterior remains cool and safe to touch. Such efficient heat utilization keeps the running cost down. Some competitors complain of wood's tendency to warp and absorb odors, but R and D didn't notice these problems during their short testing period.

Inefficiently utilizing heat, both sheet metal and aluminum dryers can get as hot as an oven door after the turkey's been roasting a few hours. But on the positive side, metal won't warp, it is easy to clean, and it won't absorb odors. The difference in weights of comparable wooden and metal dryers is surprisingly insignificant.

Running Cost Choosing a dryer which makes as efficient use of expended energy as possible saves you money and saves the world some energy. To make that choice you need to know the current price of electricity, the number of watts the dryer draws, and the amount of drying space the dehydrator provides. Here's a formula which puts all dryers on a fairly equal basis for comparison, assuming drying time is the same:

watts × 3.75¢/kwh ÷ sq. ft. of drying space = ¢/hr./sq. ft.

Using this formula, you will see that a dryer's size doesn't automatically tell you how much it will cost to dry your food. Though a small dryer may cost one-fourth as much to run per hour as a larger dryer, it may take so many more hours to dry the same amount of food that this one will actually cost more to use. Dryers will differ this way, so check the formula on all of them.

If you're going to get an electrically run dryer, we suggest that you try to buy one with a manually operated thermostat. Not only does a thermostat let you dry at your favorite temperature, it lets you keep it there, and that is a help. One dryer R and D worked with didn't have one. The temperature was controlled by leaving the dryer drawers open a bit. Twice the temperature soared to a scorching 176°F. and burned the food. Even after the testers got used to the unit they still found they couldn't walk away from the dryer for more than 2 hours at a time.

Fans and Air Flow For fast, efficient drying, a fan to improve ventilation is a must. Your aim in drying is not to heat the food but remove moisture from it. The way to do that is to get warm, dry air moving across as much surface area of the food as possible. You will find basically two air flow systems to choose from: vertical and horizontal. The horizontal system is preferred from R and D's point of view because it keeps air moving across both the top and underside of the food (provided you don't use solid shelves) instead of up through the dryer between the pieces of food.

You'll find manufacturers emphasizing air flow recycling systems as a big selling point on their dryers. Such recycling systems are supposed to cut down on the amount of air allowed to leave the unit, retain more heat, put less strain on the heating unit, conserve energy, and improve drying efficiency.

Shelf Construction Shopping for shelves you'll find almost as many different types available as there are dryers. A popular style uses an aluminum frame to support a thin mesh nylon screening without support underneath. Easy to clean and lightweight, these are hot to handle fresh out of a hot dryer. If consistently overloaded with moist heavy fruits, they will stretch. Wooden frames are a bit harder to clean.

Some shelves are made of expanded metal or heavy plastic and have a pattern such as you'd find on a radiator cover. The great thing about these is their sturdiness, but the holes are so big that some foods will fall through. Covering the shelf with plastic wrap or nylon netting is a big help in such cases. Plastic wrap can be hard to work with, so plan to have an extra set of hands handy and tape it to the shelf.

Other shelves are made of solid wood or metal and are the least desirable. Air has no way of getting to the underside of the food so the drying process is slowed and the food needs to be turned. Most shelves of this type don't have turned up edges to keep things like peas and grapes from rolling off. It's a trick to balance rolling peas on an edgeless shelf, too!

Warning: When buying any kind of plastic shelf, make sure it is FDA food quality approved and that it will withstand high temperatures. Hardware cloth and aluminum screening will lose some of its metallic finish to the food.

Miscellaneous Don't let cute little extras cloud your evaluation. On-off switches mean nil. Timers are nice, but remember, you are drying by the hour, not the minute. Warranty times are not indicative of the product's quality. A new manufacturer will put a 1-year warranty on its unit to make it look as good as the rest, while a big money company will gladly replace a few rejects for the added sales a 5-year warranty will bring in.

This is one of the old-fashioned stove top dryers used 50 and more years ago. It's designed to fit on top of a wood stove, but as you see here, it works well on top of a modern electric range.

This simple solar dryer constructed at the new Rodale Organic Farm can double as a cold frame.

Demand a safe product. Check for a yellow stamp marked UL where the electrical description is given. If it's there, you know this unit passes a safety inspection by Underwriter's Laboratory. Grounded plugs are required on appliances by law for a reason, so use a grounded outlet.

Stove Top Dryer

Primary Heat Source Required   This is the kind of dryer that was popular about 50 years ago. It was originally designed to sit on top of the old wood cook stove and make use of the ever-present heat it generated. Unless you're lucky enough to have a wood stove, you'll have to use some other primary source of heat like an electric or gas burner.

A water reservoir 3 inches deep below a stainless steel drying surface is used to temper direct heat from the stove so that moisture may be gently driven from the food. The unit has 3 square feet of drying space, small compared to electric dryers but certainly enough for drying odds and ends. With regular use this dryer will provide enough dried food to fill up your pantry in no time.

It'll take a little time until you know just how to regulate heat under the dryer. On R and D's first trial run, the testers forgot to grease the top and applied enough heat to cook the food in an hour. After that disaster it didn't take them long to catch on.

No gadgets are provided with stove top dryers to improve ventilation, but then the food isn't confined within a box either. Left exposed, moisture is carried away by air currents rising from the heat of the stove and by the dry breeze coming in a nearby window.

Solar Dryers

Most solar drying units commercially available, including collectors, racks, trays, ventilated cold frames, and hoods, are relatively inexpensive, especially compared to electrical units. There's a reason—a simple design. If you are the least bit handy, you can build your own.

A hood dryer is little more than a plastic frame placed over your drying food to intensify the sun's radiant energy. It keeps dust, rain, and insects away from the food, but the open sides let dampness in if left outside overnight. Use opaque plastic to keep the food out of direct sunlight.

Enclosed solar collector dryers with controlled ventilation are about the best you can do going 100 percent solar. The balance between temperature and air flow is a delicate one, and success depends a lot on the operator. The frame should be airtight except for vents so you can direct air flow through the dryer. It will be useful to have a thermometer installed inside. A regular outdoor thermometer doesn't usually go up high enough, so see if you can find one to handle at least 150°F. (not that you want it that high, but just to keep track if it does). Again, a black cover will encourage warmer temperatures and keep food out of the sun. Make sure that vents can be securely closed overnight so you aren't bothered moving trays in and out.

Don't expect solar dryers to do things they weren't meant to. You may live in a hot climate, but if it's humid, your food has a good

Here's a drum-type solar dryer that you can build using plans available from Solar Survival, Cherry Hill Road, Harrisville, New Hampshire 03450

Rodale R&D's cardboard box dryer costs almost nothing to make. The size of the box should be determined by the size of the cookie sheet (or vice versa) so that the cookie sheet fits on top of the box exactly.

chance of mildewing before it is dry. If you don't insist on a 100 percent solar unit, install a light bulb and maybe even a small fan. It all depends how important successful drying is to you. Despite popular belief, it does rain in Southern California on occasion, so don't be caught short.

Homemade Dryers

A Low-Cost Indoor Dryer

If you only want to dry small batches of food at a time or wish to experiment with preparing and using dried foods, a simple, inexpensive dryer built by the Rodale Research and Development Group may interest you.

The simple materials you'll need are a topless cardboard or wooden box at least 8 inches deep; one or more 60-watt bulbs; a socket base and cord; a stainless steel cookie sheet or piece of a stainless steel sheet cut to fit the box; some tinfoil; and a few brushfuls of black paint.

Start by painting the bottom of the cookie sheet or metal sheet black for maximum heat absorption. While it's drying, line your box with tinfoil, shiny side up. Then place your bulb setup in the center of the box, angling the bulb at 45°. (To help diffuse heat evenly, you might also put a little tinfoil "shade" on the top of the bulb.)

After notching the top corner of the box so the cord can exit, place the tray over the box (black side down) so it is suspended a few inches over the light bulb. Then coat the tray with a little vegetable oil to prevent the food from sticking to it (or put a layer of nylon or fiberglass netting or cheesecloth over the tray), fill the tray with a layer of sliced fruit or vegetables, and plug it in. In about 12 hours (more time on a high humidity day; much less time for herbs and foods cut into fairly small pieces; and perhaps longer for high-moisture foods like most fruits), you'll have a trayful of dried goodies for storage or snacking.

R and D's prototype dryer features a cardboard fruit box 12 × 18 inches and 8¼ inches deep. The 1-inch-deep cookie sheet, which fits the top exactly, was purchased at a hardware store and will hold about 1½ pounds of raw prepared food.

To dry larger amounts at one time, simply increase the size of your box and tray. For every 2 to 3 square feet of tray you add, use one additional 60-watt-bulb setup, taking care to space the bulbs carefully for even heating. This should keep your surface tray temperature at about 125° to 130°F.—"cool" enough so you can just bear to touch it, but hot enough to dry your harvest slowly and surely without scorching or experiencing the kind of nutrient loss that begins at around 150°F.

A Low-Cost, Kerosene-Heated Dryer

California organic grower Elmer Kulsar made a more elaborate dryer than the one put together by the Rodale R and D group, but his is still pretty simple and inexpensive. Bigger than the cardboard

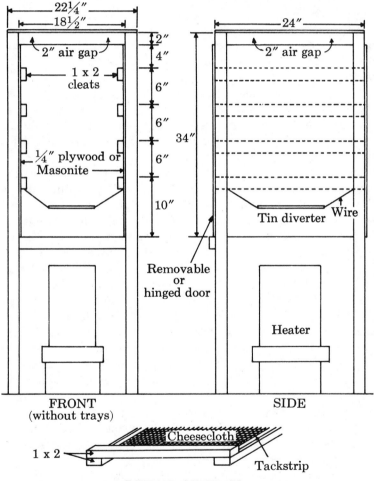

FRONT
(without trays)

SIDE

DETAIL OF TRAY

This 22-by-24-by-34-inch dryer holds 4 drying trays. The trays slide in and out easily so that those with the driest food can be shifted closest to the heater.

dryer, Kulsar's unit is capable of drying a 5-acre crop of fruit; it uses controlled heat from a kerosene heater placed under the unit.

Mr. Kulsar began by building four rectangular frames made of 1-by-2 lumber that he nailed together with thin, cement-coated 6-penny nails. He tacked cheesecloth over the frames to hold the fruit. He then cut four legs from 2-by-2 stock, each 60 inches long, which allowed him to set the trays about 12 inches above the heater. He cut two pieces of ¼-inch plywood for the sides, each 24-by-34 inches, and nailed them flush to the legs, 24 inches apart and 2 inches down from the top, to allow the warm air to escape. He nailed eight 24-inch-long strips of 1-by-2 cleating to the sides 6 inches apart above the plywood to hold the trays securely. He made the 22-by-34-inch back out of ¼-inch plywood, and nailed it in place, again leaving a 2-inch gap at the top and spacing the sides 18½ inches apart so the tray could slide easily. The top, cut 22 by 24 inches, also allowed an 18½-inch space between the sides.

To complete the hastily made but effective dryer, he added a 22-inch spacer strip to secure the front legs, and hung a 22-by-34-inch front door on a pair of nails. He then set the heater in position and hung sheet metal to serve as a heat diverter.

Any type of heater can be used providing it supplies a steady flow of heat at a moderate and fairly even temperature. An old gas plate, electric hot plate, electric heat lamp, or kerosene heater (clean and adjusted so as not to smoke) will dry food satisfactorily. Too much heat is bad, so the distance under the bottom tray will have to be adjusted. Drying time can be cut down and production stepped up if the shrinking food is periodically moved closer together on the trays and shifted downward from the upper trays. Prepared food is then added to the top tray.

Preparing Foods for Drying

Drying is especially popular with fruit. Apricots, peaches, pears, prunes, grapes, nectarines, figs, cherries, and apples are commonly dried. Although most vegetables can be dried, they have limited uses, and except for some varieties of beans and peas, most vegetables show a significant loss in taste, color, and form when they are dried. All herbs can and are commonly preserved this way.

Both fruit and vegetables should be perfect for drying. Blemished or bruised fruit will not keep as well and may turn a whole tray of drying fruit bad. Fruit must be fully ripe so that its sugar content is at its peak. However, it should not be overripe. Overripe fruit can be saved for making fruit leathers—see directions later in this chapter.

Try to cut uniformly all fruit or vegetables that are to be dried in the same batch so that each piece requires about the same amount of drying time as the rest. Don't peel unless the skin is especially thick or the food has been sprayed with chemicals, since peeling will remove many of the nutrients concentrated right under the skin surface.

Fruits

In most commercial drying operations, apricots, apples, and peaches are usually sulphured, but while sulphured fruit will retain more of its original color than unsulphured fruit and feel less leathery, the sulphuring process requires some special equipment and extra time and work. More importantly, sulphured fruit may impart a slightly sour or acid taste and a questionable chemical to the diet.

If you wish to pretreat fruits like apricots, apples, and peaches for drying we suggest that you dip them in an ascorbic acid solution. You can prepare a solution by crushing three or four 500-milligram ascorbic acid (vitamin C) tablets, or 1,500 to 2,000 milligrams of powdered ascorbic acid, into 1 quart of water. You may also dip the fruit into unsweetened lemon or pineapple juice instead if you prefer.

You'll probably be slicing most and peeling some fruits before you dry them. But there are some fruits—perhaps grapes, prunes, berries, cherries, apricots, and rose hips, to name a few—that you'll only be pitting or slicing in half. Before you put these uncut fruits out to dry, we suggest that you crack their skins so that moisture inside the fruit can readily escape. There are three ways to do this:

1. Blanch them in either water (for ¼ to 1 minute) or steam (1 to 2 minutes), depending upon the thickness of the skin and the size of the fruit. As in any other blanching procedure (see page 45), blanch only a small amount of fruit at a time and cool it immediately by plunging it in cold water.
2. Make several nicks in the skin with a sharp knife.

These two above methods are best for fruits you'll be leaving whole or pitting. For fruits that you'll be pitting and halving, try the next method:

3. Take one of the halves in both hands and place both thumbs in the middle of the skin side. Then pull up on either end so that you are in effect turning the half inside out. The strain should crack the skin sufficiently to allow moisture to escape.

When first setting fruit on the drying tray, put skin-side down (if fruit has skin). By the time you're ready to turn the fruit, the moister exposed side should have dried out a bit and lost some of its stickiness, and there's less chance that it will stick to the tray.

Vegetables

Vegetables (except onions, garlic, and leeks, which are used primarily for seasoning, and mushrooms) should be blanched or precooked in steam or boiling water after they are sliced for drying. This blanching sets the color, hastens drying by softening the tissues, checks the ripening process, and prevents undesirable changes in flavor during drying and storage. Vegetables blanched before drying require less soaking before they are cooked for eating and have a better flavor and color when served. Blanching vegetables by steaming is preferable to blanching by boiling because nutrients are dissolved in the boiling water.

A pressure cooker or a large, heavy pot makes a good steamer. Place a shallow layer of vegetables, not over 2½ inches deep, in a wire basket or stainless steel or enamel colander. Have 2 or more inches of boiling water in the pot, set the basket on a rack above the water, cover tightly, and keep the water boiling rapidly. Heat until every piece of vegetable is heated through.

If there is no convenient way of steaming, boiling is second best. Use a large amount of boiling water and a small amount of food so that the temperature of the water will not appreciably lower when the food is added. About 3 gallons of water to every quart of vegetables is good. Place the vegetables in a wire basket and immerse them in the boiling water for the required time, as suggested below.

PREPARING VEGETABLES FOR DRYING

Asparagus	Use only the top 3 inches of the spear. Blanch until tender and firm, about 10 minutes.
Beans, lima and snap, Soybeans	Shell and blanch 15 to 20 minutes. Also see the discussion of drying beans, peas, and corn that comes later in this chapter.
Beets	Remove tops and roots and blanch about 45 minutes or until cooked through. The time will depend upon the size of the beets. Cool, peel, and cut into ¼-inch cubes or slice very thin.
Broccoli	Trim and slice into small (½-inch) strips. Blanch 10 minutes.

Brussels sprouts Cut into lengthwise strips about ½-inch thick; blanch 12 minutes and dry until crisp.

Cabbage Cut into long, thin slices and blanch 5 to 10 minutes.

Carrots Wash and slice thinly. Blanch 8 to 12 minutes.

Celery Remove leaves and cut stalks into small pieces. Blanch about 10 minutes.

Corn Husk and remove the silk, then blanch the whole cob 10 minutes to set the milk. Cut the kernels deep enough to obtain large grains, but be careful not to cut so deeply as to include any cob. Also see page 115 for directions for drying corn on the cob without blanching.

Herbs See page 124.

Mushrooms Peel and cut off stems if they are tough. Leave whole or slice, depending upon their size. Do not blanch, but dry while still raw.

Onions, Garlic, Leeks Peel and slice into small strips, or peel and grate. Blanch onions and leeks 5 to 10 minutes if you plan to use whole, as in a cream sauce or casserole. If for seasoning, do not blanch.

Peanuts Hang vines in well-ventilated shelter or attic, away from cold temperatures, or spread them on a dryer, outdoors or inside. When dry and crunchy, they should cure for about 2 months before they can be eaten or roasted. Also see page 461 for directions on roasting.

Peas Shell peas and blanch (15 minutes if steamed and 6 minutes if boiled). Also see directions for drying peas in the pod on page 114.

Peppers, sweet Clean and slice into thin strips. Blanch 10 minutes.

Peppers, hot If possible, do not pick until they are mature and fully red. However, if frost threatens, harvest your crop even if some are still green; many should ripen while drying. String the peppers by running a needle and thread through the thickest part of the stem. Hang them outdoors or in a sunny window to dry. They will shrink and darken considerably and will be leathery when they are dry. Although dried hot peppers can be kept in storage con-

	tainers, they are best left hanging in a dry place.
Potatoes	Wash and slice into ¼-inch rounds. Peeling is optional. Blanch for 5 minutes in steam and then soak in ½ cup lemon juice and 2 quarts cold water for about 45 minutes to prevent the potatoes from oxidizing during drying.
Pumpkin, Winter squash	Clean and cut into 1-inch strips and then peel. Blanch about 10 minutes, until slightly soft.
Rhubarb	Cut into thin strips (about 1 inch wide) and blanch 3 minutes.
Spinach, Swiss chard, Kale	Cut very coarsely into strips. Blanch spinach and Swiss chard about 5 minutes and kale about 20 minutes. Spread not more than ½ inch thick on trays.
Summer squash, Zucchini	Do not peel, but slice into thin strips and blanch about 7 minutes.
Tomatoes	Wash, quarter, and blanch for about 5 minutes. Run through a food mill to remove skins and seeds. Strain out the juice through a jelly bag or several layers of cheesecloth. Use a little hand pressure to extract more water, then spread the remaining pulp on glass, cookie sheets, or pieces of plastic. Turn the drying pulp frequently until it becomes dry flakes.
Turnip	Wash and cut into thin slices. Blanch 8 to 12 minutes.

PREPARING FRUIT FOR DRYING

Apples	Pare, core, and cut into thin slices or rings. Don't peel unless the apples have been heavily sprayed.
Apricots	Cut in half, remove pit, and leave in halves or cut into slices or pieces.
Bananas	Peel and slice thinly.
Berries	Halve strawberries and leave other, smaller berries whole. Crack skins by quick blanching or nicking with a knife.
Cherries	Pit and remove stems. (If you don't pit them, the dried cherries will taste like all seed.) Let drain until no juice flows from them.

Grapes	Remove stems and crack skins by blanching quickly or nicking with a knife. Drain until no juice flows.
Peaches	Cut in halves and remove pits. Skin if desired, or just remove fuzz by rubbing briskly with a towel. Then slice.
Pears	Skin, remove core. Cut into slices or rings.
Plums	May be pitted or left whole. Crack skins by quickly blanching or nicking with a knife.
Prunes	May be pitted or left whole. Like plums, they may be blanched quickly or nicked with a knife to crack skins.
Rose hips	Cut off blossom ends and stems. Crack skins by quickly blanching or by nicking with a knife.

Special Techniques for Drying Peas, Beans, and Corn

Drying right in the garden on the vine or stock is often the easiest way of drying peas, beans, and corn, provided you have the following:

1. A long growing season, so that the food can remain on the vine or stock until the pods are thoroughly dry, but before frost hits.
2. Dry weather, not wet, which might cause some beans, peas, and corn to sprout right in their pods and shatter.
3. Good timing, so that you get out and collect your dried food before the pod splits open and scatters what it's holding all over the ground. Limas and soybeans should be watched quite carefully as they're near the dry stage because their pods split easily once dry.

Bean and pea pods can be picked when mature and spread out in shallow layers in an attic, a covered porch, or in a spare room to dry. Green or wax beans can be strung on heavy thread or string, about one-third of the way down from one end. Then blanch the whole string in steam for 20 minutes, and hang in a dry, warm place, like attic rafters, to make "leather britches." They can be kept there until you're ready to use them so long as the air is dry and warm enough.

You can also cut the whole bean or pea plant when most of the pods are mature and hang it upside down in a dry, well-ventilated place, allowing the beans or peas to dry. Pinto beans are hard to dry this last way because they cling so tightly to the poles on which they climb.

To shell dried beans and peas, place them in cloth bags and beat them with a mallet, or else stomp on them. The beans are then easily sorted from the shells. Soybeans and chick-peas are much tougher pods and must be shelled by hand—a lot of work, but these high-protein legumes are well worth the effort.

Before storing beans or peas, place on shallow trays in a 175°F. oven for 10 to 15 minutes to kill any insect eggs they may contain. Oven heating also assures that the beans or peas are thoroughly dry. If they aren't completely dry when they are put into storage, they will build up a great deal of heat and either smolder or crumble. Store in sterilized glass jars, or you can put them in paper bags which are then packed in plastic bags and sealed with wire twists or in metal cans with snap-on lids. For extra protection from weevils and other insects try putting a dried hot pepper in each jar or plastic bag of dried beans.

The husk covering corn won't allow the kernels to dry completely on the stalk, so you'll have to pick the corn and remove this outer husk. Then sun or oven-dry the corn cob until the corn is hard and cannot be squeezed. Corn can be stored on the cob, but it will take up much less space and be more convenient for later use if it is shelled, pasteurized or heat treated, and stored like beans and peas. The easiest way to shell dried corn is to hold the cob between both hands and twist in opposite directions, allowing the kernels to fall into a container underneath.

When Your Food Is Dry

Drying is finished when fruit feels dry and leathery on the outside, but *slightly* moist inside; beans, peas, and corn should be very hard; leafy and thin vegetables should be brittle; and larger chunks or slices of vegetables should be leathery. If in doubt, leave the food on the trays a little longer, but reduce the temperature if you're drying with an oven or indoor dryer. Fruit seems to be moister when it is hot, so remove a few pieces from the tray occasionally and allow them to cool before you determine if they are dry. Since some pieces of food will dry faster than others, it is important to remove pieces as they dry rather than wait until every piece of fruit or vegetable is totally dehydrated to stop the drying process. Food that overheats near the end of drying will scorch easily. As we mentioned earlier, you should try to cut up or slice a particular fruit or vegetable as uniformly as possible, so the pieces will take approximately the same time to dry.

Pasteurizing

Once you think the food is dry, pasteurize it to insure that no insect eggs or harmful spoilage organisms will develop; low heat may have dried the food, but it probably didn't kill all such contaminants. Spread dried food 1-inch thick on cookie sheets or trays, and heat for 10 to 15 minutes in a 175°F. oven. Then cool thoroughly.

After you have pasteurized the food, store it in open glass or other open containers in a warm, dry area free of insects and animals. For four successive days stir the contents thoroughly each day to bring the drier particles in contact with some that are more moist. In this way the moisture content will be evenly distributed. If at the end of the four days the food seems too moist, return it to the dryer or leave it in the sun for further drying.

Storing

When thoroughly and uniformly dry, the food should be packed in airtight, sterilized glass jars, or in plastic bags or metal cans with snap-on lids that are lined with new brown paper bags to keep food out of contact with potentially harmful compounds in the plastic or metal. If you are storing in clear glass it is very important that the storage area be dark, and if storing in paper bag-lined plastic bags, make sure the area is rodent-proof! It's generally preferable to package small quantities, enough, say, for one meal. Then if one package spoils, only a small amount of food will be wasted. Several brown paper bags of dried food can be stored inside one larger plastic bag.

After your food is packaged, place a label on each indicating the kind of food and the date it was packaged. Then store in a dark place in a cool (below 60°F.) dry basement or pantry. During warm, humid weather dried foods retain their quality best if they are kept under refrigeration. It is a good idea to examine dried food occasionally for mold. The danger of mold is prevented if the dried product can be stored at freezing temperatures or below. Dried foods keep well for six months to a year if they are stored properly, and longer if kept under refrigeration or in the freezer. If you discover bugs or worms in your food in late winter or early spring, don't throw it out. Spread it in shallow pans and put it in your oven for about 20 to 25 minutes at 300°F. The heat will take care of the vermin and sterilize your food at the same time. You can also heat dried food in a low oven if it becomes too limber and moist during storage.

Rehydrating

Water is taken out of fruits and vegetables for preservation, and, in most cases, you will want to put the water back in before you eat the food. Dried fruits are quite good eaten just as they are—by themselves or chopped up in cereals and desserts—but vegetables, and fruits intended for baked products and compotes, should be rehydrated or given back the water lost during drying.

To rehydrate fruits and vegetables pour 1½ cups of boiling water over each cup of dried food. Do not add salt to rehydrating vegetables or sugar to fruits; both cause food to absorb less water than they normally would. Let the mixture stand until all the water is absorbed. Vegetables, except for dried beans and peas, generally absorb all the water they are capable of retaining in about 2 hours. Fruits require a longer soaking time—anywhere from 2 to several hours. Overnight soaking may be necessary for complete rehydration of some fruits and dried peas and beans. The amount of water dried foods will absorb and the time it takes for complete rehydration varies according to the size of the food and its degree of dryness. If the water is absorbed quickly, add more—a little at a time—until the food will hold no more. Avoid adding more water than the food can absorb, since nutrients will be lost to this extra water.

Cooking Vegetables

Rehydrated fruits need not be cooked (unless, of course, you prefer them that way), but vegetables are always cooked after they have been soaked. To cook the vegetables, put them and any water they did not absorb in a pot. Add only enough extra water to cover the bottom of the pot. Cover and quickly bring the vegetables to a boil. Reduce the heat and simmer until the vegetables are plump and tender. If the vegetables are still tough after about 5 minutes of cooking or all the water is quickly absorbed by the cooking vegetables, they have not been soaked long enough. Next time, extend the soaking time so that the vegetables are fully rehydrated before cooking. You'll find rehydrated dried vegetables taste more like fresh vegetables than do those that have been canned. Fruits are cooked in the same manner.

Soak soybeans overnight in the refrigerator, throw out their soaking water, then cook for 1 to 2 hours in fresh liquid before using in recipes. Soybeans contain an antinutritional enzyme and should be eaten only after they are completely cooked.

When measuring out dried foods to use in place of fresh in recipes, keep in mind that one part dried fruits or vegetables equals about four parts of the same fresh food.

Freeze-Drying

Freeze-drying seems to be an excellent way to store foods. The foods are substantially reduced in weight and volume and they will keep for about 2 years without much loss in nutrients, color, or flavor. Unfortunately, freeze-drying is a sophisticated process that requires special equipment not available to most people. It is not a technique that can be carried out under normal home situations.

Freeze-drying is, simply, a drying method in which water is removed from frozen foods. The food is first sliced, diced, powdered, granulated, or liquefied. Then it is frozen. Once frozen it is spread out on trays and placed in a vacuum cabinet. The door is closed and the pressure is lowered, creating a vacuum. Heat is applied, and the ice within the food disappears in the air and is taken out of the cabinet with a pump. Drying takes about 10 hours (during drying the food is kept frozen) and almost all of the water is removed from the food. The moisture content is usually 2 percent or lower. The food is taken from the drying chamber and tightly packaged in a can so it will stay dry until used.

Fruit Leather

A variation on dried fruit slices is fruit leather. Basically, fruit leather is the pulp from juicing peaches, apricots, prunes, and apples that is dried to form a naturally sweet, confectionlike food that will keep in good condition for one year or more. Fruit leather can be made from almost any fruit or any combination of fruits. Here are a few recipes to get you started.

Apricot, Peach, or Nectarine Leather

1 gallon pitted apricots, peaches, or nectarines	honey
1½ cups unsweetened pineapple juice	3 teaspoons almond extract (optional)

Place the pitted fruit and pineapple juice in a large, heavy pot. Cover the pot and set it over low heat. Cook the fruit until it is soft. Drain off the juice well, lifting the fruit from the sides of the strainer

To make fruit leather, spread the sweetened fruit pulp ¼ inch thick on a lightly oiled cookie sheet. After about 2 weeks at room temperature, the pulp should be dry enough to be pulled from the cookie sheet in one solid, thin layer.

to allow all the juice to run out freely. The more juice strained out, the quicker the process of "leather-making." The juice is too good to discard. Can or freeze it for later or drink it fresh.

Run the fruit through a blender, food mill, or sieve, removing the skins if you prefer a smooth product, or use the skins as part of the pulp for the leather. Sweeten the pulp to taste with honey and add the almond extract if you wish. The pulp should be as thick as apple butter or more so. Spread it on lightly oiled cookie sheets or cookie sheets covered with preferably freezer paper, although clear plastic wrap will do, so that it is ¼ inch thick. If it is much thicker than this, it will take very long to dry. Cover the cookie sheets with a single layer of cheesecloth or plain brown paper to keep out dust and insects, and place them in a warm dry place to dry. Depending upon the weather, the pulp will dry in 1 to 2 weeks. Drying can be hastened by placing the cookie sheets in a low oven or food dryer. If using an oven, turn the control to warm (120°F.) and leave the oven door slightly open to allow moisture to escape.

When the leather is dry enough to be lifted or gently pulled from the cookie sheets, put the leather on cake racks so that it can dry on both sides. Dust the leather lightly with cornstarch or arrowroot powder when all the stickiness has disappeared, then stack the leather in layers with freezer paper, wax paper, or aluminum foil between each sheet. Cover the stack with freezer paper, wax paper, or aluminum foil and store in a cool, dry place.

Prune Leather

1 gallon pitted prunes
1½ cups water

3 teaspoons almond extract
honey (optional)

Place the fruit and water in a large, heavy pot and cook over low heat until the fruit is tender. Drain off the juice and save it for a breakfast drink or punch. (It can be diluted with water, half-and-half or more.) Run the pulp through a food mill or sieve. Discard the skins—they are a little too tough to be used in the leather. Add the almond extract and honey to taste. Then spread the pulp on oiled cookie sheets, or on cookie sheets lined with freezer paper or plastic wrap, about ¼ inch thick. Dry the pulp as the apricot or peach pulp above until it can be lifted out of the cookie sheet. Then place it on cake racks so that it can dry on both sides. Dust the leather with cornstarch or arrowroot powder when all the stickiness has disappeared, and wrap and store as apricot and peach leather above.

Apple Leather

1 gallon apples
1½ cups apple cider
honey

cinnamon, cloves, and/or
   nutmeg, added to taste
   (optional)

Peel and core apples, cut them in pieces, and run them through a food chopper. Catch the juice which runs from the food chopper and return it to the ground apples. (A blender may also be used. This mashes a limited number of apple pieces at a time, but there is no need to worry about escaping juice.)

Place the ground apples and their juice in a large, heavy pot and add 1 cup of apple cider. Apples are drier than other fruit and will scorch as they are heated if no liquid is added to the pot. Place the pot over low heat and bring the apples to a boil. Add more cider if needed to prevent the apples from sticking to the bottom of the pot. If the apples are tart, add honey when the mixture looks somewhat clear and is boiling well. Then add the spices if you wish.

When the mixture reaches the consistency of a very thick apple butter, remove it from the heat and spread the pulp on oiled cookie sheets, or on cookie sheets lined with freezer paper or plastic wrap, about ¼ inch thick. Dry this apple pulp just as the other pulps were dried until it can be lifted from the cookie sheets. Then place it on cake racks so that it can dry on both sides. Dust the leather with cornstarch or arrowroot powder when all the stickiness has disappeared, and wrap and store the apple leather as the other leathers above.

Quince Leather

5 quinces
water to cover
½ cup honey

Wash, core, and chop quinces. Cover with water. Bring to a boil and simmer until quinces have turned a deep pink orange color, 45 minutes to an hour. Drain off juice. Reserve for jelly.

Purée pulp, add honey, and cook in a heavy-bottom saucepan over low heat, stirring frequently, until it is as thick as possible.

Line a cookie sheet or flat pan or platter with cheesecloth, spread the purée on it ½ inch thick, and cover with more cheesecloth. Leave in a cool, airy place for 1 to 2 weeks to dry, turning it every few days to enable it to dry on both sides. When the paste peels off of the cheesecloth easily, lay it between two sheets of freezer paper. Roll out as thin as possible with a rolling pin. Roll up paste in the paper into a tube and store in cool, dry place.

Old-Fashioned Sun-Cooked Preserves

Years ago, when Americans were mostly rural dwellers, sun-cooked preserves were often made by country and farm women. The sweetening (they used sugar, but we'll use honey) and the fruit were cooked together, and the mixture spread out thinly on platters or shallow trays. The trays of preserves were set out in direct sunlight for about 7 sunny days. The sun's gentle warmth gradually tenderized the fruit. Evaporation made the mixture thicker. The result was a thick preserve.

In areas where days are sunny and hot, preserves can still be successfully "sun-cooked" in 3 to 7 days. But if weather conditions aren't favorable, the indoor dryer or stove can take the place of the sun. Indoor drying will probably take considerably less time than 3 to 7 days, but you'll have to experiment and find out for yourself.

To make sun-cooked preserves wash and prepare fresh fruit such as berries, cherries, peaches, or pears. Fruits may be combined, if desired. Berries should be hulled. Peaches and pears should be pared. All pits or seeds should be removed. Berries and cherries, unless very large, can be left whole. Cut other fruit into pieces about the size of medium strawberries.

Cut 4 cups fruit into thin slices and put into a saucepan. Add 1 cup mild-flavored honey. Bring the mixture to a boil, stirring constantly. Remove from heat and pour mixture into shallow trays. It should be about ½ inch deep. Spread out pieces of fruit so they are not on top of each other.

Cover the trays with cheesecloth or plastic wrap, stretched taut so it does not touch the fruit, or put a piece of glass over the top. If plastic wrap is used, fold back a small corner of the wrap, taping it securely, to assist in evaporation. If glass is used, leave a ½-inch strip of the narrowest side of the pan uncovered by the glass. Put the trays out in the direct sun for 3 to 7 sunny days. Tilt the trays slightly towards the sun. Move the trays during the day, if necessary, so that the direct rays of the sun fall upon the fruit. A sunlit window sill or even the backseat of a car will do—as long as there is direct sun and it is very hot weather. Bring in the trays each night and whenever it rains.

When the preserves are as thick as you want them to be, pack into sterilized jars and refrigerate or freeze.

Either wild or cultivated berries are excellent for making sun-cooked preserves. This is a good recipe for using up small berries. The smaller berries can be left whole, giving the preserves a nice appearance. Fruit selected for these preserves should be ripe but firm, not mushy.

Dried Pumpkin Flour

Pumpkin flour, made by grinding up dried pumpkin slices, can be used along with regular wheat flour to add moisture, color, and extra food value to baked goods.

Prepare the pumpkin for drying by following the directions given earlier in this chapter. When thoroughly dry, pasteurize the slices in a 175°F. oven for about 15 minutes. Then pulverize them in a blender.

Store the flour in airtight containers in a cool, dry place, or preferably in the refrigerator or freezer. This flour can replace small quantities of wheat flour in cookie, pie crust, cake, muffin, bread, and pancake recipes.

Vegetable Powders

These are made just like pumpkin flour. Dry your vegetables and then grind them up in a blender. Package separately to combine later, or make your own dried mixes, with or without dried herbs.

Powdered vegetables, added to boiling water and allowed to dissolve for about 1 minute, make instant vegetable broth. They are also good to add flavor to stews, soups, meat loaves, and casseroles.

Dried Tomato Paste

Drying tomato paste in the sun is a simple and satisfying way of turning a bountiful harvest into a concentrated, flavorful paste. Any tomato sauce recipe can be used, but here's a suggestion:

For every 2 pounds of ripe, cut-up tomatoes, add:

½ large onion, chopped	1 sprig thyme, chopped
1 stalk celery with leaves, chopped	¼ bay leaf
	5 crushed peppercorns
1 garlic clove, minced	2 whole cloves
several sprigs chopped parsley	1 teaspoon salt
1 sprig oregano, chopped	

Italian plum tomatoes are best, but other types can be substituted. Simmer everything together (don't add water—just mash the tomatoes a little so they make their own juice) for an hour or so, stirring it every time you pass the stove. Now, 2 or 3 cups at a time, purée the sauce in the blender, and then put it through a sieve if you wish.

Put the pot of strained pulp back on the stove, but turn the heat to low so that it doesn't scorch. Simmer very slowly, uncovered, stirring occasionally, until the pulp is reduced by one-half and quite thick. This will take several hours.

Next, spread the pulp ½ inch thick on plates or stainless steel cookie sheets and put out in the sun. As it starts to dry, cut through the paste in a criss-cross pattern, to allow air to penetrate as much as possible. Protect it from insects with a storm or screen window, a piece of cheesecloth, or netting. A day or two of hot sun will dry the pulp to the stage where you can scrape it off the plates and form it into nonsticky balls.

If you don't live in a sunny climate, you can dry the paste in a warm oven or even in a light bulb-heated box. (See Drying Indoors, earlier in this chapter.) After the paste is rolled into balls, let them dry for a day more at room temperature, and then store them in a tightly lidded jar. To use, dilute with a little boiling water or stock, or add a couple to a batch of minestrone or spaghetti sauce.

Herbs

Leaves, seeds, flowers, and even roots of many plants can be collected and dried in the time-honored way, to add distinctive flavor and aroma to many foods all year round.

Harvesting and Drying Herb Leaves

Herb leaves are cut when the plant's stock of essential oils is at its highest. In the leafy herbs—basil, savory, chervil, and marjoram—this occurs just before blossoming time. Lemon balm, basil, parsley, rosemary, and sage can be cut as many as four times during the outdoor growing season. Cutting should be done on the morning of a day that promises to be hot and dry. As soon as the dew is off the plants, snip off the top growth—perhaps 6 inches of stem below the flower buds.

If the leaves are clean, do not wash them; some of the oils will be lost in the rinsing process. If the leaves are dusty or have been thickly mulched, however, wash them briefly under cold water. Shake off any excess water and hang the herbs, tied in small bunches, in the sun, just until the water evaporates from them.

Before the sun starts to broil them, take them in and hang them in a warm, dry place which is well ventilated and free from strong light. Traditionally, herbs were hung above the mantles of kitchen fireplaces or in attics. Herbs are tied and hung leaves down so that the essential oils in the stems will flow into the leaves. To prevent dust from accumulating on the drying leaves, place a brown paper bag with many holes punched in it for circulation around the leaves. Gather the bag together at the stems and close it with string, a rubber band, or wire tie. A paper bag will also shade the leaves from direct light which would otherwise darken the leaves unnecessarily. Sage, savory, oregano, basil, marjoram, mint, lemon balm, and horehound are best dried in this fashion.

The leaves from thyme, parsley, lemon verbena, rosemary, and chervil may be removed from the stems and spread in a single layer to dry in indoor or outdoor dryers used also for fruits and vegetables. Be certain, though, that the screen or netting on the tray is very fine so that no leaves can fall through its openings. For best flavor, the temperature inside the dryer should stay under 105°F. When thoroughly

Traditionally, bunches of fresh herbs were hung by the hearth where the fire's heat would gently dry them. Herbs are hung upside down so that the aromatic oils in the stems will flow into the leaves as they dry.

dry, remove leaves from stems, but don't crush them unless you plan to use them right away.

Harvesting and Drying Herb Seeds

To harvest seeds, gather anise, coriander, cumin, caraway, dill, and fennel plants when the seed pods or heads have changed color, but before they begin to shatter. You may spread the pods one layer thick on drying trays just as you do the leaves. When they seem thoroughly dry, rub the pods between the palms of your hands, and the seeds should fall out easily. You can also dry the seeds by hanging the whole plant upside down inside a paper bag to dry. As the seeds dry and fall from the pods, the paper bag will catch them.

Harvesting and Drying Herb Flowers

Flowers to be used in cookery are cut on the first day they are opened. If petals alone are used, they are removed from the calyx and spread on a tray. Rose petals should have their claws—the narrow white portion at the base of each petal—removed. Flower heads used for tea, like camomile, are dried whole.

Harvesting and Drying Roots

The roots of certain plants, particularly angelica, burdock, comfrey, ginseng, ginger, and sassafras, are highly aromatic and can be dried and cut for candy, teas, and cold beverages. These roots are much thicker than the delicate seeds, flowers, and leaves of other plants, and the drying process is a long one.

So as not to injure the plant, roots should be dug out and cut off during the plant's dormant stage, when there is sufficient food stored in the plant cells. This is usually during the fall and winter months. Cut off tender roots—never more than a few from each plant. Scrub them with a vegetable brush to remove all dirt. If the roots are thin, they may be left whole, but for quicker drying, thick roots should be sliced lengthwise. Place one layer deep in any indoor or outdoor dryer.

Storing the Herbs

Leaves may be crushed before being stored away, but they retain their oils better if they are kept whole and are crushed right before they are used. Seeds should not be ground ahead of time because they deteriorate quickly when the seed shell is broken. If the seeds are kept whole and stored properly, they will keep several years. Store leaves, seeds, flowers, and roots in tightly sealed jars in a warm place for about a week. At the end of that time, examine the jars. If there is moisture on the inside of the glass or under the lid, remove the contents and spread it out for further drying. Checking the jars is especially important if you are storing dried roots because it is difficult to know when they are completely dry. If further drying is not done, there is a strong chance that mold will develop, and the leaves, seeds, flowers, or roots will not be fit to use. Do not store herbs in cardboard or paper containers, because these materials absorb the oils and leave the dried herbs tasteless. Ideally, herbs should be stored in a cool place, out of strong light, either in dark glass jars, in tins, or behind cabinet doors.

Storing herbs in dried form is so popular because it is a simple means of preservation which produces a product that can be conveniently used in the kitchen. Dried herbs can be taken from their jars just as they are needed, to be mixed with foods while they are cooking or just before they are served. However, the fresh quality of the just-picked herb is lost in the drying process. This is particularly true of chervil, borage, burnet, chives, parsley, and basil. For this reason, some people prefer to preserve some of their herbs by other methods, like freezing (see page 53) and preserving in vinegar (page 248).

While herbs are usually stored separately, there are advantages to storing some herbs in combination. During storage, the distinct aromas and flavors of the different herbs are given a chance to blend together and form delightful herb mixtures that can be sprinkled on foods or tied in cheesecloth and plunged in soups and stews while they are simmering. Below you'll find some recipes for popular herb combinations. It is important that you do not mix fresh herbs with dried herbs when measuring out for recipes unless you remember that 1 tablespoon of fresh herbs equals ½ teaspoon of dried herbs or ¼ teaspoon of dried powdered herbs.

Fines Herbes

Fines herbes are a mixture of finely ground fresh or dried herbs that is sprinkled on food in the last few minutes of cooking or added to the food just before it is placed on the dinner table. Fines herbes can be made from equal parts of any of the following herbs: basil, celery seed, chervil, chives, marjoram, mint, sweet savory, parsley, sage, tarragon, and thyme. Fines herbes are usually served in or on sauces, soups, cheese dishes, and egg entrées.

Bouquets Garnis

Bouquets garnis are a mixture of dried herbs which are tied in a cheesecloth sack or put in a spice ball and plunged into a stew or soup to be left in the food while it is cooking. The sack and the herbs are removed when the stew or soup is taken from the heat. Bouquets garnis can be made from almost any spices, but we recommend the combinations below for starters.

For meat and vegetable soups and stews, mix together:

1 part thyme
2 parts rosemary
1 part sage

2 parts savory
2 parts marjoram

For fish stews and soups mix together:

½ part dill
1 part basil
¼ part oregano

1 part lemon balm
½ part thyme
1 part savory

Root crops keep well when left right in the ground where they grew. Cover the rows with mulch to prevent the ground from freezing, and mark them so that you'll know where to place your shovel when you want to dig up some vegetables.

Underground Storage

Underground storage is perhaps the easiest method for storing large amounts of food crops, for once the storage area is constructed, there is very little effort and expense involved in storing a good many fruits and most vegetables. But store only sound fruits and vegetables of good quality this way. Diseased or injured ones may be used early in the fall or preserved in some other way by cutting out bad sections and freezing, canning, or drying the rest. Harvesting, in most cases, should be delayed as long as possible without danger of freezing. All vegetables and fruits to be stored must be handled with great care to avoid cuts and bruises. Food that needs to be cleaned, usually just the root crops, should be rubbed lightly with a soft cloth or glove or rinsed under gentle running water. (Let excess water evaporate before storing.)

VEGETABLES IN UNDERGROUND STORAGE

Vegetables not listed here do not keep well in underground storage. Refer to the chapters on canning, freezing, drying, and juicing for other methods of preservation.

Beets, Carrots, Turnips, Rutabagas These root vegetables should be harvested in late November, after 30°F. nights. Root crops of this type can be stored by removing the tops—do not wash—and placing them in an area just above freezing, with 95 percent humidity. They can be packed in cans, boxes, or bins, surrounded by straw, or they can be placed in moist sand, or in any outdoor storage pit or root cellar. They may also be left in the garden where they grew. In the middle of the winter you can go out, brush away the snow, dig them up and use them. If you want to make the digging easier, cover the rows with leaves or straw. Make the cover a foot deep, and weight it down with chickenwire or rocks.

Cabbage, Chinese cabbage   Prepare these for storage by removing loose outer leaves. If produce is to be wrapped with newspaper, burlap, or some other material, roots and stem should be removed; otherwise leave these in place. Wrapped cabbages should be stored in boxes or bins at a just above freezing temperature in a cool, damp area. When stems and roots are left on, any of the outdoor storage areas that are made all or in part of damp sand or soil are effective. Cabbage emits a strong odor during storage that is usually not welcome in the house, so most people prefer to store it in one of the many outdoor storage arrangements.

Celery   Celery is best maintained by pulling the crop. Leave the tops dry, do not wash. The roots should be placed in slightly moist sand or soil, and the plant maintained at 32° to 34°F. To avoid odor contamination, do not store with cabbage or turnips. In areas without very severe winters, celery may be left in the ground, covered with a thick layer of leaves or straw.

Endive or Escarole   This lettuce, which goes under both names, will keep 2 or 3 months in storage if kept at 32° to 34°F. Like celery, endive roots should be kept in slightly moist sand or soil.

Garlic   Garlic must be cured before storing. Dry garlic thoroughly, making sure bulbs are not in the direct rays of the sun. For large quantities, garlic may be cured in the garden with its tops covering the bulbs. (Small quantities may be bunched and tied or braided and hung in a well-ventilated cool room to store and dry.) Remove tops and roots with knife or shears, leaving 1 inch of root on the bulb, and store as onions in cool, slightly humid (60 to 75 percent) area.

Kohlrabi   Remove leaves and roots and store at 32° to 34°F. in an area with about 95 percent humidity. Root cellars and basement storage rooms are ideal locations.

Onions   Onions must be cured before storing. Leave the vegetable on the ground after pulling for at least 2 or 3 days so that several layers of onion skin dry out and form a protective layer. Then place in crates in an open shed for several weeks to complete curing. Remove the tops and store in bins or string bags at temperatures ranging from 33° to 45°F. in an area with about 60 to 75 percent humidity. Attics often prove to be good storage areas.

Parsnips, Salsify, Jerusalem artichokes   These can be left in the ground throughout the winter. To make digging easier, cover the rows with about 1 foot of leaves or straw before the ground has frozen. Jerusalem artichokes are thin skinned and do not keep well once dug up, so dig up no more than a 2 weeks' supply at a time.

Peas, Soybeans, Beans   All should be shelled and dried. To eliminate fumigation, which is practiced by commercial growers in order to kill weevils, simply heat the crop in an oven for 30 minutes to an hour at a constant temperature of 135°F. Spread the vegetable in pans for this treatment, and do not let the temperature drop or rise significantly. After drying thoroughly, place in jars or bags for storage. The temperature of the storage area is not important, but it must be dry.

For more on drying these vegetables, see page 114.

Peppers   If green and mature, they should be picked just before frost. They may be kept for 2 to 3 weeks at temperatures between 45° and 50°F. in moderate humidity. Hot varieties of peppers store best if they are dried first and then stored in a cool, dry place. Do not store them in cellars. To dry hot peppers, pull the plants from the ground and hang them up until dried, or harvest the peppers and string them up on a line to dry.

Potatoes These tubers must be stored in the dark. For several months after harvest they can be held in almost any storage location, as this is their normal resting cycle. After this period, temperatures between 34° and 41°F. are necessary to prevent sprouting. The lower temperatures tend to turn starch to sugar and sweeten the vegetable. Only experience with the crop will enable you to determine proper storage in your area. During the storage period, moisture should remain high. Never store with apples.

Pumpkins, Winter squash Both must be cured before storage. This helps to toughen their skins so they will keep better. Leave them in the field for 2 weeks after picking. If weather is near freezing, cure squash in a room with a temperature of about 70°F. for several days. Leave a partial stem on the fruit and take exceptional care to prevent bruising, storing only the best undamaged produce. After curing, place them gently on shelves, separated from each other, in a 50° to 60° F. dry place. Examine them every few weeks for mold. If you find some, wipe the squash carefully with a cloth made slightly oily with a good vegetable oil. Treated this way, they will keep 5 to 6 months.

Sweet potatoes Sweet potatoes should be free from injury, and need to be cured before final storage. Lots of air circulation and high temperature over a period of 10 days to 3 weeks are necessary to eliminate excess water, change some starch to sugar, and cause "corking over" of cuts in the skin. After curing, sweet potatoes should be placed in a warm, 50° to 60°F. room which is well ventilated, with moderate humidity (up to 75 percent).

Tomatoes Ripe tomatoes do not store well, but green or slightly red ones can be held in storage and be encouraged to ripen there. Harvest all tomatoes that are of good size, be they ripe or still green, just before the first killing frost. Remove tomatoes from plants, wash, and allow to dry before storing. Separate green tomatoes from those that show red, and pack green tomatoes no more than two deep in shallow boxes or trays for ripening. Green mature tomatoes will ripen in 4 to 6 weeks if held at 55° to 70°F. in moderate humidity.

FRUITS IN UNDERGROUND STORAGE

Many major fruits do not store well for extended periods of time. Of the ones that do—most notably apples and pears—the varieties vary in keeping quality, and it is best to plan to grow good-keeping varieties if you know that you will be storing many of them.

If the fruit you wish to store is not listed here, refer to the chapters on canning, freezing, drying, and juicing for alternative preservation methods.

Apples These are among the better-keeping fruit. The Winesap and Yellow Newton are among the best apples, frequently lasting from 5 to 8 months satisfactorily. Next in keeping quality are the Stayman-Winesap, Northern Spy, York Imperial, Arkansas Black Twig, Baldwin, Ben Davis, and Rome Beauty. Normal storage ranges from 4 to 6 months with these varieties. Jonathan, McIntosh, Cortland, and Delicious (red or golden) can be kept for shorter periods. There are other factors that influence keeping qualities of different apple varieties such as locality (McIntosh apples grown in New England store better than those grown in the Middle Atlantic states), seasonal conditions, maturity when picked, and length of time between picking and storing.

Good keeping qualities are increased with careful handling to prevent bruising. Storage at between 30° to 32°F. and 85 to 90 percent humidity is preferred for most varieties. Yellow Newton, Rhode Island Greening, and McIntosh are better stored at 36° to 38°F. Wrapping in oiled paper or in shredded paper helps prevent scald, acknowledged to be the most serious disorder, particularly with Cortland and Rhode Island Greening.

Grapes Grapes should be cooled to 50°F. as soon as possible after picking and spread out in single layers. Allow the fruit to remain in this condition until the stems shrivel slightly. Then place the grapes in trays no more than 4 inches deep in a cellar which is slightly humid and has a temperature of about 40°F. Stored this way, grapes will keep several months.

Oranges Florida oranges may be kept 8 to 10 weeks at 30° to 32°F. with 85 to 90 percent humidity. California oranges can be kept 6 to 8 weeks at 35° to 37°F., but are subject to rind disorders at lower temperatures.

Quinces This fruit will keep 2 to 3 months if picked before it is thoroughly ripe and if held in a cool, moist storage area.

Peaches Peaches are fairly perishable and may be stored only several days to 2 weeks in a cool cellar.

Pears These can be held from 8 weeks to several months, depending upon variety. Winter Nelis, Anjou, and Easter Beurre are the most hardy, with Bosc, Kieffer, Bartlett, Comice, and Hardy in the lower range.

Pears should be harvested in a condition that would seem to the amateur to be immature. If allowed to begin to "yellow" on the tree, pears develop hard, gritty cells in the flesh. They should be harvested when the dark green of the skin just begins to fade to a yellowish green and the fruit begins to separate more or less readily from the tree.

Pears ordinarily do not ripen as satisfactorily at storage temperatures as apples. For highest eating quality they should be removed from storage while they are still comparatively hard and green, and ripened at room temperature with a high relative humidity. Pears often keep somewhat better in home storage if wrapped in newspaper or other paper. This fruit, like apples, should be stored at a temperature as close to 32°F. as possible and with a high relative humidity, ranging from 85 to 90 percent.

Where and in What to Store

While many of the storage ideas that follow are suitable for fruit storage, fruit should never be stored with potatoes, turnips, or cabbage. The gases released from apples during respiration can cause potatoes to sprout. Cabbage and turnips can transmit their odor to apples and pears. (Wrapping apples and pears in paper or packing them in maple leaves in barrels is recommended if these fruits must be stored with cabbage and turnips, because these materials will prevent absorption of such odors.) Constant air circulation in fruit storage is essential to remove gaseous substances, such as ethylene and volatile esters, which, if they linger, can speed up the ripening process.

Storage Containers

In most storage areas there is a need for small containers. Here are some suggestions:

Wooden boxes, which are originally designed to store and ship

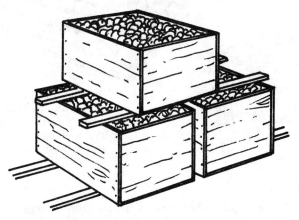

When stacking boxes, place furring strips between them, the floor, and other tiers to permit full air circulation.

apples and other fruits, make ideal storage units for root cellars or larger storage areas. Interior packing for stuffing between food may be leaves (dry and crisp), hay, straw, string-sphagnum moss, or crumpled burlap.

Gardeners Helen and Scott Nearing store their Maine garden crops this way with great success. By alternating layers of dried leaves with layers of produce in wooden boxes, they have firm and edible potatoes, apples, rutabagas, carrots, and beets for as long as 50 weeks after they stored them, well into the next growing season.

Pails, baskets, and watertight barrels are used just as boxes are. Layer packing material and produce alternately, finishing with 2 inches or more of packing at the top. These containers are used in pit storage areas as well as in larger units.

Metal tins are adaptable for storage, providing you patch-paint raw metal or use galvanized metal to keep rusting at a minimum. Leave them open topped or cover produce with leaves, sphagnum moss, or straw.

Bins are used primarily in larger storage units. They should be constructed for permanent use 4 inches off the floor. They are good for potato and other root crops.

Onions can be safely stored in crates or mesh bags, providing they are placed in a cool, dry place, slightly above the freezing point.

Watertight barrels and pails are excellent for storing vegetables. Since vegetables need ventilation, cover tops with a few inches of moist crumpled burlap or string-sphagnum moss.

Orange crates and mesh bags are excellent for onion storage and other foods that need good air circulation. Old, clean stockings are also good for storing onions and garlic. Stuff cured onions or garlic bulbs right down in the toe and up to the top, leaving a little room for twisting and knotting closed and hanging on a nail or hook.

In the Storage Room

Most vegetables shrivel rapidly unless stored in a moist atmosphere. Shriveling may be prevented: 1. By keeping the air quite moist throughout the storage room. 2. By protecting the vegetables either by wrapping them or putting them in closed containers. 3. By adding moisture directly to the vegetables now and then.

If the first method is used, provide a layer of coarse, well-washed gravel, 3 inches thick, on the basement floor. The pebbles should be about equal in size. The shelves and all other equipment are then placed on the gravel. Whenever the cellar air gets too dry (a hygrometer will tell) the gravel is sprinkled with a watering can. Care must be taken, however, that no surplus water is left on the floor.

The wet gravel on the floor will evaporate much more water than it would if it were on the floor alone, the gravel having a much greater surface area. This sprinkling has to be repeated when the relative humidity of the air falls below 80 percent, and for this reason a watering can should always be kept in the cellar. It must be said that one sprinkling usually goes a long way, especially when the cellar is filled to capacity. Garden-fresh produce releases a considerable amount of humidity, any surplus of which can usually be removed with moderate ventilation. However, harsh and prolonged ventilation, open doors, insufficiently insulated walls and doors, or a sudden change of outdoor humidity can disturb this balance and make water sprinkling necessary.

Only those vegetables which require moist conditions may be stored successfully this way. Vegetables like onions, pumpkins, and squash must be stored in some other location where the air is drier. Moist storage rooms are not well suited for canned foods because can lids and metal cans or other metal containers will rust readily.

A dry storage room is more satisfactory for canned foods and also vegetables which prefer a dry place. Root crops like carrots, which shrivel easily, may be stored in a dry room by adding water directly to the vegetables when needed or by placing them in closed containers. Large crocks, metal cans, tight wooden boxes, and barrels are all suitable. Closed containers should be clean, dry, and lined with paper before the vegetables are packed; a layer of paper may also be placed between each layer of vegetables. If stored in this way, sand or other materials are not needed to prevent shriveling.

Those gardeners who prevent excessive shriveling by adding water directly to the vegetables commonly store such crops as carrots in crates, boxes, or baskets which are kept covered with burlap or a piece of old rug or carpet. Water is added by sprinkling as needed, and the covering itself is kept moist.

Regardless of the method, stored vegetables should be carefully watched to avoid loss from decay, growth, or excessive shriveling. Decaying vegetables should be taken out as soon as noticed. If vegetables start to sprout and grow, the temperature is too high. Vegetables which begin to shrivel a great deal should be wrapped, placed in closed containers, or sprinkled with water. In a moist storage room extra moisture may be provided by sprinkling the floor.

Cleanliness is one of the first things to be observed in storage cellars. Walls and ceiling should be whitewashed and the floor must be kept clean at all times. Dead leaves, stalks, and the like must be removed from under shelves and planking. Spring cleaning is obligatory, and it is best done during dry summer days. Shelves, crates, baskets, and other containers must be brushed in the open air or with all windows open.

Basement Storage Ideas

The cellars in old houses didn't have dirt floors just because they were cheap and easy to construct. These dirt floors made cellars excellent food storage areas; they helped to keep food cool and moist. These old root cellars have long been used for storage in the colder parts of our country, and some houses without central heating on farms and homesteads are now being built with dirt floor cellars for just this reason. These cellars usually have an outside entrance which can be opened to ventilate the cellars and regulate the temperature inside. Many have insulating material on the ceiling to prevent cold air in the cellar from chilling the whole house.

Centrally heated homes with concrete floor basements are generally too warm to be used just as they are for food storage. But with a little ingenuity and a minimal expense, part of just about any basement can be converted into a storage room.

Essentially the storage room is a place where temperature and humidity are held to the proper level for keeping produce. This means lower-than-usual household or basement temperatures, ranging from 30° to 40°F.

With few exceptions, the most desirable temperature is at or very near 32°F.—the freezing point of water. Except for potatoes, vegetables are not injured at this temperature. It is difficult, however, to keep the temperature as low as 32°F. without danger of it going low enough to cause actual freezing during exceedingly cold weather. It is suggested, therefore, that the storage room temperature be kept between 35° and 40°F. Such temperatures cannot be reached and kept there except in a room separated from the rest of the basement, reasonably well insulated, and having adequate ventilation.

The size of the basement storage room will vary with the space available and the family's needs; 8 by 10 feet is suggested for most families who plan to store both vegetables and other foods in the same room; a room this size will hold 60 bushels of produce. A storage room, if properly constructed and managed, will be suitable for nearly all foods commonly preserved. Where practical, the storage room should be located either in the northeast or northwest corner of the basement and away from the chimney and heating pipes.

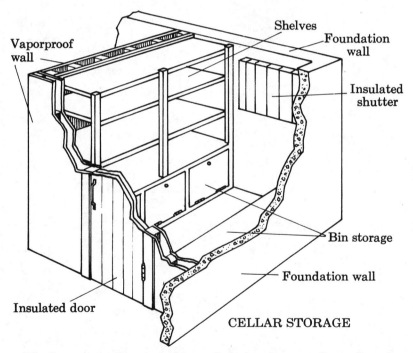

The basement often provides needed storage space for fruit and vegetable storage, but too much heat and too little humidity frequently found in basements of centrally heated homes often cancel out its usefulness. A cellar storage room can be constructed rather easily by blocking off and insulating a corner of a basement. See the accompanying plans.

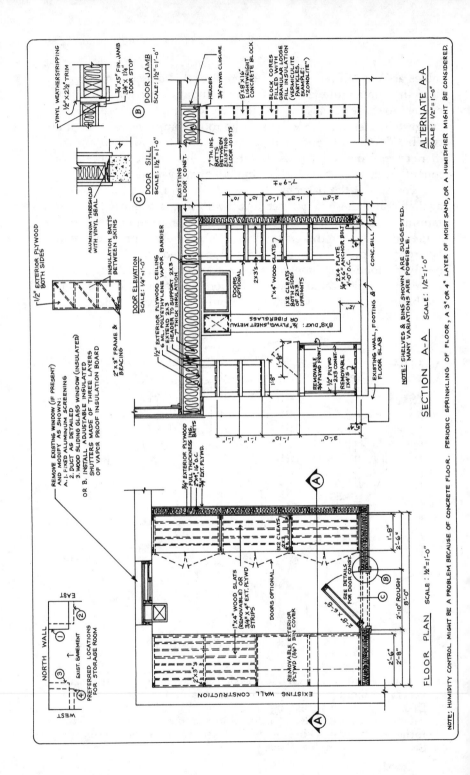

NOTE: HUMIDITY CONTROL MIGHT BE A PROBLEM BECAUSE OF CONCRETE FLOOR. PERIODIC SPRINKLING OF FLOOR, A 3" OR 4" LAYER OF MOIST SAND, OR A HUMIDIFIER MIGHT BE CONSIDERED.

At least one wall having outside exposure should be used, preferably the one with the least sunlight, on the north side and with a window that is easily reached. The other walls can be made of wood which will do a good job of keeping the storage area cool, providing their construction is tight.

Two types of insulating material can be used: board or loose fill. It is important to keep them dry or their insulating properties will be reduced, so moisture-vapor barriers are used, inside and out, such as dampproof paper, tar, and asphalt.

Board insulation can be nailed to the walls and ceiling. Two thicknesses should be used to prevent leakage through joints.

Loose-fill insulation includes planer shavings, cork dust, and minerals. Stagger the studs so they are exposed on one surface only. Sheathing and dampproofing should be done when the wall space is being filled.

Planer shavings are very satisfactory insulation for side walls and ceiling; they are dry and do not tend to settle. Add hydrated lime to the shavings, from 20 to 40 pounds per cubic yard, to help keep the shavings dry and as a repellent to vermin and rodents. Tamp the shavings as they are filled between studs until a density of 7 to 9 pounds per cubic foot is obtained.

Loose-fill insulation is not used where floor insulating is needed. Use board-type insulation—wood-fiber insulating wall-board to floor insulation. (The floor to side wall jointing is the same as for wall corners.) The insulating board is then mopped with hot tar, and cement or other flooring is laid on top. Where chinks occur at rough wall or floor surfaces, caulking compound, which fills the space and protects the exposed insulating material from dampness, is used.

Temperature of the surrounding basement will determine thickness of the walls. A prevailing temperature of 60°F. in the rest of the basement outside the storage room calls for the following thickness of insulating materials:

3 inches: wood-fiber insulating board, cork board, granular cork, fibrous rock, rock wool,
4 to 5 inches: planer shavings,
6 inches: compressed peat,
8 to 9 inches: white pine and soft woods.

Lower basement temperatures will require less insulation.

The walls to close off a corner of the basement are easily constructed with a 2-by-4-inch framework, sheathing on both sides, and 3-inch insulation batts between the studding. Leave an opening in one wall for a door. This may be framed with 2-by-2-inch studs, faced on each side with ¼-inch plywood, and the center filled with insulation. Fit the door tightly and secure it with a type of latch that holds it firmly closed.

To insulate the ceiling of the storage space, sheathe underneath the ceiling joists and apply 4 inches or more of insulation between the joists, extending the insulation out over the walls of the storage space.

Ability to maintain a desirable temperature range of 35° to 40°F. in the basement storage room depends largely on outside weather conditions. In both early fall and late spring, day temperatures are likely to be higher than those desired in the storage area. Therefore, it is important during these times to keep windows closed. As a general rule, windows should be opened whenever the outside temperature is lower than that in the storage room and the inside temperature is above 40°F. When the temperature in the storage room drops to 35°F., windows should be closed. Place one reliable thermometer inside and another outside a window for guidance.

Light must be excluded from stored vegetables and fruits, so cover the windows with opaque material. Wide, wooden louvers

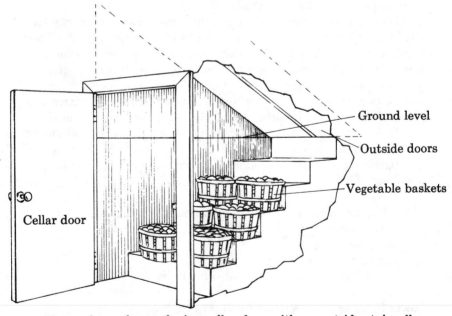

Ground level

Outside doors

Vegetable baskets

Cellar door

If your house has a sloping cellar door with an outside stairwell into the basement, you can adapt this area into storage with relative ease and little expense.

fitted to the outside of the window frame aid in excluding light if the window is opened for ventilation in the daytime. Cover louvers, or open windows if louvers are not used, with screening to keep out insects and animals.

Using the Cellar Steps

A small, but simple and inexpensive storage area can be made by utilizing the cellar steps of an outside cellar entrance. Install an inside door to keep out basement heat at the bottom of the steps. If you want to create an even larger storage area, build inward into the basement, but take care to insulate extra wall space. Temperatures in the stairwell will go down as you go up the steps, and a little experimenting will help you determine the best levels for the different crops you are storing. If the air is too dry, set pans of water at the warmest level for extra humidity.

Window Wells

In a pinch, window area wells can hold bushels of food over part or all of the winter. Because they are adjacent to the house and below ground level, the temperatures inside them should remain fairly constant throughout the winter. Cover the wells with screening and wood. To raise the temperature in very cold weather, open the basement windows and allow some house heat to enter. When the temperature in the window well gets too high, remove wood from the

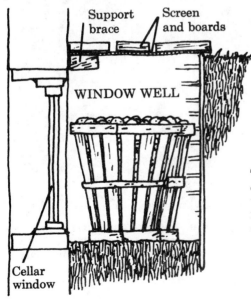

The insulating capabilities of the ground should keep a window well cold, but above freezing except in the colder parts of the country. Cover the wells with screens and wooden boards to keep the ground's heat in and the cold air out.

top of the well to permit the cold, outside air to cool the area. If basement windows open inward or are the sliding type, access can be convenient and simple during the cold winter months.

A Root Cellar Adjacent to the House

In Myerstown, Pennsylvania, Enos Hess, with the help of his county farm agent, has constructed an extensive and unique storage facility in the form of a modern "old-fashioned root cellar." He has constructed a large garage-type building attached to the house in which he stores packaged foods. At the outside wall of the structure, and near the garage door, a door opens onto a stairway and conveyor belt which leads downward for about 12 feet.

Behind the door at the base of the steps is a 12-by-18-foot room with 12-inch-thick concrete block walls and a 10-inch-thick reinforced concrete ceiling, covered by 4 solid feet of earth. The cellar is built over a natural spring which is covered by a 4-inch-thick concrete slab. Hydrostatic pressure forces the cool water up through the slab,

Enos Hess has successfully stored a variety of crops in open boxes in his storage room, which is something of a modern "old-fashioned root cellar." The entire chamber is built of wooden siding over 12-inch-thick concrete block walls and a 10-inch-thick reinforced concrete ceiling—all under a solid 4 feet of earth.

wetting the surface of the concrete floor. Wood slats lie over the concrete, and any excess water is pumped out of the cellar automatically.

Air circulation is most effective. Intake air entering through a 12-inch-round concrete overhead flue is directed out at the floor through an adjustable grill. The air is cooled as it passes over the water and flows through the room.

Asked why he installed a concrete floor when the water condition was as indicated, Enos explained there could be a muddy condition if he hadn't used concrete. He is so well satisfied that when he next builds the adjacent 12-by-30-foot room, he will not change this proven method one bit.

The flat ceiling created one problem he had not anticipated which will be solved in his new room: condensation on the ceiling causes dripping all over the room. A tentlike installation of plastic sheeting sloping to either side of the room catches the water and directs it to the walls, and thence to the floor.

Enos Hess has very successfully stored potatoes, beets, and carrots in open boxes, and celery and cabbage in plastic bags well into spring of the following year.

A Homemade Storage Unit

Do you know of a supermarket going out of business or remodeling? With a little luck, some effort, time, and determination, you may

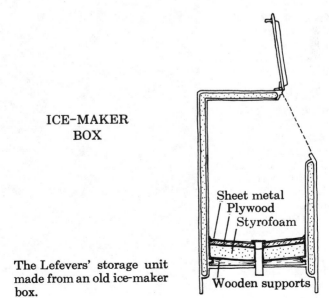

ICE-MAKER
BOX

Sheet metal
Plywood
Styrofoam

The Lefevers' storage unit made from an old ice-maker box.

Wooden supports

be able to have a 6-cubic-foot cold storage unit sitting in your house for a meager price. Grace and Tim Lefever of Sonnewald Farm near Spring Grove, Pennsylvania, do. Although they had to chop up and cart away a 6-inch concrete slab in the process, the Lefevers got an insulated ice-maker box from a supermarket for free. They put it in an unheated area in their house and built a new floor for the box by placing 5-inch-deep Styrofoam between wooden supports that were cut to a uniform slope to the drain in the center. This understructure was covered with plywood and finally faced off with a sheet metal floor.

A Concrete Storage Room

Root cellars can be made from scratch by making a concrete structure and then burying it below ground level. For details, see the plans and material specifications that follow.

But storage areas need not be so intricate. Many gardeners and small-scale farmers have found that the ground itself can protect vegetables and fruits from winter's freezing temperatures. They have taken advantage of the ground's insulating qualities to make small, efficient storage space for their garden products.

John Keck, of the Rodale Organic Experimental Farm, didn't have a cellar to work with, so he constructed an outdoor, underground storage area from an old steel "REA" Express truck body. He supported the roof of the truck body with pipe columns and light beams, imbedded it into a bank and sodded it down with earth. (He found that the steeper the bank and the more earth coverage, the better.) A small stairway down to the truck acts as a door entry. Ventilation is provided through the roof in two places.

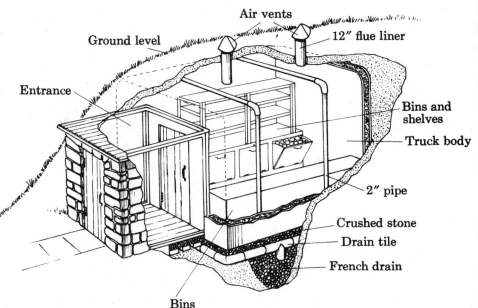

The truck body rests on a bed of crushed stone and is covered with earth for insulation. An entrance made from concrete blocks makes the storage area easily accessible. See plans on page 148.

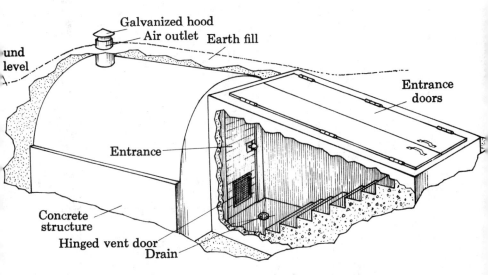

For proper insulation, make the depth of the earth covering this storage structure 12 inches deeper than the local frost line depth. This means that if the frost line is 30 inches deep, the earth cover should be 42 inches in depth. See plans on page 149.

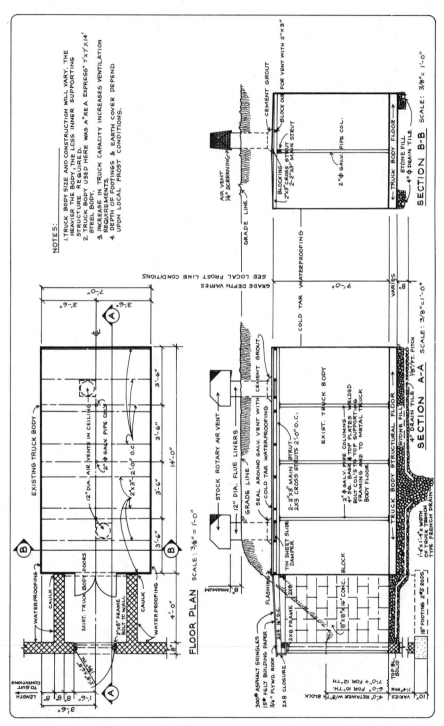

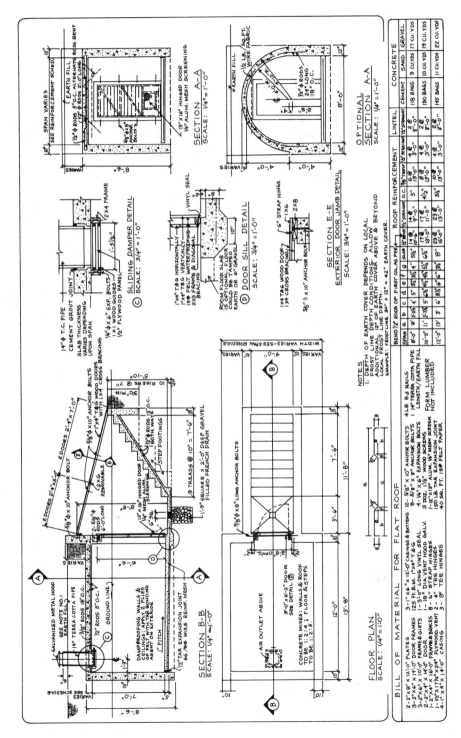

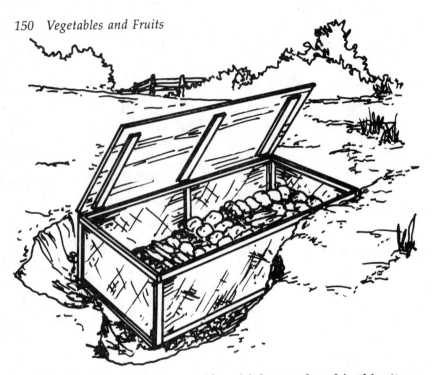

Separate layers of root vegetables with layers of sand in this pit-type storage area to retain moisture and keep the temperature constant.

Soil-Pit Storage

The glacial plain of northwestern Pennsylvania provides excellent conditions—moist soil with a winter temperature around 52°F.—for soil-pit storage like the kind Jane Preston devised:

> Our pit is neatly cut out, 3 feet wide, 6 feet long, and 2 feet deep. It is on sloping ground, so that excess water promptly drains away. We lined the inside entirely with ¼-inch hardware cloth, carefully kept tight—as it *must* be—to keep out rodents. The top is finished with a 2-by-4-inch frame and a neat wooden lid. (If mice do get in, they are in "clover" and you are in deep trouble!)
>
> At harvest time—late autumn—select carrots, beets, parsnips, potatoes, turnips, and the like. Wash them free of soil, but be careful not to bruise the skin. Put a layer of clean, sharp builder's sand (washed) on the pit bottom. Then place a neat layer of root vegetables. Do not "dump" vegetables into the pit.

After the first layer of root crops is in place, cover with a layer of sand, then continue as you fill the pit. A "map" is mighty helpful in keeping account of where different roots are to be found. Finish the pit with a final layer of sand.

Four bales of straw can be laid on the cover, and this then is covered with a plastic sheet to keep off snow. The insulation provided by the straw bales not only helps keep things "warm" in winter (avoiding freezing), but also cool in the warm days of early spring. Good-quality vegetables have been taken from the pit as late as May—even early June. Then it is time to clean everything out and let in sunshine and ventilation over the summer.

Our pit has been in steady use for over 10 years. We have had to replace the wooden parts or mend wire on occasion, but that's about all. And we've been able to supply a homegrown table with good varieties of all root vegetables, plus cabbage and celery as well. The secret, if there is one, resides in 100 percent humidity at a very uniform temperature all winter.

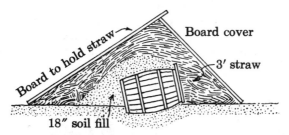

A variation of the Prestons' soil pit storage is the barrel pit, which is made by partially burying a barrel with earth and covering it with straw and used lumber.

Hay-Bale Storage

In Ohio, gardener John Krill stores his apples and root crops in on-top-of-the-ground storage areas. Building these areas is quite simple, and the materials used may be turned into mulch or compost after the storage season. Here John describes how he built them:

Bales of spoiled hay are commonly available nearly everywhere at low cost. Bales average in size about 42 inches long by 18 inches high and 14 inches wide. I lay two bales end-to-end. About 14 inches across from them, I place two more bales end-to-end parallel to the first two. This is done on a well-drained spot, of course. The ends are closed off by placing a bale across each end. A thin layer of hay is then placed in the bottom of the resulting "box."

Apples, pears, potatoes, or other root crops are then placed inside of this hay-bale enclosure. Do not dump the produce in. Careless handling will cause bruising which will quickly cause rot. Place carefully by handfuls until within several inches of the top. Cover the top with hay. Place hay bales across the opening. Place a stone under each bale on top to keep it from completely sealing the opening.

Wait until freezing weather sets in, then remove the stones so that the top bales may completely seal the opening. This allows for natural ventilation of the stored crop. When unseasonably warm weather comes, lift one bale and place a stone under it to permit ventilation until severe cold again sets in. Then remove the stone to again seal the top.

The great thickness of the hay bales makes excellent insulation against cold. Enough air seeps through to create ventilation without permitting freezing cold to enter. Dig a shallow trench alongside of the bottom bales to carry away rain.

These hay-bale enclosures are so quickly and simply made that a number of them may be constructed. No soil is heaped up

HAY-BALE STORAGE

Top-of-the-ground storage is suitable for most root crops. Here, a rectangle is made from bales of hay. A lid made from additional hay bales covers the food in the center. A stone can be placed under the top bales for ventilation during mild weather and can be removed when freezing weather prevails.

around them as in most other methods of storage. Access to the stored produce is quickly and easily gained. A single top bale is removed to expose the interior. It is easily replaced after removal of a quantity of the contents.

After the contents have been used up, the hay bales are converted either into mulch or added to the compost pile. Thus the bales play two important parts in the scheme of the organic gardener. But the best appreciated part is the lack of hard physical labor and trouble in constructing the hay-bale storage enclosures.

Hay Storage in the Garden

I have kept tomatoes ripening on their vines long after frosts invaded our area. The method is quite simple and highly effective. I stake my tomatoes and the stake plays an important part in the method. All small tomatoes that have no chance of ripening are removed. Only those showing signs of even the faintest blush color are kept on the vine. The vine is then tied as compactly as possible to the stake.

Next, old hay is pushed up around and over each tomato plant like the skirt around the waist and legs of a hula dancer. Keep this hay wrap loose and 3 or 4 inches thick, and secure by wrapping with twine around the tomato plant and its stake. The stake makes a support for the entire arrangement. To pick tomatoes, carefully part the hay without pulling it free. Remove the ripened fruit and replace the hay.

Tomatoes kept in this fashion are far more trouble-free than by other methods commonly used. Best of all, the tomato tastes like a garden-ripe fruit. They will slowly ripen in this fashion until freezing weather arrives to stay.

Storing in Garbage Pails

Gardener Elmer L. Onstott, of St. Louis, Missouri, had a problem keeping his late-summer vegetables garden-fresh over the winter because he had no room for basement storage. A good part of his storage problem was solved when he discovered that 10-gallon garbage pails can be converted into storage bins by burying them in 16-inch holes, leaving their rims above ground. In addition to being inexpensive and readily accessible, the garbage-pail storage bins are water- and rodent-proof and store easily over the summer as a compact stack of pails.

Here is how Mr. Onstott got his project of year-round fresh vegetables started:

Late in November or in early December, I wait for several 30°F. nights and then start harvesting my carrots, beets, and turnips. All I have to do is pull them, shake off the loose dirt, cut the tops off, and place them in cans *without washing them*. Then I put the lid on the cans and cover everything with 6 inches of straw, adding more when the ground freezes. I add nothing to the cans, unless the vegetables are dry at the time of storage. If their skins are dry, I sprinkle the vegetables with water as I put them in the cans, being careful not to form a pool of water at the bottom of the cans. During the winter, especially if it is dry, it is a good idea to sprinkle a little water over the vegetables once and a while. But I have found little dehydration; almost all the vegetables remain firm and fresh until the middle of March.

Instead of emptying one pail at a time, I take a little from each in turn, so that the level in each can is lowered as the winter progresses. The lower the contents are, the less chance of damage from frost. If the vegetables are 6 to 8 inches below the bin cover, with even a little straw, it will take a very severe freezing to cause frost damage because the heat from the earth below the frost line will feed into the bins.

While I cannot equal or even duplicate the airtight, atmospherically controlled and refrigerated rooms that large commer-

A more permanent version of Mr. Onstott's garbage pail idea is the tile pit storage. Place large (18-by-3-inch or 24-by-24-inch) tiles in a pit on well-drained soil. Positioning should be near the kitchen and shaded from the sun. After the pit is filled with boxes or baskets of food, place wire screen over the top, and cover the screen with a large rock or concrete tile.

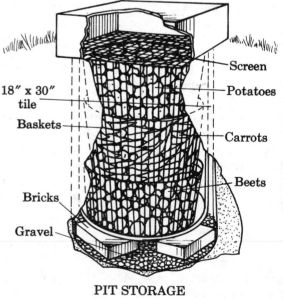

PIT STORAGE

cial operations boast of, my simple, inexpensive garden storage bin gives me results. All through the winter months and into early spring, long after the average gardener has all but forgotten his garden, I bring in those garden-fresh, organically grown vegetables from the storage bins of my garden. Red beets, the best keeper of them all, I have enjoyed into May—weeks after the new crop was up and growing.

According to Gordon Morrison, of Michigan, garbage cans or similar containers don't even have to be underground to make good storage bins. He keeps apples, potatoes, carrots, and the like perfectly crisp from fall until early spring just by using watertight wooden or metal barrels, watertight butter tubs, and candy pails. He puts them in a cool basement room, a sheltered back porch or garage—anywhere that they can safely be protected against freezing.

He provides enough ventilation by topping off the contents of the barrel with a well-moistened, but not drippy, 4- or 5-inch head cover of something like burlap or moistened sphagnum moss, which can be rewetted when necessary. By this method he's kept many kinds of garden crops in good condition in a cement-floored basement room next to the furnace room, where temperatures range quite often between 40° to 50°F.

Mound Storage

Other gardeners have had success storing root crops and pears and apples right on the ground in a mound storage construction. To

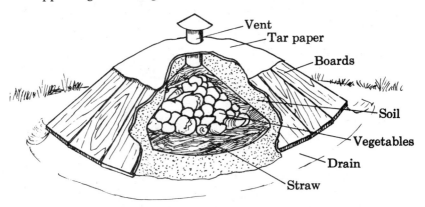

MOUND STORAGE

Once the mound is opened, all the food inside should be removed. Place just enough fruit or vegetables to last your family 1 or 2 weeks in each storage area.

make a mound storage, place straw, hay, or dry leaves on the ground, place fruit or vegetables on top of this, and cover with more mulch. Cover the mound with soil, then boarding. A ventilating pipe can be added by placing stakes or a pipe through the center of the pile. This should be capped in freezing weather to prevent cold temperatures from entering. A trench is then dug around the mound for drainage.

Once the pit is opened, all the food should be removed. It is therefore a good idea to build a few of these, each containing a small number of different vegetables. Separate the various vegetables in the mound with mulch. Do not mix fruits with vegetables, but make separate mounds for them. Mounds should be made in different places every year, because leftovers in used mounds are usually spoiled.

Pickles
and Relishes

Pickles and relishes have been enjoyed by millions throughout the centuries. The Chinese are said to have invented pickling—Chinese laborers who worked on the Great Wall of China carried salted vegetables to the Wall with them. In the United States, perhaps the ethnic group most famous for its pickles is the Pennsylvania Dutch. Every Pennsylvania Dutch cookbook contains dozens of recipes for both fruit and vegetable pickles, and every meal served by the Pennsylvania Dutch is supposed to contain the "seven sweets and seven sours." (There doesn't have to be seven of either; the cook picks out the best pickles and relishes to go with the meal he or she is serving.)

There are actually four different kinds of pickled products:

Brined pickles, which usually include sauerkraut and pickling cucumbers, go through a curing process of about 3 weeks. Most brined pickles are made in a low salt brine (3 to 5 percent salt) and do not require desalting before they are used. Cucumbers may also be cured in a high salt (10 percent salt) brine; these should be soaked in water before they are processed any further.

Fresh-pack or quick-process pickles are the easiest to prepare. They are either soaked in a low salt brine for several hours or overnight, then drained and processed with boiling hot vinegar, spices, herbs, or other seasonings, or they are cooked with the spiced vinegar and packed and processed right away.

Fruit pickles are usually prepared from whole fruits—pears, peaches, and watermelon rind are good choices—and simmered in spicy, sweet-sour syrup, then packed and processed.

Relishes are mixed fruits and vegetables which are chopped, seasoned, and then cooked, packed, and processed. Relishes may be hot and spicy or sweet and spicy. Familiar ones include piccalilli, chutneys, corn relish, catsup, and Pennsylvania Dutch chowchow.

Ingredients

Fruits and Vegetables   Use only tender fruits or vegetables that are in prime condition. Produce should not be more than 24 hours old and should be refrigerated or cooled immediately after picking. This is especially important for cucumbers, which deteriorate rapidly after picking at room temperature. You may, if you wish, sort the produce for uniform size. Some canners—those who win prizes at county fairs—always sort for uniformity. By using vegetables or fruits of the same size you can be sure that all the food will cook and cure evenly.

Do not use any fruits or vegetables for pickling whole that have mold damage or are injured. Damaged ones can be cut up for relishes or, in the case of cucumbers, sliced for bread and butter pickles, with the injured or moldy part being cut out and discarded. Although proper processing does kill the spoilage organisms in moldy parts, an off-flavor will develop from the mold growth that cannot be masked by spices or herbs.

Fruits and vegetables should be washed thoroughly under running water. Scrub the food with a soft brush or with the palms of your hands. Rinse well so that all soil drains off. It is important that the food be handled gently so that it does not become bruised; this is especially important for cucumbers or for soft fruits like peaches or pears (although the latter two may be picked slightly underripe for pickling). Be sure to remove blossoms from cucumbers—the blossoms may be a source of spoilage enzymes.

Produce should never be picked for canning after a heavy rain because it will be waterlogged. Wait 12 hours after the rain to pick your fruits and vegetables.

When picking cucumbers, *cut* them from the vines, leaving a short bit of stem on the fruit. If pulled off the vines, they are likely to rot where the stem was broken from the skin.

Drain your produce on a tea towel or on a dish drain. Wipe them dry if you must, but be careful not to bruise them.

Vinegar   As practical and self-satisfying as it might be to use your own homemade vinegar in your pickles, *don't*. Vinegar should be a good grade of 40 to 60 grain strength (4 to 6 percent acetic acid). The vinegar you make on your homestead may be some of the best you've ever had, and it is fine for your salads and other cooking. But homemade vinegars vary in acidity, so you'll never know what the acidity of your vinegar is unless you test it, or have it tested. It is much safer to rely on commercial vinegar for your pickles.

Cider vinegar has a good flavor and aroma, but it is not good for white pickles such as onion or cauliflower because it may discolor them. Distilled vinegar is clear so there is no chance of discoloration. The fact that it is slightly more acidic than cider vinegar makes little difference.

Honey Light honeys, such as clover or alfalfa, are very mild in flavor and are good to use for canning or pickling for that reason. Dark honeys are strongly flavored. If they are used in pickling, you might find that their flavor will overpower the flavors of the other ingredients. Therefore, no matter what kind of honey you are using, we suggest that you taste your syrup as you add the honey to it. If you think that the syrup is sweet enough, stop adding honey, although you may not have reached the amount suggested in the recipe. Those amounts are intended to be guidelines only, not hard and fast rules. Your best guides are your tastebuds and a knowledge of your family's likes and dislikes.

Pickling recipes from most cookbooks suggest that you boil the syrup, consisting of vinegar, spices, and sugar, for a certain amount of time. This allows the sugar to dissolve completely and also allows the spices to flavor the syrup. If you are altering a recipe that you've found in another book because you want to use honey in it rather than sugar, remember that the heating of honey tends to break down the sugars in honey and cause a change in flavor and a darkening in color. Honey can be heated to high temperatures for short periods without causing too much damage, but it will not stand sustained boiling. For this reason, never add the honey to the syrup *before* it is boiled. Instead, boil the vinegar and spices together for the stated time; then add the honey, tasting the syrup as you add it to determine sweetness. (As a general rule of thumb, if you are substituting honey for sugar, cut the amount of sweetener by one-half.) Bring the syrup to a boil and pour it over the pickles. Process as directed.

Spices Always use whole, fresh spices or herbs. Whole spices or herbs should be tied in a bag (cheesecloth will do) or a stainless steel spice ball and removed before the pickles are packed. Never use ground herbs or spices; they tend to darken pickles. Whole spices, if left in the jar after the pickles are canned, may cause an off-flavor in the product.

If garlic is used, it should first be blanched for 2 minutes or removed from the jar before it is sealed.

The liquid should be tasted first before canning. Spices vary considerably in strength, and you can correct the seasoning by adding more spices. Unused spices should be kept in airtight jars in a cool place, as heat and humidity tend to sap their quality.

Water   Soft water is recommended for the best-looking pickles. Iron or sulphur in hard water will darken pickles; calcium and other salts can interfere with the fermentation process.

Salt   You should use plain salt. Iodized salt may cause darkening, and table salt sometimes causes a cloudy brine. Rock salt, dairy salt, or pickling salt is best. You can also use sea salt.

Equipment

You shouldn't need any more specialized equipment for pickling than you already have to do your other canning chores. Use unchipped enamelware or stainless steel for heating liquids you may be using in your pickling. Do not use copper, brass, galvanized iron, or aluminum utensils. These metals react with acids and salts in the liquids and may cause undesirable color changes in the finished product.

For fermenting or brining use a crock, stone jar, plastic container, unchipped enamelware pan, or large glass jar, bowl, or casserole. Use a heavy plate or large glass lid that fits right inside the circumference of the crock and a weight to hold the vegetables in the brine. Clean rocks or a glass jar filled with water can be used as a weight, or a large plastic bag filled with water can serve as both weight and cover (so long as it completely covers the food and is tight against the circumference of the brining container).

If you are using recipes that specify ingredients by weight, a small household scale is a necessity. It is best to use a scale to make large quantities of sauerkraut to insure the proper proportion of salt and cabbage.

Canning Pickled Foods

Jars should be filled, as recommended in the recipes, leaving the necessary headspace. Pack jars firmly and uniformly, but avoid packing so tightly that the brine or syrup is prevented from filling spaces around and over the product. Wipe the rims and threads of the jar with a clean hot cloth, and cap. Pickles are processed in a boiling-water bath. (See the instructions on boiling-water bath processing on page 76.)

Processing times in the following recipes are given for altitudes less than 1,000 feet above sea level. At altitudes above 1,000 feet, add

1 minute of processing time for each 1,000 feet. For example, at 2,000 feet, process for 2 minutes more than the required time; at 3,000 feet, process for 3 minutes more; at 4,000 feet, process for 4 minutes more.

Storing Pickles

Pickled products should be stored in cool, dark, dry places. Extreme fluctuations of temperature may cause a breakdown of texture, resulting in an inferior product. It might also cause enough expansion of the product to break the jar or the seal. Light causes products to fade and become less appetizing in appearance. This does not mean that the product is spoiled, however.

A storage area in a basement is fine if it is cool, dry, and dark. Do not store canned products with vegetables or fruits that require high concentrations of moisture. Dampness may rust enclosures and cause spoilage.

Brine Curing

Just about any vegetable can be cured this way: snap or string beans, cucumbers, cauliflower, onions, broccoli, green tomatoes, Jerusalem artichokes, and carrots.

By covering such foods with a brine and keeping them in a moderately warm room, you can create ideal conditions for the lactic acid forming bacteria existing on the food surface to feed upon the sugar naturally present in the food. The lactic acid will continue to grow (or ferment) until enough has formed to kill any bacteria present that would otherwise cause the food to spoil. Lactic acid, which aids digestion and helps to kill harmful bacteria in the digestive tract, gives the brined food a slightly acid or mildly vinegar flavor.

The following is a general recipe for brine curing that can be used with any one vegetable or mixture of the vegetables mentioned above:

1. Choose fresh, perfect vegetables, wash them carefully to avoid bruising, and drain well.
2. Pack vegetables in a crock or other wide-mouthed container, leaving about a 3- or 4-inch headspace. Cover with a 10 percent brine solution, made by dissolving 1 cup salt in 2 quarts water. *Brine in which a fresh egg floats is approximately 10 percent.* Make sure that the salt is completely dissolved in the water before pouring it over the vegetables. The amount of brine you'll need will be approximately one-half the volume of

your container. In other words, if you are packing a 10-gallon crock with vegetables within 3 or 4 inches of the top, prepare about 5 gallons of brine.

3. Cover vegetables with a plate or something similar, and place a clean rock or other heavy object on the plate to weight down the vegetables so they are underneath the brine.

4. Place the crock or other container in a moderately warm area—70° to 80°F. is best.

5. Then, or not later than the following morning, add salt at the rate of 1 cup for each 5 pounds of vegetables. This is necessary to maintain a 10 percent brine solution. *Salt should be added on top of the plate so that it can dissolve slowly in the water and work its way to the bottom of the crock.*

6. Remove scum when it forms on top of the brine. This, if left on, will destroy the acidity of the brine and result in spoilage of the product.

7. At the end of a week and for each succeeding week, add ¼ cup salt for each 5 pounds of vegetables. Add in same manner as in step number 5.

8. Fermentation resulting in bubble formation should continue for about 2 to 4 weeks. Fermentation time varies depending upon temperatures, so after about 10 days check for bubble formation. If no bubbles rise to the surface after you've gently tapped the side of the container, fermentation has stopped.

9. When fermentation is complete, you can either remove the food, desalt it, and then pickle it (see recipes later), or keep the food in the brine in the container for future use.

If you plan to leave the food in the brine, transfer the container to a cool cellar or store room. There's no need to add additional salt, but you should cover the surface of the liquid with either a thin layer of hot melted paraffin or ¼ inch of good vegetable oil. This is done to seal the surface so that scum which would otherwise spoil the food cannot form.

Make sure fermentation has stopped before using the paraffin, or the bubbles that rise to the surface will crack the wax and you'll have to remove it, melt it down, and re-cover the surface. Remove the paraffin before using the brined vegetables. It can be saved, melted, strained, and used again and again.

Although the oil can be eaten, many prefer to skim it off the surface before using the food underneath.

Vegetables may be added during the curing process if enough brine is added to cover them, *and* if salt is added in definite amounts to maintain a 10 percent brine.

Some recipes suggest using alum or lime for crisper pickles. It really isn't necessary to add either of these if proper procedures are followed. If you do want to add something to your pickles to crisp them, try using grape or cherry leaves during the brining process.

To desalt these pickles for further use, soak them in cold water for a few hours. You can hasten the soaking process by using large amounts of water (three to four times volume of water as volume of pickles), changing the water often and stirring often, being careful not to bruise the brined food.

Brine-Cured Dill Pickles

Euell Gibbons, nationally known author and expert on wild foods, wrote in ORGANIC GARDENING AND FARMING about his own adventures with pickled vegetables, using his "dill crock." This method has great appeal because it not only brines the vegetables, but flavors them as well. Gibbons wrote:

Naturally, I got started at this tasty sport with wild foods. A nearby patch of wild Jerusalem artichokes had yielded a bumper crop, and I wanted to preserve some. I used a gallon-size glass jar, getting all of these jars I wanted from a nearby school cafeteria.

Packing a layer of dill on the bottom of the jar, I added several cloves of garlic, a few red tabasco peppers, then some cored and peeled Jerusalem artichokes, plus another layer of dill. With room still left, I looked around for other things to add. The winter onions had great bunches of top sets, so I peeled a few and made a layer of them. Then I dug up some of the surplus onions and used the bottom sets—shaped like huge cloves of garlic—to make still another layer. I then put in a layer of cauliflower picked apart into small florets, and added some red sweet pepper cut in strips, along with a handful or so of nasturtium buds.

This was all covered with a brine made by adding three-fourths of a measure of sea salt to 10 measures of water. I added some cider vinegar too, but only ¼ cup to the whole gallon. I topped the whole thing with some more dill, set a small saucer weighted with a rock on top to keep everything below the brine, and then let it cure at room temperature.

After 2 weeks I decided it must be finished. The Jerusalem artichokes were superb, crisp and delicious. The winter onions, both the top and bottom sets, were the best pickled onions I ever tasted. The cauliflower florets all disappeared the first time I let my grandchildren taste them, while the nasturtium buds make better capers than capers do.

The next summer I determined to get started early and keep a huge dill crock running all season. Any size crock can be used, from 1-gallon up. I use a 10-gallon one and wish it were bigger. Never try to use a set recipe for a dill crock, but rather let each one be a separate and original "creation." I plant plenty of dill, and keep planting some every few weeks so I'll always have some on hand at just the right stage.

What is good in a dill crock? Nearly any kind of firm, crisp vegetable. Green beans are perfect, and wax beans also very good. These are the only two things cooked before being added to the brine, and they should be cooked not more than about 3 minutes. And small green tomatoes are great. Nothing else so nice ever happened to a cauliflower. Just break the head up into small florets, and drop it into the dilled brine. In a week or two—the finest dilled cauliflower pickle ever tasted.

If you have winter onions, clean some sets and put them in the crock. It's a tedious job, but the results are worth it. Not only do they add to the flavor of all the rest of the ingredients in the jar, but the little onions themselves are superb. If you don't have winter onions, you can sometimes buy small pickling onions on the market and use them. If not, just take ordinary onions and slice them crosswise into three or four sections. These will come apart after curing, but so what? They are simply great pickled onion rings. I've even cut off the white part of scallions and thrown them in the brine, with some success, and one late-fall dill crock was flavored with white sections of leek, which did it wonders.

To preserve these pickles, pack them in hot, sterilized jars along with some dill. Strain the brine, bring to a boil, and pour over pickles, leaving ½-inch headspace. (You can also make new brine using ½ cup salt and 4 cups of vinegar to 1 gallon of water, but the old brine is much more flavorful.) Seal and process in boiling water for 15 minutes.

Spiced Vinegar

This is for use with brined pickles after they have been soaked in water for a few hours to desalt them. Warning: You have to make this 3 weeks before you want to soak brined pickles in it.

To 1 gallon of vinegar add a spice bag containing:

½ ounce allspice   1 stick cinnamon
½ ounce cloves   1 piece mace

Boil vinegar and spices for 15 minutes. Add 1 cup honey for sour pickles, or 2 cups honey for less tart pickles, or 4 cups honey for sweet pickles.

Heat slightly, and set aside for 3 weeks before removing spice bag.

Sweet Cucumber Pickle

Put desalted cucumbers into sterilized jars. Bring spiced vinegar—about 1 pint vinegar per quart of pickles—to a boil. Pour over cucumbers and let them set until following day. Drain off vinegar and add honey at the rate of ¼ cup per pint of vinegar. Bring to a boil and pour over cucumbers. On the following day again drain off vinegar and add honey in the same proportion. Seal jars, and process 10 minutes in boiling-water bath.

Sour Cucumber Pickle

Select *small* brined cucumbers and soak to desalt in cold water for a few hours.

To 6 quarts of cucumbers use 1 gallon of plain vinegar. Prepare spice bag as directed in spiced vinegar recipe above.

Bring vinegar and spices to boil. Add the cucumbers, 1½ quarts at a time, and let them boil for 2 minutes. Don't allow them to get soft. Place them in a large stone crock as they are taken from the kettle. When all the cucumbers have been packed, cover them with the boiling vinegar. If a sweeter pickle is desired, add 2 cups of honey to the vinegar just before it is poured over the pickles. Place the spice bag in the crock.

Cover the top of the stone crock with layers of thick paper tied tightly to exclude air. Let pickles remain in the vinegar solution for

3 weeks. Then pack the pickles in hot, sterilized jars. Remove the spice bag from the vinegar, bring to a boil, and pour vinegar over the pickles, leaving a ½-inch headspace. Seal and process for 10 minutes in a boiling-water bath.

Pickled Nasturtium Buds
(False capers)

If you pride yourself on your gourmet cooking, one recipe you can't afford to pass up is the following. As Euell Gibbons already noted, "Nasturtium buds make better capers than capers do." Nasturtium buds should be gathered while they're still green—yellow ones are useless.

Place the nasturtium buds in a 10 percent brine—made by adding 1 cup salt to 2 quarts water—to cover. Weight them, if necessary, to hold them in the brine. Allow them to cure in brine for 24 hours.

Remove from brine, and soak in cold water for an hour. Drain the buds. Bring vinegar to a boil. Pack nasturtium buds in hot, sterilized jars, cover with boiling vinegar, leaving a ½-inch headspace. Seal and process 10 minutes in a boiling-water bath.

It is best to let your "capers" stand for 6 weeks before you use them, as they will be more flavorful. You can, however, use them immediately if you wish.

Sauerkraut

Sauerkraut has been made and enjoyed by millions for hundreds of years. For our ancestors, it was an important winter source of vitamin C and was used as a cure for scurvy on sea voyages. In addition, like other cured vegetables, sauerkraut contains the beneficial lactic acid.

Sauerkraut, like the brined vegetables discussed earlier, is cured in salt. But it is packed in a dry salt, not covered in a salt water solution, because the cabbage contains a great deal of water and forms its own brine when the salt draws out water in its shredded leaves.

After fermentation, sauerkraut can either be canned in a boiling-water bath or stored in the container in which it was made. If you're just keeping it in the container, be sure to keep it in a cool place. Temperatures just above freezing are best. Low temperatures will discourage the growth of surface scum. Still the kraut should be checked periodically and scum should be removed.

Tight-forming head lettuces can be used instead of cabbage to make a milder form of "sauerkraut," and although we've never made it ourselves, we have heard of people who have used shredded carrots and turnips instead of cabbage.

Sauerkraut can be made in either large quantities or in small batches. We give instructions for both methods. The recipe below for making small batches, suggested by natural foods writer Beatrice Trum Hunter, has the added advantage of being made with less salt than most other recipes, which might interest those on low-sodium diets.

Saltless Sauerkraut
(small batches)

Assemble a few simple items. Locate a bowl, pot, or other wide-mouthed container that will hold 1 gallon of liquid measure. Glass, well-glazed clay, or other impermeable material is suitable. A stoneware crock and cover are ideal for the purpose. If you consider purchasing a used stoneware crock, check its inside. If the crock has been used to store surplus eggs in waterglass, it will have a permanent whitish stain. Reject such a crock for sauerkraut production.

Next, find a flat plate slightly smaller in diameter than the inner surface of your container. If you plan to use a sloping bowl, measure the plate against the diameter near the top of the bowl. Then locate a few large, smooth stones. Plan to keep the container, plate, and stones reserved exclusively for "Project Sauerkraut."

Depending upon the amount of kraut you want to make, shred a whole or even a half of a solid head of cabbage. Place it at the bottom of the container. For each head of cabbage, pulverize a half teaspoonful each of dill, celery, and caraway seeds in an electric seed grinder or with mortar and pestle. These seeds can be added whole, but the ground seeds release more flavor. Sprinkle this blend on top of the shredded cabbage. If the flavor of such seeds doesn't appeal to you, omit any or all of them.

Whether or not you use the seeds, you should sprinkle over the shredded leaves:

½ teaspoon ground kelp
½ teaspoon salt

Now pour cold water over this mixture, so that the cabbage is completely covered: 2 quarts of water, more or less, for each cabbage.

The liquid should reach no higher than 2 or 3 inches from the top of the container to prevent overflow during fermentation.

Put the plate over the shredded cabbage, seasoning, and water. Press down gently, so that the liquid flows over and submerges it. Next, weight it down with the freshly scrubbed stones. Cover the container, place it in a warm room, and let nature take its course. As a precaution against overflow, which rarely occurs with this method, place the crock in a glass pie plate to catch any drippings.

Within a few days you will begin to sniff the pleasant fermentation process. Take a peek now and then: Be sure that the plate keeps the cabbage submerged under the liquid. Skim off any scum that develops on the surface, because, as mentioned earlier, it will otherwise grow and cause the cabbage to spoil. One little trick to avoid more scum is to sprinkle a bit of additional salt on the surface of the kraut after you've skimmed off the scum. The length of fermentation will be determined by the room's temperature. Sauerkraut may be ready in as few as 7 days. Sometimes it takes a little longer. Toward the end, check daily.

When you're convinced that the product is well fermented, remove the stones and plate. Using a slotted spoon, transfer the drained sauerkraut to a bowl. Strain the remaining liquid. The flavor of both the low-salt raw sauerkraut and its juice will be subtle and delicate. If you're unaccustomed to low-salt, you may find the flavor "flat." Nothing prevents you from adding a dash of salt. Or, if you wish, simply make the sauerkraut as described, adding salt instead of the kelp.

After your family eats its fill of sauerkraut, any remainder can be refrigerated. Although raw, it keeps well, due to its natural preservative, lactic acid. If you have prepared a large quantity you can pack it hot in canning jars, leaving ¼-inch headspace, and process in a boiling-water bath 15 minutes for pints and 20 minutes for quarts.

Other raw vegetables can be added to ferment with the cabbage. Good additions are thinly sliced onions, carrots and turnip strips, cauliflower segments, and radish slices. Do your own experimenting.

Sauerkraut
(6 quarts)

If your grandmother made sauerkraut, it is likely that she used this recipe or one very close to it. Huge stoneware crocks of sauerkraut were a common sight in springhouses and cellars on farms in the past, and farm children were assigned the task of skimming bad kraut and scum off the tops of these crocks. This recipe has been adapted from one sent us by Susan Ferris of the Maine Organic Food Association.

Select about 15 pounds of firm, green cabbages. Let stand at room temperature for 1 day. Remove any bruised outer leaves, wash, quarter, and remove cores. Cabbages should be dry before grating for sauerkraut. Shred or cut about the thickness of a dime.

Thoroughly mix 3 tablespoons plus 2 teaspoons salt (use pickling/canning salt) with each 10 quarts of shredded cabbage. As each batch is salted, get ready your crock or crocks. Pack the cabbage firmly, but not tightly, into the crocks, pressing down with a wooden spoon or paddle.

Lay a clean cloth over the cabbage with a plate on top that fits snugly inside the crock. It is important that the cabbage is covered by the tight-fitting plate; it may spoil otherwise. Weight with a stone or a gallon jar filled with water. The weight should be heavy enough so that the liquid just reaches the bottom of the cover. To vary the weight, use heavier or lighter stones or fill or empty the jar as needed as fermentation increases.

Allow cabbage to ferment at room temperature (68° to 72°F.) for 9 to 14 days or more. (The lower the temperature, the slower the fermentation.) Change and wash the cloth, adjust the weight, and skim off the scum daily. Fermentation has ended when bubbles stop rising to the surface. Taste at the end of a week and can or place in a cool cellar or storeroom when taste suits you.

To can your kraut, use hot, sterilized quart jars. Bring kraut to a boil with 3 quarts water. Pack lightly into jars, filling spaces with brine from the crock. Leave a ½-inch headspace. Process in a boiling-water bath for 20 minutes for quarts and 15 minutes for pints.

Sauerkraut
(16 to 18 quarts)

This recipe is from a booklet prepared by the U.S. Department of Agriculture.

about 50 pounds of cabbage
1 pound (1½ cups) salt

Remove the outer leaves and any undesirable portions from firm, mature heads of cabbage; wash and drain. Cut into halves or quarters; remove the core. Use a shredder or sharp knife to cut the cabbage into thin shreds about the thickness of a dime.

In a large container, thoroughly mix 3 tablespoons salt with 5 pounds shredded cabbage. Let the salted cabbage stand for several minutes to wilt slightly; this allows packing without excessive breaking or bruising of the shreds.

Pack the salted cabbage firmly and evenly into a large, clean crock or jar. Using a wooden spoon or tamper, or your hands, press down firmly until the juice comes to the surface. Repeat the shredding, salting, and packing of cabbage until the crock is filled to within 3 or 4 inches of the top.

Cover cabbage with a clean, thin, white cloth (such as muslin) and tuck the edges down against the inside of the container. Cover with a plate, round paraffined board, or other clean cover that just fits inside the container so that the cabbage is not exposed to the air. Put a weight on top of the cover so the brine comes to the cover but not over it. A glass jar or heavy-duty plastic bag filled with water makes a good weight.

The amount of water in the glass jar or plastic bag can be adjusted to give just enough pressure to keep the fermenting cabbage covered with brine.

Formation of gas bubbles indicates fermentation is taking place. A room temperature of 68° to 72°F. is best for fermenting cabbage. Fermentation is usually completed in 4 to 6 weeks.

To store the sauerkraut, heat to simmering (185° to 210°F.). Do not boil. Pack hot sauerkraut into sterilized, hot jars and cover with hot brine to ½ inch of the top of the jar. Adjust jar lids. Process in a boiling-water bath, 15 minutes for pints and 20 minutes for quarts.

Sauerkraut by the Quart

Here's a simple, sure-fire way to make sauerkraut: pack it and let it ferment right in the canning jar.

Simply use quart glass jars with rubber rings and zinc lids. After washing the jars, place them upside down in a large pan of water with the rubber rings and lids, and slowly bring to a boil.

When enough cabbage is cut for several jars, pack and press the cabbage into each sterilized jar.

When the juice begins to show as it is squeezed out of the cabbage, continue to fill and press until 1 inch of space is left at the top of the jar. Then add to each quart jar 1 teaspoon of salt and ½ teaspoon of honey. Fill slowly with boiling water, allowing it to settle down into each jar. Insert a knife blade at intervals to allow air bubbles to escape. Leave about ½ inch of space at the top, put the rubber ring in place, and screw the zinc lid down tight.

Wipe the jars, set them in an old dishpan or a small tub in a cool place (65° to 70°F.) outdoors or in the corner of the garage, and wait. Inspect the jars every few days. Don't be alarmed when the zinc lids begin to bulge—that shows the kraut is fermenting properly. If (and it frequently happens) the juice spews between the rubber ring and lid, wipe it off and tighten the jar lid some more. That is why the jars are set in a container outside—there may be leaks and there will be plenty of kraut odor! This will continue as long as the kraut is working. Never under any circumstances loosen the lids.

After about 6 weeks the kraut will have cured and the jars can be washed and brought inside. If you're going to use the kraut within a month, there's no reason to process the cans; just store them in a cool place. But if you do plan to keep them longer, place the filled jars in a large pot and cover the jars with water. Put the lid on the pot and bring the water to a boil. Once boiling, process for 20 minutes.

What You Did Wrong If Your Sauerkraut Spoiled

Spoilage of sauerkraut is indicated by undesirable color, off-odors, and soft texture. If your kraut has spoiled, here's what you might have done wrong:

If your **sauerkraut is soft,** it may be due to insufficient salt. Try using more salt the next time. High temperatures during fermentation may also cause softness. Uneven distribution of salt may also be a cause of softness—be sure that your salt is well mixed with the kraut next time. Air pockets caused by improper packing may also make your kraut soft. Your crock or jar may have had air spaces that caused poor fermentation; this can be remedied by packing the jar or crock tightly and being sure to weight it properly.

Pink kraut is caused by the growth of certain types of yeast on the surface of the kraut. These yeasts may grow if there is too much salt, if there is an uneven distribution of salt, or if the kraut is improperly covered or weighted during fermentation.

Rotted kraut is usually found at the surface where the cabbage has not been covered sufficiently to exclude air during fermentation. This scum does not cause trouble as long as you skim it off before it stops fermentation. Remove every day or two.

Darkness in kraut may be caused by unwashed and improperly trimmed cabbage. It may also be caused by insufficient brine in the fermenting process. Be sure that the brine completely covers the fermenting cabbage. Exposure to air or a long storage period in the crock after fermentation is complete may also result in darkened kraut. Another cause of darkening may be high temperatures during fermentation, processing, or storage.

Fresh Vegetable Pickles

These pickles are different from the brine-cured ones discussed earlier in that they are not packed in a salt-water solution for several days or weeks, but take, with few exceptions, from 1 hour to 4 days in all to make. Some traditionalists claim these pickles can't match the long-cured ones, but many prefer them because they are simpler to make and quite good, in their opinion.

Sweet Gherkins

Gherkins, immature cucumbers, are often made into sweet pickles.

5 quarts (about 7 pounds) cucumbers, 1½ to 3 inches in length
½ cup salt
¾ teaspoon turmeric
2 teaspoons celery seed
2 teaspoons whole mixed pickling spice

8 1-inch pieces stick cinnamon
½ teaspoon fennel (optional)
6 cups (1½ quarts) vinegar
4 cups honey
2 teaspoons vanilla (optional)

First Day

Morning: Wash cucumbers thoroughly; scrub with vegetable brush. Drain cucumbers; place in large container, and cover with boiling water.

Afternoon: Six to 8 hours later, drain, and cover with fresh boiling water.

Second Day

Morning: Drain, cover with fresh boiling water.

Afternoon: Drain, add salt, and cover with fresh boiling water.

Third Day

Morning: Drain. Prick cucumbers in several places with table fork. Add turmeric and spices (including fennel, if you are using it) to 3 cups of vinegar and bring to a boil. Add 1½ cups of honey. Pour over cucumbers (they will be partially covered at this point).

Afternoon: Drain syrup into pan; add 2 cups of the vinegar to syrup; bring to boil. Add 1 cup of honey. Pour syrup over pickles.

Fourth Day

Morning: Drain syrup into pan; add ½ cup of the vinegar to syrup. Heat to boiling. Add 1 cup honey and pour over the pickles.

Afternoon: Drain syrup into pan. Add remaining ½ cup honey (plus vanilla if you are using it) to the syrup. Bring to quick boil. Pack pickles into clean, hot pint jars. Cover with boiling syrup, leaving a ¼-inch headspace. Adjust seals and process for 5 minutes in a boiling-water bath.

Yield: 7 pints

Fresh-pack Dill Pickles
(unsweetened)

17 to 18 pounds cucumbers, 3 to 5 inches in length (pack 7 to 10 per quart jar)
2 gallons 5 percent brine (¾ cup salt per gallon water)
6 cups (1½ quarts) vinegar
¾ cup salt
9 cups water
2 tablespoons whole mixed pickling spice
2 teaspoons whole mustard seed per quart jar
1 or 2 garlic cloves per quart jar (optional, but if added, they turn the pickles into the kosher dill type)
3 heads fresh dill per quart jar (you may substitute dried dill) or 1 tablespoon dill seed per quart jar

Wash cucumbers thoroughly, scrub with vegetable brush, and drain. Cover with brine. Let set overnight. Drain.

Combine vinegar, salt, water, and mixed pickling spices tied in a clean, white cloth bag or spice ball. Heat to boiling. Pack cucumbers in hot, sterilized quart jars. Add mustard seed, garlic, and dill plant or seed to each jar. Cover with boiling liquid to within ¼ inch of top of jar. Adjust seals and process in a boiling-water bath for 15 minutes.

Yield: 7 quarts

Crosscut Pickle Slices

4 quarts cucumbers, medium-sized, sliced (about 6 pounds)
1½ cups onions, sliced
2 large garlic cloves
⅓ cup salt
2 quarts ice cubes or chips
2 cups honey
1½ teaspoons turmeric
1½ teaspoons celery seeds
2 tablespoons mustard seeds
3 cups white vinegar

Wash cucumbers thoroughly and scrub with a vegetable brush. Drain on rack. Slice unpeeled cucumbers into ⅛- to ¼-inch slices; discard ends. Add onions and garlic.

Add salt and mix thoroughly, cover with crushed ice or ice cubes, and let stand 3 hours. Drain thoroughly and remove garlic.

Combine honey, spices (in a spice bag), and vinegar. Heat just until boiling. Add drained cucumber and onion slices and pack hot pickles loosely in clean, hot pint jars, leaving ¼-inch headspace. Adjust seals and process in a boiling-water bath for 5 minutes.

Yield: 7 pints

Bread and Butter Pickles

30 medium-sized cucumbers
10 medium-sized onions
4 tablespoons salt

Slice cucumbers and onions and sprinkle with salt. Let stand 1 hour. Drain in cheesecloth bag. Make a spiced vinegar using the following ingredients:

5 cups vinegar
2 teaspoons celery seed
2 teaspoons ground ginger

2 cups honey
1 teaspoon turmeric
2 teaspoons white mustard seed

Let spiced vinegar come to a boil, add cucumbers and onions, and bring to boiling point. Pack in jars, leaving ¼-inch headspace. Adjust seals and process 10 minutes in a boiling-water bath.

Yield: 8 pints

Old-fashioned Cucumber Chunks

1 gallon cucumbers, cut into
   1-inch pieces
1½ cups salt
9 cups vinegar

water
2 tablespoons mixed pickling
   spices
2 cups honey

Wash, dry, and cut cucumbers into 1-inch pieces before measuring. Put in crock or large container. Dissolve salt in 1 gallon water. Pour over cucumbers. Cover with plate and weight so that the cucumbers remain submerged in the brine. Let stand 36 hours.

Drain. Pour 4 cups vinegar over the cucumbers; add enough water to cover. Simmer 10 minutes. Drain and discard liquid. Add spices (tied in bag) to 3 cups of water and 5 cups of vinegar. Simmer 10 minutes. Add 1 cup honey. Pour over cucumbers. Let stand 24 hours.

Drain syrup into kettle. Add remaining honey. Heat to boiling. Pour over cucumbers. Let stand 24 hours.

Pack pickles into hot, sterilized jars. Heat syrup to boiling and pour over pickles, leaving ¼-inch headspace. If there is not enough liquid to cover pickles, add more vinegar. Adjust seals and process in a boiling-water bath for 10 minutes.

Yield: 4 quarts

Cucumber Oil Pickles

100 medium-sized cucumbers
3 onions
2 cups salt
1 gallon plus one cup water
2 tablespoons peppercorns

4 tablespoons mustard seed
4 tablespoons celery seeds
4 cups vinegar
2 cups honey
1 cup olive oil

Wash, dry, and thinly slice unpeeled cucumbers and peeled onions. Dissolve salt in 1 gallon cold water. Add cucumbers and onions. Let stand 12 to 18 hours. Drain. (Taste the cucumbers. If they are too salty, rinse well in cold water.)

Place spices in a spice bag, and put it with 1 cup water and the vinegar in a pot. Boil 1 minute. Add honey, cucumbers, onions, and oil. Simmer until cucumbers change color. Then bring to boiling. Pack, boiling hot, into hot, sterilized jars, leaving ¼-inch headspace. Adjust seals and process for 10 minutes in a boiling-water bath.

Yield: 15 quarts

Pickled Beets

1 gallon small beets
Water to cover beets
1 tablespoon whole allspice

1 long stick cinnamon
1 quart vinegar
1 cup honey

Cook beets with roots and about 2 inches of stem left on in water to cover. Cook until tender; dip beets into cold water and slip off skins. Put beets in large preserving kettle. Combine the spices and vinegar, pour over beets and simmer 15 minutes. Then add the honey. Pack hot into sterilized jars. Cover beets with boiling syrup, leaving ¼-inch headspace. Adjust seals and process pint and quart jars 10 minutes in a boiling-water bath.

Yield: 3 to 4 quarts

Dilled Brussels Sprouts

2 pounds Brussels sprouts
2½ cups water
2½ cups vinegar
3 tablespoons salt

1 teaspoon cayenne pepper
4 heads dill
4 garlic cloves (optional)
4 heads dill (optional)

Cook Brussels sprouts until just tender, leaving whole. Combine water, vinegar, salt, pepper, and dill, and boil about 5 minutes. Pack Brussels sprouts into hot jars. Pour vinegar solution boiling hot over the sprouts, leaving ¼-inch headspace. (A clove of garlic and 1 extra head of dill can be placed into each jar if desired.)

Adjust caps and process pints 15 minutes in a boiling-water bath.

Yield: about 4 pints

Sweet and Sour Cabbage

4 quarts finely shredded red
    cabbage
4 tart apples, diced
1½ quarts cider vinegar
1 or 2 cups water (sufficient
    amount for juice)
4 teaspoons sea salt

½ teaspoon peppercorns
2 teaspoons caraway seed
½ teaspoon mace (optional)
½ teaspoon whole allspice
¼ teaspoon cinnamon log
    pieces
1 cup honey

Place spices in a spice bag and simmer with all the ingredients except honey in a large pot for 20 to 25 minutes. Remove spice bag and add honey. Pack into hot pint jars to within ¼ inch of the top. Process in a boiling-water bath for 15 minutes.

This cabbage will be much too stout for most people to eat from the jar, but it is delectable when the juice is drained off and it is simmered in a small amount of water.

Yield: 4 quarts

Dutch Spiced Red Cabbage

2 heads red cabbage
½ cup salt
1 gallon vinegar
½ cup water
1 teaspoon celery seed

1 teaspoon pepper
1 teaspoon each mace, allspice,
   cinnamon
½ cup honey

Shred the cabbage, sprinkle with the salt, let stand 24 hours. Press moisture out, stand in sun for 3 hours. Boil the vinegar for 8 minutes with water and spices. Add honey.

While hot pour over the cabbage. Keep in large bowl or earthen jar or can, as for sauerkraut (see page 84).

Yield: about 6 pints

Carrot Pickle

2 or 3 bunches carrots
2 cups vinegar
1½ cups water
½ tablespoon whole cloves

½ tablespoon allspice
½ tablespoon mace
½ stick cinnamon
½ cup honey

Pare carrots and cut in strips that are the desired size and the length of your canning jars, if possible. Boil in water until just heated through. Pack hot carrots lengthwise in hot, sterilized pint jars. Make a syrup of vinegar, water, and spices (in spice bag), bring to a boil, and simmer for 5 minutes. Add honey, bring to a boil, and pour over carrots. Allow ¼-inch headspace. Adjust seals and process pint jars 10 minutes in a boiling-water bath.

Yield: about 3 pints

Pickled Cauliflower

1 quart vinegar
2 tablespoons mustard seed
½ cup honey

8 whole cloves
4 sticks cinnamon
2 heads cauliflower

Simmer all ingredients except cauliflower together for 15 minutes.

Meanwhile wash cauliflower, cut away all leaves, and break into uniform flowerets. Blanch them in boiling water for 2 minutes. Drain and put flowerets into jars. Strain out spices from the syrup and pour hot over the cauliflower. Leave ½-inch headspace. Seal and process pint and quart jars for 10 minutes in a boiling-water bath.

Yield: 3 to 4 quarts

Pickled Eggplant

6 eggplants
2 medium or 1 large onion, chopped
½ teaspoon allspice
½ teaspoon cloves
½ teaspoon whole white peppercorns

1-inch cinnamon stick
1½ cups cider vinegar
2½ cups honey
1 cup water

Peel and chop the eggplants, then put them and the chopped onion in a pot. Cover with lightly boiling salted water, boil quickly for 5 minutes; drain, cover with cold water; drain again.

Place the spices in a spice bag and cook it with the vinegar, honey, and water to boiling. Add eggplant and onion mixture; simmer until syrup is thick and vegetables tender; remove spice bag; fill hot, sterilized jars, leaving ¼-inch headspace. Adjust lids and process pints 15 minutes in a boiling-water bath.

Yield: 4 to 5 pints

Dilled Green Beans

4 pounds whole green beans (about 4 quarts)
¼ teaspoon crushed red pepper per pint jar
½ teaspoon whole mustard seed per pint jar

½ teaspoon dill seed per pint jar
1 garlic clove per pint jar
5 cups vinegar (1¼ quarts)
5 cups water
½ cup salt

Wash beans thoroughly; drain and cut into lengths to fill pint jars. Pack beans into clean, hot jars; add pepper, mustard seed, dill seed, and garlic.

Combine vinegar, water, and salt; heat to boiling. Pour boiling liquid over the beans, filling jars but leaving ¼-inch headspace. Seal and process in a boiling-water bath for 5 minutes.

Yield: 7 pints

Jerusalem Artichoke Pickles

1 peck (8 quarts) artichokes
vinegar to cover artichokes
2 cups salt
4 tablespoons turmeric
1 gallon vinegar
2 tablespoons turmeric

1 box mixed pickling spices tied
   in a spice bag
6 cups honey
medium-sized pod red peppers
onions

Wash and cut artichokes. Pack in a large crock or enamel pot. Cover with vinegar. Add 2 cups salt and 4 tablespoons turmeric. Soak for 24 hours.

In the meantime, make spiced vinegar by combining in a pot 1 gallon vinegar, 2 tablespoons turmeric, and spice bag with pickling spices. Boil the mixture for 20 minutes. Remove the spice bag, add 6 cups honey, and bring the mixture to a boil.

Drain the artichokes. Pack in pint jars, covering with the boiling spiced vinegar. Allow ¼-inch headspace. To taste to each jar add 1 medium pod red pepper and onions. Process the jars for 10 minutes in a boiling-water bath.

Yield: 16 pints

Mixed Mustard Pickles

¼ cup mild yellow mustard
4⅔ cups distilled white
   vinegar
1¾ cups honey
½ cup salt
3 tablespoons celery seed
2 tablespoons mustard seed
½ teaspoon whole cloves
½ teaspoon ground turmeric

1½ teaspoons powdered alum
4 pounds 3- to 4-inch pickling
   cucumbers, cut into chunks
2 pounds small onions, peeled,
   quartered
1 quart (1½ inch) celery pieces
2 cups chopped sweet red
   peppers
2 cups cauliflower flowerets

In a pot blend mustard with a little vinegar; stir in remaining vinegar, honey, and next 6 ingredients; heat to boiling. Add cucumbers and remaining vegetables; heat just to boiling.

Simmer while quickly packing one clean, hot pint jar at a time. Fill to within ½ inch of top, making sure vinegar solution covers vegetables. Cap each jar at once. Process 5 minutes in boiling-water bath.

Yield: 9 to 10 pints

Pickled Mixed Vegetables

2 medium heads cauliflower
2 medium green peppers
2 medium sweet red peppers
1½ pounds onions (6 to 8
    medium)
2½ cups distilled white vinegar

1½ cups water
3 tablespoons salt
1 tablespoon mustard seed
1 tablespoon celery seed
¼ teaspoon ground turmeric
¾ cup honey

Break cauliflower into small flowerets; cook in unsalted boiling water 5 minutes; drain. Cut peppers into ¼-inch strips; quarter onions. Combine vinegar and all remaining ingredients except honey in a pot; heat to boiling. Add vegetables; simmer 2 minutes, then add honey.

Continue simmering while quickly packing one clean, hot pint jar at a time. Fill to within ½ inch of top, making sure vinegar solution covers vegetables. Cap each jar at once. Process 5 minutes in a boiling-water bath.

Yield: 5 to 6 pints

Okra Pickle

3½ pounds small okra pods
1 garlic clove for each jar
3 small hot peppers, if desired,
    for each jar

2 cups white vinegar
4 cups water
2 teaspoons dill seed
⅓ cup salt

Wash okra, and pack firmly in hot sterilized jars. In each jar put a clove of garlic and hot peppers if you wish. Make a brine with the vinegar, water, dill seed, and salt. Boil. Pour boiling brine over okra, leaving ¼-inch headspace at top. Seal.

Process for 10 minutes in a boiling-water bath. Let ripen several weeks before using.

Yield: 6 pints

Onion Pickles

1 quart small white onions,
    peeled
¼ cup salt
1 tablespoon mustard seed
2 teaspoons' prepared

horseradish
3 cups white vinegar
1 tablespoon honey
small hot red peppers

To make peeling easier, drop onions in boiling water, and after about 2 minutes, remove and plunge them into cold water, then peel. Drain; sprinkle onions with salt; add cold water to cover. Let stand 12 to 18 hours in a cool place.

Drain; rinse and drain thoroughly.

Combine mustard seed, horseradish, and vinegar; simmer 15 minutes, then add the honey.

Pack onions into sterile, hot jars, leaving ¼-inch headspace, adding 1 hot red pepper to each jar. Heat pickling liquid to boiling. Pour, boiling hot, over onions, leaving ¼-inch headspace.

Adjust caps and process half-pints and pints 10 minutes in a boiling-water bath.

Yield: 4 half-pints or 2 pints

Pickled Peppers

4 quarts long red, green, or
    yellow peppers
    (Hungarian, banana, or
    other varieties)
1½ cups salt
4 quarts plus 2 cups water

2 tablespoons prepared
    horseradish
2 garlic cloves
10 cups vinegar
¼ cup honey

Cut two small slits in each pepper. You may want to wear gloves to prevent burning hands. Dissolve salt in 4 quarts water. Pour over peppers and let stand 12 to 18 hours in a cool place. Drain, rinse, and drain thoroughly. Combine 2 cups water and all remaining ingredients except honey; simmer 15 minutes, then add honey. Remove garlic.

Pack peppers into hot jars, leaving ¼-inch headspace. Pour boiling hot pickling liquid over peppers, leaving ¼-inch headspace. Adjust caps. Process half-pints and pints 10 minutes in a boiling-water bath.

Yield: 16 half-pints or 8 pints

Pickled Sweet Peppers

Wash, stem, and core peppers. Slice them lengthwise into thin strips. Blanch them in steam for 2 minutes, then plunge them into ice water to cool them quickly. Drain.

Pack the cooled strips into hot, sterilized pint or half-pint jars and cover them with a hot syrup made from ½ part honey to 2 parts vinegar. Leave ¼-inch headspace. Process half-pints and pints for 10 minutes in a boiling-water bath.

Pickled Pimiento Peppers

16 to 20 large pimiento
  peppers
3 cups vinegar

1 cup honey
½ teaspoon salt

Wash, stem, and seed peppers, cut into strips, cover with boiling water, and let stand 3 minutes. Drain well; pack in hot, sterilized jars.

Boil remaining ingredients 5 minutes and pour over peppers, filling jars to within ¼ inch of the top of the jar. Adjust lids and process pints and half-pints 10 minutes in a boiling-water bath.

Yield: 8 half-pints or 4 pints

Green Tomato Pickle

4 quarts sliced green tomatoes
6 large onions, sliced
½ cup salt
6 cups vinegar
6 sliced green peppers
3 diced sweet red peppers
6 garlic cloves, minced

1 tablespoon dry mustard
1 tablespoon whole cloves
1 stick cinnamon
1 tablespoon powdered ginger
½ tablespoon celery seed
2¼ cups honey

Combine sliced green tomatoes and onions. Sprinkle with salt. Let mixture stand for 12 hours. Wash in clear water and drain. Heat vinegar to the boiling point, and add the green and red peppers and garlic. Then add the tomato-onion mixture. Tie the spices in a square of cheesecloth and drop into the mixture. Simmer for about 1 hour, or until tomatoes are transparent, stirring frequently. Then add the honey.

Pour into hot jars, leaving ¼-inch headspace, and process quarts 15 minutes and pints 10 minutes in a boiling-water bath.

Yield: about 6 pints or 3 quarts

Sweet Pickled Tomatoes

1 gallon green tomatoes
water
1 cup salt (approximately)

2 cups honey
about 2 cups vinegar

Slice green tomatoes. Put them in a large crock or glass container. Pour over them enough water to cover. Sprinkle ¼ inch of salt on top of the water—about 1 cup. Let stand for 24 hours. Drain. Put the tomatoes into a large kettle. Add 2 cups honey and enough vinegar to cover. Bring to boil, take off the heat, and pour into hot, sterilized jars. Leave ¼-inch headspace at top. Adjust seals and process in a boiling-water bath for 10 minutes.

Yield: 5 to 6 pints

Pickled Zucchini

4 pounds small zucchini
1 pound small white onions
water to cover vegetables
½ cup salt
1 quart cider vinegar

1 cup honey
2 teaspoons celery seed
2 teaspoons turmeric
2 teaspoons dry mustard
2 teaspoons mustard seed

Cut unpeeled zucchini into very thin slices, like cucumbers. Peel onions and slice thin. Cover vegetables with water and add salt. Let stand 1 hour; then drain.

Combine remaining ingredients, bring to a boil, and pour over vegetables. Let stand 1 hour.

Return to heat, bring to a boil, and cook 3 minutes. Pack in sterilized jars, leaving ¼-inch headspace. Adjust lids and process pints and quarts for 10 minutes in a boiling-water bath.

Yield: 4 pints or 2 quarts

Fresh Fruit Pickles

Spiced Sweet Apples

1 quart vinegar
2 cups water
1 ounce allspice
1 ounce cinnamon stick

4 cloves
4 cups honey
7 pounds sweet apples,
    quartered and cored

Put vinegar, water, and spices in pot. Bring to a boil. Add honey and apples, bring to a boil, and simmer gently until the fruit is tender.

Place the apple quarters in hot, sterilized jars, bring syrup to a boil, and pour over the apples, leaving ¼-inch headspace. Adjust seals and process 10 minutes in a boiling-water bath.

Yield: 4 pints

Minted Sweet Apples

Use the above recipe, but substitute 1 cup of mint tea for the spices. Make the tea by simmering 1 cup of fresh green mint in a pint of water. Strain before adding to the syrup. Reduce water to 1 cup.

Pickled Cantaloupe

4 cups cantaloupe, cubed
   (about 2 pounds)
2 cups water
⅔ cup vinegar
¼ cup honey

¼ teaspoon mace
¼ teaspoon allspice
8 whole cloves
2 sticks cinnamon

Cook cantaloupe in water until tender. Drain. Do not discard water.

Combine cantaloupe with 1⅓ cups cooking water, vinegar, honey, and spices. Simmer over low heat about 30 minutes or until desired flavor is reached and liquid has cooked down.

Pour hot into sterilized jars, wipe rim clean, adjust sterilized caps, leave at room temperature to cool, and check to be sure each is sealed before storing.

Yield: about 2 pints

Spiced Crabapples

12 pounds of crabapples
2 cups honey
1 quart vinegar
1 stick cinnamon

1 tablespoon cloves
1 teaspoon allspice
1 teaspoon mace

Wash apples well; be sure to remove blossom ends. Place the spices in a spice bag and add it to the honey and vinegar. Bring the mixture to a boil.

When this syrup is cool, add the crabapples and heat slowly so as not to burst the fruit. Sometimes it is well to prick each apple to avoid bursting. Bring to a boil. Allow to cool overnight.

Remove spice bag. Heat to boiling point. Pack crabapples into hot, sterilized jars, fill to within ¼ inch of the top with syrup. Adjust seals and process 10 minutes in a boiling-water bath.

Yield: 12 to 15 pints

Lemon Pickles

Use only organically grown, unsprayed lemons for this recipe, as for any recipe that calls for fruit rind (marmalades, relishes, etc.).

12 large lemons	1 tablespoon nutmeg, grated
½ cup salt	1 teaspoon red pepper
8 garlic cloves, peeled	4 tablespoons dry mustard
1 tablespoon mace	½ gallon vinegar
1 tablespoon allspice	

Wash and dry the lemons. Cut each lengthwise into 8 sections. Put these lemon sections into a pan with salt, garlic, and spices (tied in a spice bag). Add vinegar and bring to a boil. Simmer 30 minutes. Pour into large stone jar or crock. Stir daily for 1 month.

At the end of a month, drain the liquid into a pot. Place the lemons in hot, sterilized glass jars. Bring the liquid to a boil, remove spices, and pour over the lemons, leaving ¼-inch headspace. Adjust seals and process for 10 minutes in a boiling-water bath.

Pickled Peaches

7 2-inch pieces stick cinnamon	6 cups honey
2 tablespoons whole cloves	7 pieces stick cinnamon
2 quarts vinegar	14 to 21 whole cloves
16 pounds small or medium-sized peaches	

Place spices in a spice bag. Combine vinegar and spices in a large kettle. Bring to a boil, cover, and let simmer about 30 minutes.

Wash the peaches and remove skins. (Dipping the fruit in boiling water for 1 minute, then quickly in cold water makes peeling easier.) To prevent pared peaches from darkening during preparation, immediately put them into cold water containing 2 tablespoons salt and 2 tablespoons vinegar per gallon. Drain just before using.

Add honey to the syrup and bring to a boil. Add peaches, enough for 2 or 3 quarts at a time, and simmer for 5 minutes. Pack hot peaches into clean, hot jars. Continue heating in syrup and packing peaches. Add 1 piece stick cinnamon and 2 or 3 whole cloves (if desired) to each jar. Cover peaches with boiling syrup, leaving ¼-inch headspace. Adjust seals and process in a boiling-water bath for 10 minutes.

Yield: 7 quarts

Pear Pickle

8 pounds pears
10 2-inch pieces stick cinnamon
2 tablespoons whole cloves
2 tablespoons whole allspice
1½ quarts vinegar
2 pounds honey

For Seckel pears: Wash the pears and remove the blossom ends only. Boil the pears for 10 minutes in enough water to cover. Drain. Prick the skins. Put spices in a bag, boil with vinegar for 5 minutes. Add the honey and bring to boil. Add pears, simmer for 10 minutes or until the pears are tender. Do not overcook. Let stand overnight.

In the morning, remove the spice bag. Drain syrup from the pears and heat syrup to boiling. Pack pears in clean, hot, sterilized jars. Pour hot syrup over the pears, leaving ¼-inch headspace. Adjust seals and process 10 minutes in a boiling-water bath.

Yield: about 8 pints

For Kieffer pears: Prepare the pears, and reduce vinegar to 5 cups. Wash the pears, pare, cut in halves or quarters, remove hard centers and cores. Boil 10 minutes in enough water to cover. Drain. Proceed as for Seckel pears.

Watermelon Pickle

2 pounds watermelon rind
  (with outer green rind and
  pink meat removed)
¾ cup vinegar
¾ cup honey
3 sticks cinnamon
½ teaspoon nutmeg
½ teaspoon mace
15 whole cloves
¼ cup lemon juice

Cut watermelon rind into pieces about 1 inch long and ½ inch wide. Cover with cold water and leave in a cold place overnight.

Next morning, bring to a boil in soaking water. Simmer, covered, until tender, about 10 minutes. Drain, reserving ¾ cup of cooking water. Combine ¾ cup cooking water with the vinegar, honey, and spices which should be put into a bag made of cheese-cloth. Boil syrup for 15 minutes. Add lemon juice. Pour over rind and bring to a boil. Simmer about 5 minutes.

Pack hot into sterilized pint jars, adjust jar lids and process in a boiling-water bath for 5 minutes.

Yield: about 3 pints

If Your Pickles Fail

If your pickles, for some reason, don't turn out the way you'd like them to, learn from your mistakes and don't do the same thing wrong next year. Here are some common causes of pickle failure and how you can correct them:

If your pickles are **shriveled,** you may have used too strong a vinegar or salt solution at the start of the pickling process. In making very sweet or very sour pickles, it is best to start with a dilute solution and increase gradually to the desired strength. Shriveling may also be caused by overcooking or overprocessing.

Hollowness in cucumber pickles usually results from one or several of the following: poorly developed cucumbers, holding the cucumbers too long before pickling, too rapid fermentation, or too strong or too weak a brine during fermentation.

Soft or slippery pickles are spoiled pickles. Do not use them. This condition is generally a result of microbial action which caused the spoilage, and it is irreversible. Proper processing should halt microbial activity, but if it results, here are some things you might have done wrong: used too little salt or acid, failed to cover your cucumbers with brine during fermentation, allowed scum to scatter through the brine during fermentation, processed the pickles for too short a time, did not seal the jar airtight, or used moldy garlic or spices. Also, if you failed to remove the blossoms from the cucumbers before fermentation, they may have contained fungi or yeasts responsible for the softening action.

Dark pickles are not spoiled pickles; however, if you are one of those people who prides himself on the looks of his home-canned products, darkening can be annoying. Darkening may result from the use of too much spice, iodized salt, overcooking, iron in the water you used, or the use of iron utensils.

Relishes, Chutneys, and Spicy Sauces

Chowchow

2 small heads cauliflower, cut
  into small flowers
1 large bunch celery, cut into
  slices
2 pounds onions, sliced, or 2
  pounds pearl onions
2½ quarts fresh lima beans
1 quart sliced green peppers
2 quarts yellow string beans,
  cut into small pieces
2 dozen ears corn, kernels
  removed

2 quarts carrots, chunked
2 to 3 quarts small cucumbers
  (may be brined)
2 quarts kidney beans
salt to taste
2 quarts cider vinegar
5 cups honey (less, if desired)
1 tablespoon mustard seed
1 tablespoon peppercorns
1 tablespoon whole cloves
cinnamon stick

Sprinkle vegetables with salt and let stand 24 hours. Drain.
Make a syrup from the rest of the ingredients. Bring syrup to boil, add vegetables, and simmer until the vegetables are soft. Pack in hot, sterilized jars and adjust seals. Process for 10 minutes in a boiling-water bath.

Cherry Relish

2 cups pitted cherries
1 cup seedless raisins
1 teaspoon cinnamon
¼ teaspoon cloves

½ cup honey
½ cup vinegar
1 cup broken pecan nut meats

Combine all ingredients but the nuts in a pot. Cook slowly for 1 hour. Add nuts and cook 3 minutes longer. Pour into hot, sterilized jars, leaving ¼-inch headspace. Adjust seals and process in a boiling-water bath for 10 minutes.

Corn Relish

2 quarts whole kernel corn (use
  16 to 20 medium-sized ears
  of fresh corn or 6 10-ounce
  packages of frozen corn)
1 pint green peppers, diced

1 pint sweet red peppers, diced
1 quart (1 large bunch) celery,
  chopped
8 to 10 small onions, chopped or
  sliced

¾ cup honey
1 quart vinegar
2 tablespoons salt
2 teaspoons celery seed

2 tablespoons powdered
mustard
1 teaspoon turmeric

For fresh corn: Remove husks and silks. Cook ears of corn in boiling water for 5 minutes; remove and plunge into cold water. Drain and cut corn from cob. Do not scrape cob.

For frozen corn: Defrost overnight in refrigerator or for 2 to 3 hours at room temperature.

Combine green peppers, red peppers, celery, onions, honey, vinegar, salt, and celery seed. Cover pan until mixture starts to boil, then simmer uncovered for 5 minutes, stirring occasionally. Mix dry mustard and turmeric and blend with liquid from above mixture; add, with corn, to mixture. Heat to boiling and simmer for 5 minutes, stirring occasionally.

Bring to boil. Pack loosely while boiling into clean, hot pint jars. Allow a ¼-inch headspace. Adjust seals and process in a boiling-water bath for 5 minutes.

Yield: 7 pints

Horseradish Relish

1 cup grated horseradish
½ cup white vinegar
¼ teaspoon salt

Wash horseradish roots thoroughly and remove the brown outer skin with a vegetable peeler. The roots may be grated or cut into small cubes and put through a food chopper or a blender.

Combine ingredients. Pack into clean jars. Seal tightly. Store in refrigerator.

Pepper Relish

1 pint coarsely ground onion
2 cups coarsely ground sweet
green pepper
2 cups coarsely ground sweet
red pepper

1½ cups honey
1 quart vinegar
2 teaspoons salt

Drain onions and pepper, then combine all ingredients and bring slowly to a boil. Simmer until slightly thickened, about 25 minutes. Pour into clean, hot sterilized jars. Fill jars, leaving a ¼-inch headspace. Adjust seals and process 10 minutes in a boiling-water bath.

Yield: 2 to 3 pints

Pepper-Onion Relish

1 quart onions (6 to 8 large), finely chopped
1 pint sweet red peppers (4 or 5 medium), finely chopped
1 pint green peppers (4 or 5 medium), finely chopped
½ cup honey
1 quart vinegar
4 teaspoons salt

Combine all ingredients and bring to a boil. Cook until slightly thickened (about 45 minutes), stirring occasionally. Pack the boiling hot relish into clean, hot jars; fill to top of jar. Seal tightly. Store in refrigerator.

If extended storage without refrigeration is desired, this product should be processed in a boiling-water bath. Pack the boiling hot relish into clean, hot jars to ¼ inch of top of jar. Adjust seals and process in a boiling-water bath for 5 minutes.

Yield: 5 half-pints

Plum Relish

2 quarts pitted and coarsely ground ripe plums
1 orange, ground
½ cup vinegar
dash of nutmeg
1 cup chopped nuts (optional)
2 cups honey

Combine all ingredients except nuts and honey; cook until thick. Remove from heat and stir in nuts and honey. Pour into hot, sterilized pint jars, leaving ¼-inch headspace. Adjust caps and process in a boiling-water bath for 15 minutes.

Yield: about 4 pints

Green Tomato Mincemeat

6 cups green tomatoes, cut up
6 cups apples, cut up
½ cup suet, ground up
½ pound seedless raisins

1 tablespoon grated lemon rind
1 tablespoon grated orange rind
1½ cups honey
½ cup vinegar
¼ cup lemon juice in ¼ cup
   water

½ tablespoon ground
   cinnamon
¼ teaspoon mixed allspice and
   cloves

Put the green tomatoes, apples, suet, and raisins through a meat chopper. Then combine these with all the other ingredients and bring them to a boil in a large enamel or stainless steel kettle. Reduce the heat and simmer for 2 or 3 hours, using an asbestos pad and stirring frequently to prevent scorching.

Pour into hot pint jars, allowing 1-inch headspace, and process 25 minutes at *10 pounds pressure*.

Yield: 3 to 4 pints

Currant and Green Tomato Chutney

1½ cups currants
2¼ cups green tomatoes,
   chopped
2¼ cups tart apples, peeled and
   chopped
1 lemon, seeds removed,
   quartered, sliced thin
1 cup onions, minced

1 garlic clove, minced
½ cup honey
½ cup vinegar
½ cup water
1 tablespoon mustard seed
¾ teaspoon salt
¼ teaspoon cayenne
1 teaspoon ginger

Combine all ingredients. Simmer for 20 minutes or until fruit is soft. Pack into sterilized jars, leaving ¼-inch headspace. Adjust seals and process for 5 minutes in a boiling-water bath.

Yield: 2 pints

Honey Chutney

2 quarts sour apples
2 green peppers
⅓ cup onions
¾ pound seedless raisins
½ tablespoon salt
1 cup honey

juice of 2 lemons and the grated
   rind of one
1½ cups vinegar
¾ cup tart fruit juice
¾ tablespoon ginger
¼ teaspoon cayenne pepper

Wash and chop fruit and vegetables. Add all other ingredients and simmer until thick. Pour into hot, sterilized jars and adjust seals. Process in a boiling-water bath for 10 minutes.

Mango Chutney

2 cups cider vinegar
½ cup lime juice
5 tablespoons honey
4 large ripe mangoes, peeled
 and sliced
¾ cup dried currants
½ cup raisins
½ cup onion, minced
1 cup green pepper, chopped

⅔ cup almonds, chopped
3-inch piece ginger root (fresh
 or dry), peeled and cut fine
1 tablespoon crushed mustard
 seed
2 teaspoons salt
2 tablespoons fresh hot pepper,
 minced, or 1 ground dried
 chili pepper

Combine all ingredients. Bring to a boil, then turn heat down and simmer for 30 minutes. Drain off juice and boil it down to half of its volume. Add it to chutney and ladle into hot, sterilized jars, leaving ¼-inch headspace. Adjust seals and process for 5 minutes in boiling-water bath.

Yield: about 3 pints

Peach Chutney

1½ cups onions, sliced
1 garlic clove, minced
2 tablespoons oil
1½ teaspoons turmeric
1½ teaspoons coriander
1 teaspoon cumin
½ teaspoon ginger
½ teaspoon cayenne
¼ teaspoon black pepper

¼ teaspoon dry mustard
¼ teaspoon cardamom
⅛ teaspoon nutmeg
⅛ teaspoon cloves
¾ cup wine vinegar
¼ cup water
⅓ cup honey
2 pounds peaches, sliced

Sauté onions and garlic in oil, adding spices as they cook. Add vinegar, water, and honey. Simmer about 10 minutes. Add peaches and simmer until they are tender.

While still hot, pour into hot, sterilized jars, leaving ¼-inch headspace. Adjust lids and process for 5 minutes in a boiling-water bath.

Yield: about 2 pints

Tomato Chutney

2 pounds ripe, firm tomatoes
2 pounds tart green apples,
 peeled, cored, and sliced

2 onions, sliced
2 cups cider vinegar
2 teaspoons powdered ginger

2 dried chili peppers, crumbled   ½ cup raisins
1 teaspoon mustard seed   ¾ cup honey

Peel and slice tomatoes into a bowl. Add the apples, onions, vinegar, ginger, chili peppers, and mustard seed. If you have a blender, you can grind up the tomato and apple skins and add them too. Stir, cover, and put in a cool place for overnight.

In the morning put this mixture in an enamel or stainless steel pot; add raisins and honey and bring to a boil. Reduce heat and simmer uncovered until thick and rich colored.

Put in clean pint jars, leaving ¼-inch headspace. Adjust seals and process in a boiling-water bath for 5 minutes. Do not eat until it has aged at least 10 days.

Tomato-Apple Chutney

3 quarts chopped tomatoes (18 to 20 medium-sized)
3 quarts chopped apples (12 to 15 medium-sized)
1 cup chopped green pepper
3 cups chopped onion
2 cups seedless raisins
4 teaspoons salt
4 cups vinegar
⅓ cup whole mixed pickle spices

Combine tomatoes, apples, green pepper, onion, raisins, salt, and vinegar. Put spices in a spice bag and add to tomato mixture. Bring to a boil and simmer 1 hour, stirring frequently. Remove spices. Pack boiling hot chutney into clean, hot, sterilized jars, leaving ¼-inch headspace. Adjust seals and process 5 minutes in a boiling-water bath.

Yield: 7 pints

Tomato-Apple Chutney—2

1 seeded, chopped lemon
2 cups peeled, chopped apples (a hard, tart variety)
2 cups skinned, chopped tomatoes (6 medium-sized; choose ripe but firm ones)
2 seeded, chopped red peppers (sweet)
1 cup minced onion
2 cups raisins
1 cup cider vinegar
1⅓ cups honey
1½ teaspoons salt
2 teaspoons ground ginger
dash cayenne

Combine all ingredients. Bring to a boil, cover, turn down heat, and continue to simmer until fruit and vegetables are just tender but not mushy. Stir occasionally to prevent sticking and so that every-thing cooks evenly. Pour into hot, sterilized pint jars, leaving ¼-inch headspace. Adjust seals and process 5 minutes in a boiling-water bath.

Yield: about 6 pints

Tomato-Pear Chutney

2½ cups tomatoes, quartered, fresh or canned
2½ cups pears, diced, fresh or canned
½ cup seedless white raisins
½ cup green pepper, chopped (1 medium)
½ cup onions, chopped (1 or 2 medium)

½ cup white vinegar
1 teaspoon salt
½ teaspoon ground ginger
½ teaspoon powdered dry mustard
⅛ teaspoon cayenne pepper
¼ cup canned pimientos

When fresh tomatoes or pears are used, remove skins; when canned pears are used, include syrup.

Combine all ingredients except pimientos. Bring to boil. Cook slowly until thickened (about 45 minutes), stirring occasionally. Add pimientos, bring to boil, and cook 3 minutes longer.

For refrigeration storage: Pack the boiling hot chutney into clean, hot jars, filling to the top. Seal tightly.

For canning: Pack the boiling hot chutney into clean, hot jars, leaving ¼-inch headspace. Adjust seals and process in a boiling-water bath for 5 minutes. Remove jars, complete seals if necessary. Set jars upright on rack to cool.

Yield: 3 to 4 half-pints

If a slightly sweeter chutney is desired, you may add honey to taste just before canning. Stir in honey, tasting until desired sweet-ness is reached, just before pouring into jars. The cayenne pepper may be reduced or eliminated if a less spicy chutney is desired. The best test, again, is a taste test.

Piccalilli

1 quart green tomatoes, about 16 medium tomatoes, chopped
1 cup sweet red peppers, chopped (2 to 3 medium peppers)

1 cup green peppers, chopped (2 to 3 medium)
1½ cups onions, chopped (2 to 3 large)
5 cups cabbage, chopped (about 2 pounds)

⅓ cup salt
3 cups vinegar

2 tablespoons whole mixed
    pickling spice
1 cup honey

Combine vegetables, mix with salt, let stand overnight. Drain and press in a cheesecloth bag to remove all liquid possible.

Combine vinegar with the spices tied in a spice bag. Bring mixture to boil. Add honey. Add vegetables, bring to boil, and simmer about 30 minutes, or until there is just enough liquid to moisten vegetables. Remove spice bag. Pack hot relish into clean, hot pint jars, leaving ¼-inch headspace. Adjust seals and process in a boiling-water bath for 5 minutes.

Yield: 4 pints

Chili Sauce

4 quarts peeled and chopped
    tomatoes
2 cups chopped sweet red
    pepper
2 cups chopped onion
2 garlic cloves, minced
1 hot pepper, chopped
2 tablespoons celery seed
1 tablespoon mustard seed

1 bay leaf
1 teaspoon whole cloves
1 teaspoon ground ginger
1 teaspoon ground nutmeg
2 3-inch pieces stick cinnamon
½ cup honey
3 cups vinegar
2 tablespoons salt

Combine the tomatoes, sweet pepper, onions, garlic, and hot pepper in a large stainless or enamel pot.

Tie the celery seed, mustard seed, bay leaf, cloves, ginger, nutmeg, and cinnamon in a cloth bag. Add to tomato mixture, and boil until volume is reduced to half—2 to 3 hours. Stir frequently to prevent sticking. Remove the spice bag.

Add the honey, vinegar, and salt. Bring to a rapid boil, stirring constantly. Allow to simmer about 5 minutes. Pack into clean, hot, sterilized pint jars, leaving ¼-inch headspace. Adjust seals and process 10 minutes in a boiling-water bath.

Yield: 8 or 9 pints

Catsup

2½ quarts sliced tomatoes (15
    to 17 medium-sized)
¾ cup chopped onion

3-inch piece stick cinnamon
1 large garlic clove, chopped
1 teaspoon whole cloves

1 cup vinegar	1 teaspoon paprika
1¼ teaspoons salt	dash cayenne pepper

Simmer together tomatoes and onion for about 20 to 30 minutes; press through a sieve. Put the cinnamon, garlic, and cloves loosely in a clean, thin, white cloth; tie top tightly; add to vinegar and simmer 30 minutes. Remove spices.

Boil sieved tomatoes rapidly until one-half original volume. Stir frequently to prevent sticking. Add spiced vinegar, salt, paprika, and cayenne pepper to tomato mixture. Boil rapidly, stirring constantly about 10 minutes or until slightly thickened.

Pour into clean, hot, sterilized jars, leaving a ¼-inch headspace. Adjust seals and process for 5 minutes in a boiling-water bath.

Yield: about 2 pints

If you want a somewhat sweeter catsup, you can add honey just before you seal. Your taste is your best guide here, so add honey, stirring it until the taste pleases you.

Catsup—2

2 cups tomato purée	2-inch piece cinnamon stick
2 cups tomato juice	½ teaspoon whole allspice (or
1 cup cider vinegar	½ teaspoon ground
½ cup honey	allspice)
1 teaspoon salt	2 stalks celery and leaves
1 teaspoon dry mustard	2 onions

In a heavy saucepan, combine tomato purée, tomato juice, cider vinegar, honey, salt, and mustard. Place over medium heat and bring to a boil.

Put cinnamon and allspice in a spice bag; add to catsup mixture; then add celery and onions. When mixture reaches boiling point, reduce heat and cook, uncovered, until catsup is thick (about 3 hours). Stir occasionally during cooking.

Remove spice bag and celery and onions. Pour into heated glass container with tight-fitting lid; cool at room temperature, then store in the refrigerator.

This catsup may also be canned. Follow directions for previous recipe.

Yield: about 1 pint

Green Tomato Catsup

6 pounds green tomatoes	1½ tablespoons salt
3 pounds onions	1 quart vinegar
1 tablespoon black pepper	1 cup honey
1 tablespoon mustard	

Slice tomatoes and onions and put in a pot with seasonings; pour vinegar over all and cook for 4 hours over slow heat, stirring occasionally. Put mixture through a sieve or food mill, return to the pot and bring to a boil again; then add honey.

Pour into hot, sterilized jars and seal at once, without processing.

Yield: about 6 pints

Red Hot Sauce

24 long hot red or green peppers (about 1½ cups chopped)	4 cups vinegar
12 large red, ripe tomatoes (about 2 quarts)	2 tablespoons mixed spices
	1 tablespoon salt
	½ cup honey

Wash and drain vegetables. Remove seeds from peppers. You may want to wear gloves if your peppers are really hot. Core and chop tomatoes. Add 2 cups vinegar to the vegetables and boil until soft. Press through sieve or food mill. Place the spices in a piece of cheesecloth and tie the corners together to form a bag. Then put this bag and the salt into a pot with the strained vegetables and boil until thick. Add remaining vinegar and the honey. Boil 15 minutes, or until as thick as wanted. Pour, boiling hot, into hot, sterilized jars; seal at once, without processing.

Yield: about 4 pints

Hot Pepper Vinegar

Wash small hot red peppers and make two or three small slits in each. You may want to wear gloves to keep the hot pepper juice off your skin; it may burn sensitive skin.

Pack into bottles, cover with cider vinegar. Cork and leave in a cool place for 3 weeks before using. As the vinegar is used, add more vinegar.

Jams, Jellies, and Fruit Butters

What's a book on food preservation without a chapter on jelly-making, you may ask, and rightly so. In fact, if you've looked at most books on food preservation, you've probably found that the chapter on making jellies and jams is the largest section in the entire book (except, perhaps, for the pickling section). This is because jelly-making offers a chance to be truly creative. As long as you follow some basic rules, you can explore some very interesting flavor combinations of fresh fruits, dried fruits, and nuts. Perhaps you remember the pride your mother had in a special jelly, or the smile on your grandmother's face as she served slabs of homemade bread spread thick with butter and her own peach jam. Jellied fruit products have been a mark of pride for farm women for years. Even today, jams and jellies are an important item of show in farm shows and county fairs throughout rural areas of the country.

Making jellied fruit products—jams, jellies, conserves, preserves, marmalades, and fruit butters—is an excellent way of using up fruits and berries that can't be used for other types of canning. When you size your fruit for canning or freezing you'll most likely have some fruit that is too small or too large. Some pieces may be rejected because they have imperfections, like soft spots or rotten portions. Other pieces may be too ripe for canning. All these fruits can be used for jelly-making, and all of them can make a definite contribution to the finished product. Ripe fruit adds to the flavor of the product and should always be used when you are using commercial pectins. And, of course, size or blemishes won't matter because all the fruit is going to be cut up and peeled anyway.

For the sake of brevity, we refer to jams, jellies, preserves, marmalades, and conserves as jams and jellies in the next several pages—but there's a fine line separating each of them.

Jellies are cooked twice: first, the juice is extracted from the fruit (you can freeze the juice, if you wish, and make fresh batches of jelly at will), and then the juice is cooked with sweetening and pectin until it gels. *Jams* are purées made with fruit, sweetening, and pectin. *Preserves, conserves,* and *marmalades* are made with bits of fruit, cooked until translucent with sweetening and pectin. *Preserves* are generally made from a single kind of fruit (strawberry preserves contain only strawberries); *conserves* are made with fresh fruits and dried fruit or nuts, or both; and *marmalades* are made most often from one or many kinds of citrus fruit.

Fruit butters fall into a category of their own, so we'll discuss them separately at the end of this chapter.

Making Jam and Jelly with Honey

As you read on in this section you'll find that the directions and recipes here are quite different from those in jam and jelly chapters of other books. Recipes for jams and jellies elsewhere call for great quantities of sugar, but you won't find any sugar in our recipes. As we said in the introduction, this book means to provide alternatives to overprocessed and highly refined ingredients. An alternative to sugar is honey, and all the recipes that follow use this natural sweetening instead of sugar.

Because of its sweetness, sugar was a highly prized commodity for centuries before its preservative powers were discovered. It was always included with trade goods for the Indians, for example, because Indians prized it so highly. When it was discovered that sugar did a good job of preserving foods, it became even more valuable on farms and homesteads. During World War II, when sugar supplies were low and what little there was had to be rationed, government researchers began experimenting with alternatives to sugar. Books on preserving written during the war tell how to substitute honey or corn syrup (another refined product) for some of the sugar in jelly and jam recipes.

It is becoming more apparent, however, that a serious effort must be made to find a real, workable alternative for sugar for many uses in the home. Evidence is mounting that points to sugar as a real danger to health. For years experts have known the role sugar plays in tooth decay, diabetes, and obesity. Now there are indications that sugar may be responsible, in part, for coronary disease, and it has even been suggested that sugar may be a cause of cancer in some people. There is, in short, growing medical evidence that we would be wise to avoid eating any sugar at all.[*]

[*] There are several articles and books that cite such evidence. One of the best references is: John Yudkin, SWEET AND DANGEROUS (New York: Peter H. Wyden, Inc., 1972).

Even though medical experts and nutritionists don't know precisely to what extent sugar does our bodies harm, they do know that sugar does us no good. Sugar, even the so-called "raw" sugar, has no significant nutritional value. The complex sugar refining process destroys almost all of the few original vitamins and minerals found in the raw sugar cane and sugar beet.

Honey, unlike sugar, is subjected to a minimal amount of refining and processing and contains its original vitamins, enzymes, and minerals. Although not a highly significant source of food nutrients, honey certainly has more food value than sugar. In its unheated, unfiltered form, it also contains some pollen, a food rich in vitamins, minerals, and proteins. What's more, honey can be produced on the homestead; sugar, for all practical purposes, cannot.

(Note: Although sorghum, molasses, and maple syrup are all natural sweeteners, we don't recommend you use them in jams and jellies. They have a strong flavor of their own and will overpower or at least affect the flavor of the fruit.)

Before we discuss making jams and jellies with honey we want to be honest and tell you that they are more difficult to make than the conventional jams and jellies made with sugar. In order for the traditional cooked-down and pectin products to gel, they have to be cooked longer than they would if made with sugar—about 8 to 10 minutes for pectin jellies and jams and even longer for the cooked-down ones that have no extra pectin added.

If you have a little patience and follow the recipes carefully, you should be able to make some fine jelly products this way. The only real problem with using pectin or the cooked-down method is that the long cooking time required will lessen the food value of both honey and the fruit and alter their flavor and color. If you're concerned about such things, you may prefer to make jellies and jams with low-methoxyl pectin, gelatin, or agar-agar, all of which don't require sweetening or long cooking to gel. Discussions of all these methods of making jams and jellies follow.

Cooked-down and Pectin Jams and Jellies

The traditional cooked-down method simply uses fruit, some liquid like water or fruit juice, and a sweetening. It should be used only with fruits that have a good amount of natural pectin in them, like tart apples and crabapples, blackberries, grapes, lemons, plums, quinces, and raspberries. The fruit must also contain a sufficient amount of acid to achieve a gel. If the fruit you're working with is low in acid, recipes will call for the addition of lemon juice or citric acid. When using this method, select your fruit carefully. A batch of fruit should contain one-fourth just-ripe fruit and three-fourths fully ripe fruit for best pectin and flavor content.

Extra pectin should be added to fruits that don't have much natural pectin in them, like strawberries and apricots. Pectin can also be added to other fruits in order to cut down on cooking time necessary to form a gel. When using honey instead of sugar, added pectin can be especially helpful, since, as we said earlier, the cooking time could be quite long. Jellies made with added pectin also require less fruit than the cooked-down type to make the same amount of finished product.

Commercial pectin is available in most supermarkets under several different brand names, and it's available in both liquid and powdered forms. You can also make your own pectin from apples and use it instead of the commercial kind.

Homemade Pectin

You can make pectin—often called apple jelly stock—ahead of time and preserve it for later use, if you enjoy making combination jellies or blending them with other fruits in season when fresh apples are not available. Or, if you are lucky enough to have the fresh apples at the same time as the other fruits, you can make use of the pectin immediately.

"Apple thinnings"—those small, immature green apples sold in the early summertime—are rich in both acid and pectin. They will make good jelly stock and give a snappy tartness to the product. However, if you prize the clarity of the jelly product, be warned that such apples will not produce as clear and transparent a jelly as pectin made from fully mature apples. If you happen to have a bumper crop of apples, you will like the idea of using some of the surplus to make apple stock. It represents one more good use for that large supply. You can use imperfect fruit, even with insect damage, bird pecking, bruises, or cuts from dropping from the trees. Merely cut away the imperfect sections and use the sound parts.

Wash the apples carefully, trim, and cut pieces into thin slices. Measure 1 pint of water for every pound of apples. Place the slices in a kettle and boil for 15 minutes.

Strain off the free-running juice through one thickness of cheesecloth, without attempting to squeeze the pulp. Return the pulp to the kettle, and add the same measure of water again. This time, cook the mixture at a lower temperature for 15 minutes. Allow it to

stand for 10 minutes, then strain the second batch of juice through one thickness of cheesecloth. Again, do not attempt to squeeze the pulp. Allow it to cool enough so that you can handle it. Squeeze out the remaining juice, and combine all you have. There should be about 1 quart of juice for every pound of apples you used.

You can use this stock immediately for blending with other fruit juices to make jelly or jam, or you can preserve it for future use. If you wish to can the stock, heat it to the boiling point and pour it immediately into hot, sterilized canning jars. Seal, and invert the jars to cool. No further processing is necessary.

If you prefer to freeze the stock, allow it to cool and then pour into freezer containers. Allow 1 inch of space at the top for expansion.

Four cups of homemade pectin replaces approximately one-half bottle or 3 ounces of commercial liquid pectin in most recipes.

Extracting the Juice for Jellies

Before you make jelly, you must first extract juice from your fruit. To extract juice, prepare the fruit as directed in the recipe and then

Making pectin from apples. (1) Wash, peel, and slice apples and boil them for 15 minutes with 1 pint of water for every pound of apple slices. (2) Strain off the free-running juice through cheesecloth without squeezing the pulp. (3) Add 1 pint of water to each pound of remaining pulp and simmer for 15 minutes. (4) Let it stand for 10 minutes and then strain through cheesecloth without squeezing the pulp. (5) Allow the pulp to cool and squeeze out all the juice by pressing it in cheesecloth.

put it in a damp jelly bag or fruit press. The clearest jelly comes from juice that has dripped through a jelly bag without pressing. If you're not too concerned with producing a perfectly clear jelly, you can get more juice by twisting the jelly bag slightly and squeezing the juice out, or by using a fruit press. Pressed juice should be re-strained through a damp jelly bag or a double thickness of cheesecloth without squeezing.

When you're finished with the jelly bag wash it out well. The last thing you want is for your next batch of fruit to pick up a sour or musty taste from a jelly bag that had bits of old fruit or juice fermenting in it.

Cooking the Fruit Mixture

Commercial pectin is commonly available in both powdered and liquid form. When using powdered pectin, mix it with the unheated fruit juice for jelly, or with the unheated, crushed fruit for jams, conserves, or marmalades *before* cooking. If you are using liquid pectin, add it to the boiling juice and honey mixture when making jelly, and to the cooked fruit and honey mixture immediately after it is removed from the heat. When making jams, preserves, conserves, and marmalades, all products (except, of course, those raw jams and jellies that use gelatin or agar-agar) should be brought to a full boil and cooked for the recommended length of time. A full boil is reached when bubbles form over the entire surface of the jams, marmalades, or conserves mixture, or when the jelly mixture reaches a full rolling boil that cannot be stirred down.

It is better to make a small quantity of jelly or jam at a time rather than a large one. Since honey foams when it boils, a large kettle is required. An 8- or 10-quart kettle with a flat, broad bottom will be fine. Do not use an aluminum pot, as acids in the fruit may react with the aluminum.

We suggest you use a light-colored, mild-flavored honey for jams and jellies unless your family likes the taste of darker honeys. Dark honeys will impart a strong flavor to the product in which they are used, and although the taste will not be unpleasant, it probably will be too strong for most people. Clover, alfalfa, or another very light honey is best for making jellied fruit products. The honey should, if possible, be fresh because the flavor of honey changes somewhat in storage.

Filling and Sealing Containers

Cooked jellied fruit products may be sealed with paraffin in jelly glasses (straight-sided containers without inner lips) or they may be sealed in glass canning jars. Although jellied foods must be cooked in an open pot, they do not need to be processed in either boiling water or pressure steam like vegetables, meats, and other fruits that are stored in canning jars. You may seal jellied products with paraffin only, if they are firm. Fruit butters, preserves, conserves, and marmalades are usually too soft to be sealed with paraffin. In warm, humid climates, where there is a chance that paraffin might melt, seal all jellied products in canning jars.

Get glasses or jars ready before you start to make the jellied product. If you are using canning jars, make sure all jars and closures are perfect. Discard cracked or chipped jars. Wash the containers in warm, soapy water and rinse with hot water, so that they will not crack or break when hot liquids are poured in them. Wash and rinse all lids and bands. Metal lids with sealing compound may need boiling or holding in water for a few minutes—follow the manufacturer's directions. If you are using porcelain-lined zinc caps, have clean, new rings of the right size for jars. Wash the rings in warm, soapy water and rinse them well.

Work quickly when packing and sealing jars. To keep fruit pieces from floating to the top, gently shake jars of jam occasionally as they cool. If you are using jars with two-piece lids, fill hot jars to ⅛ inch of the top with the hot fruit or jelly mixture. Wipe the rim clean, place a hot metal lid on the jar with the sealing compound or rubber next to the glass, screw the metal band down firmly to seal, and stand the jar upright to cool. If you are using jars with porcelain-lined zinc caps, place a wet rubber ring on the jar, fill the jar to ⅛ inch of the top, screw the cap down tightly to seal, and stand upright to cool.

If you are sealing your jars with paraffin, remember that you can only seal firm products with it. Use only enough paraffin to make a layer ⅛ inch thick. A single thin layer can expand or contract readily and makes a better seal than one thick layer or two thin ones. Prick air bubbles in paraffin. Heating it over direct heat is dangerous because wax is flammable.

Allow paraffined products to stand overnight before moving them about so that the wax can harden completely. Cover paraffined glasses with metal or paper lids.

If you make more than one batch in a day, label each jar with the type of jellied product, the day it was made, and its batch number.

For best quality, do not keep jellied fruit products for more than 6 months. Like many canned goods, they lose flavor in storage. They should be stored in a cool, dry, dark place.

Jelly Tests

Some of our recipes call for a jelly test. There are two ways you can make a jelly test. To make a spoon or sheet test, dip a cold metal spoon in the boiling jelly mixture. Then raise it at least a foot above the kettle, out of the steam, and turn the spoon so that the syrup runs off to one side. If the syrup forms two drops that flow together and fall off the spoon as one sheet, the jelly should be done. Remember that honey jellies will be slightly softer than jellies made with sugar.

You can also use the refrigerator test for your jelly. Pour a small amount of boiling jelly on a cold plate and put it in the freezer compartment of a refrigerator for a few minutes. If the mixture gels, it should be done. While you're testing the jelly remove the cooking jelly from the heat.

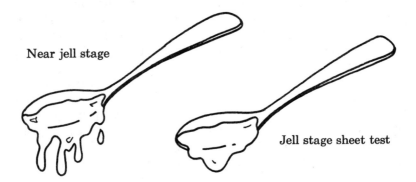

Near jell stage

Jell stage sheet test

Continue to cook your jelly until it slides off a spoon in one sheet, without dripping or breaking.

If Your Jellied Fruit Products Failed

There are many factors involved in making jellied fruit products. It is often hard to pinpoint one single factor that is responsible for a poor jelly, especially if you have little experience making jellies with honey.

If you are using honey and your products are **too runny,** try using less honey next time. Honey jellies or jams tend to be softer than jams or jellies made with sugar because honey contains so much more moisture. If your jelly was too runny and you used added liquid pectin, use the powdered kind next time. It generally gels better with honey than does the liquid kind. (Runny products can be used as toppings for ice cream, yogurt, pancakes, waffles, and cereal.)

If, on the other hand, your jelly is **too stiff,** you may have used too much pectin. If you were making a jelly with fruit that required no extra pectin, perhaps you cooked it too long. (Perhaps some underripe fruit in the batch contributed extra pectin to the product.) If the jelly is gummy, it may have been overcooked.

If jelly is improperly sealed, **fermentation** may occur or **mold** may develop. Any jelly that has mold on it should be thrown out.

Jellied fruit products made with honey are generally a bit darker than those made with sugar; however, if your jellies or jams **darken at the top of the jar,** it might be because you stored them in too warm a place or because a faulty seal is allowing air to enter the jar. The color of jellied products can **fade** if they are stored in too warm a place or if kept in storage for too long a time. Red fruits, such as strawberries or red raspberries, are especially likely to fade.

If you've stirred your jams and preserves properly before jarring them, you should have no trouble with **fruit floating to the surface.** If you find fruit floating in your preserves, stir the preserve mixture gently for 5 minutes after removing it from the heat the next time you make preserves. Also, make sure that you've used fully ripe fruit, that it was cooked long enough, and that it was properly crushed or ground.

Sometimes **part of the liquid will separate** from the jellied mass during storage. This could be the result of too much acid in the fruit or keeping the jelly in too warm a place. If there is no evidence of mold or spoilage, the jelly is still safe to eat.

Uncooked Jams and Jellies

If you miss the fresh flavor, color, and texture of fruit that is cooked away in most jams and jellies, and are concerned about destroying some of the fruit's and the honey's food value, you might like to try making some that are not cooked. Another nice thing about these jellied products is that all of them gel without sweetening, so you can use as little honey as you'd like. These jellies are not sterilized by boiling, so they must be poured into clean jars with tight-fitting lids and kept in the refrigerator for no more than about 3 weeks, or stored in the freezer. If you're going to freeze them remember to have a ½-inch headspace for expansion.

There are recipes later in this chapter for uncooked jams and jellies. Some of these recipes use natural gelatin instead of pectin. Generally, we've found that gelatin works better than pectin in recipes that use honey rather than sugar, and this is because gelatin doesn't require sweetening to make it gel.

Other recipes use agar-agar as the gelling agent. Agar-agar is a transparent dried seaweed, rich in iodine, iron, and other trace elements. To make jellies with agar-agar, soak one-quarter of a ½-ounce package of dried agar-agar in ½ cup cold water until it's soft. Then add to it 1 cup of concentrated fruit juice. Simmer the mixture for a minute or two, until the agar-agar dissolves. Sweeten to taste with honey. Pour into a jar, cool it down, and store in refrigerator or freezer.

Using Low-Methoxyl Pectin

Regular jams and jellies gel through an interaction of pectin, acid, and sugar. Those made with added regular pectin are not different in this respect from cooked-down jellies which gel with the aid of the natural pectin of the fruits. Low-methoxyl pectin differs from both the commercial and natural pectin in fruits because it requires calcium salts (or a bonemeal-lemon juice mixture explained later), not sugar, to form its gel. Euell Gibbons originally discovered low-methoxyl pectin through his brother, a diabetic, who was experimenting with it for use in his own diet, and reported on his own experiments in ORGANIC GARDENING AND FARMING. At that time, Gibbons's big problem was obtaining a small quantity of the low-methoxyl pectin since it was commercially available in large quantities only. That problem has been solved for him—and for all of us—since Walnut Acres, a natural foods store and mail-order company, began handling low-methoxyl pectin and the calcium salts used with it.

Using the low-methoxyl pectin and calcium for jelly-making is very simple. Here is how it is made:

Prepare your fruit or juice. Generally, jam fruit is simmered until soft, then put through a sieve to reduce it to a pulp and to remove skins and seeds. Fruit for jelly is usually crushed, simmered, then dripped through a jelly bag or several thicknesses of cheesecloth to obtain a clear juice. You can also use canned, bottled, or frozen juices. Measure your juice or fruit. Put it in a saucepan over medium heat and bring just to a boil.

Next measure ⅛ cup honey and ½ teaspoon of low-methoxyl pectin to each cup of fruit or juice, and mix them well in a mixing bowl. Pour this mixture into the freshly boiling juice or fruit, all at once, and stir, stir, stir until the honey and pectin are completely dissolved. Now add 1 teaspoon of the calcium solution (⅛ teaspoon calcium in ¼ cup water) for each cup of juice or fruit, and quickly stir until well mixed. The jelly or jam is done and can be immediately poured into sterilized jars and sealed.

However, one can test this jelly before sealing to see if it has exactly the right texture by putting about a tablespoonful in a metal cup and chilling it in the refrigerator. Such a small quantity chills very quickly; and as soon as it is cold, it will be the texture of your finished jelly. If it is too stiff, add more juice to the pot; if it is too runny, add another teaspoon or so of the calcium solution. It is that simple. (Remember to remove your jelly or jam from the heat while you're making this test.)

This jelly and jam cannot be merely paraffined and set on the shelf. It must be sealed in sterilized jars with sterilized lids, for even the microorganisms that cause spoilage find these products good. The half-point, straight-sided mason jars that seal with two-piece dome lids are ideal. Sterilize jars by boiling them for 10 minutes, then leaving them in hot water until they are used. Heat the jelly or jam just to boiling point and pour it, boiling hot, into the sterilized jars, immediately sealing with the sterilized lids. When opened for use, the jar will keep perfectly well in the refrigerator for several weeks.

It should be noted that since the original edition of this book came out, ORGANIC GARDENING AND FARMING food editor Nancy Albright has developed a substitute for the calcium salts used with this pectin. Her more natural and easily made bonemeal and lemon juice mixture works just as well.

To make the bonemeal and lemon juice solution, mix together ¼ teaspoon powdered bonemeal, 1 tablespoon lemon juice, and ¼ cup water. Let it stand for 30 minutes, or until the bonemeal is completely dissolved, before using.

For recipes using this bonemeal-lemon juice mixture and the low-methoxyl pectin, turn to the recipe section later in this chapter.

The beauty of using low-methoxyl pectin is, of course, that since sugar isn't necessary for gelling, you're free to add as little or as much sweetening as you want. You also don't have to worry about adding too much liquid (honey or fruit juice) because you can always adjust the gel of the mixture by adding more juice if it's too stiff or more calcium if it is too runny.

Low-methoxyl pectin and the calcium salts that you can use instead of the bonemeal-lemon juice mixture gelling action can be ordered from Walnut Acres, Penns Creek, Pennsylvania 17862.

Recipes for Jams, Jellies, Marmalades, Preserves, and Conserves

Apple Jelly

Wash apples. Remove stems and dark spots, quarter but do not pare or core apples. Add just enough water to half cover apples and cook until the fruit is soft. Drain, using a jelly bag. You'll get more juice if you squeeze the bag, but it will make a cloudy jelly. Measure juice. Add ¾ cup honey for every cup of juice. Boil until a good jelly test is obtained. Pour into hot, sterilized glasses. Cover with paraffin.

You can make mint jelly as a variation on this apple jelly. Just before removing apple jelly from the heat, add a few mint leaves which have been washed (about ¼ cup of mint leaves to 1 quart of juice) and a bit of green vegetable coloring. Stir, remove the leaves, pour jelly into hot, sterilized glasses. Cover with paraffin. This makes an attractive and delicious jelly to serve with lamb.

Apricot Jam

3 cups dried apricots*
¼ cup fresh lemon juice

7 cups mild-flavored honey
½ bottle liquid fruit pectin

Mix apricots, lemon juice, and honey. Simmer until apricot pieces are soft. Stir to prevent scorching. Remove from heat and stir in pectin. Stir and skim for 5 minutes. Pour into hot jars and seal with lids.

Yield: 3 half-pints

* You can use dried peaches instead of dried apricots.

Raw Apricot Conserve

Pit and mash fresh apricots. Stir in the desired amount of honey and thicken with blanched, ground almonds.

Raw Blackberry Jam

Mash ripe blackberries and add honey to taste. Stir it in and keep the raw jam in a covered dish.

Blackberry Jelly
(Jam can be made from contents of jelly bag.
Do not discard. See next recipe.)

2 quarts fresh blackberries
¾ cup water
3 teaspoons low-methoxyl
    pectin

¾ cup honey
5½ teaspoons bonemeal-lemon
    solution*
2 tablespoons lemon juice

Wash and remove stems from blackberries. Mash them in a pot, add water, and bring to a boil. Turn heat down to a simmer, cover and cook for about 5 minutes until berries are tender. Be careful that the water doesn't boil away, causing the berries to scorch.

Pour off as much juice as possible. Carefully put berries into a damp jelly bag or line a colander with several thicknesses of cheesecloth, dampen it, and put the berries in it. When most of the juice has drained off the berries, gather up the cheesecloth and form a bag of it, tying string around the neck securely. Let the bag hang from a faucet or some convenient handle, dripping into a bowl, for about 8 hours. Do not squeeze the bag or force the pulp through the bag in order to keep the jelly clear.

Measure juice. There should be about 3 cups. Bring it to a boil. Stir low-methoxyl pectin into honey, mixing it in well, then add it all at once to the boiling juice, stirring until it is dissolved. Add bonemeal-lemon solution and lemon juice, mixing it in well.

Pour jelly immediately into hot, sterilized jars. Wipe the rims clean, adjust sterilized caps, leave at room temperature to cool, and check to be sure each is sealed before storing.

Yield: 1½ to 2 pints

* Bonemeal-lemon solution is made up of ¼ teaspoon powdered bonemeal, 1 tablespoon lemon juice, and ¼ cup water. Mix well and leave for 30 minutes until dissolved.

Blackberry Jam
(made from contents of jelly bag after
juice for blackberry jelly has been extracted)

¾ cup blackberry pulp
¼ cup water
½ teaspoon low-methoxyl
    pectin

3 tablespoons honey
¾ teaspoon bonemeal-lemon
    solution (see above)
1 tablespoon lemon juice

Purée pulp through strainer. Add water and heat to boiling.

Stir pectin into honey and add boiling pulp, stirring until it is completely dissolved. Remove from heat. Add bonemeal-lemon solution and lemon juice, mixing well.

Pour immediately into a hot, sterilized jar. Wipe the rim clean, adjust cap, leave at room temperature to cool, and check to be sure it is sealed before storing.

Yield: approximately 1 cup

Black Cherry Conserve

4 medium-sized oranges
2 cups water
8 cups Bing cherries, pitted
1½ to 2 cups honey, according to taste

¾ cup lemon juice
1½ teaspoons cinnamon
12 cloves
2 cups pecan halves or walnut quarters

Wash but do not peel oranges. Remove seeds and slice thin. In a covered saucepan, simmer orange slices in water until tender, about 10 minutes. Watch them so that they don't boil dry.

Add cherries, honey, lemon juice, cinnamon, and cloves. Simmer, covered, for 30 minutes. Add nuts.

Pour immediately into hot, sterilized pint jars, wipe rims clean, and adjust caps. Leave at room temperature to cool, and check to be sure each is sealed before storing.

Yield: 4 pints

Raw Blueberry Jam

In the blender put the following and blend into mush:

2 cups fresh or thawed blueberries
1 stalk rhubarb
1 cup nuts

Pour the jam into a dish which can be covered tight. If too thin, stir in rice polishings to thicken and add honey to taste. Keep refrigerated.

Fresh Blueberry Jam

2 tablespoons lemon juice
2 tablespoons water

1 envelope unflavored gelatin
1½ teaspoons arrowroot flour

2½ cups washed, hulled, and sliced blueberries (about 1 quart whole berries)

6 tablespoons honey

Combine lemon juice, water, gelatin, and arrowroot in a saucepan. Heat, stirring constantly, until gelatin and arrowroot are dissolved. Add the blueberries and honey to the gelatin mixture and heat to boiling.

Pour into clean containers and store in refrigerator or, for longer keeping, in the freezer.

Yield: 3 half-pints

Pink Crabapple Jelly

3 pounds small pink crabapples (about 10 cups, quartered)
4 cups water
2 teaspoons low-methoxyl pectin

¾ cup honey
4 teaspoons bonemeal-lemon solution (see page 211)

Wash crabapples. Remove stems and blossom ends and cut into quarters. There should be about 10 cups, quartered.

Add water and bring to a boil. Simmer, covered, until crabapples are tender. Put through dampened jelly bag or several layers of damp cheesecloth and leave to drip for several hours. There should be about 4 cups of juice. Bring juice to a boil.

Soften pectin in honey and add all at once to boiling juice, stirring to dissolve pectin completely. Add bonemeal-lemon solution, mixing in well.

Pour immediately into hot, sterilized jars. Wipe rims clean and adjust sterilized caps. Leave at room temperature to cool. Check to be sure each jar is sealed before storing.

Yield: approximately 2½ pints

Currant Jelly

Pick over currants. Do not remove stems. Wash. Drain. With a potato masher mash a few in bottom of preserving kettle. Continue adding and mashing currants until all are used. Add ½ cup water to 2 quarts fruit. Slowly bring to a boil and let simmer until currants appear white. Strain through a colander, then allow juice to run through a cloth or jelly bag. Measure juice. Add ¾ cup honey for every cup of juice. Boil until a good jelly test is obtained. Pour into hot, sterilized glasses. Cover with paraffin.

Elderberry Jelly

2 pounds elderberries
  (slightly underripe berries
  have the best flavor)
1 cup water
2 tablespoons lemon juice

3 teaspoons low-methoxyl
  pectin
1 cup honey
3 teaspoons bonemeal-lemon
  solution (see page 211)

Wash elderberries and cut berries off of main stem. There should be approximately 2 quarts berries. Combine berries with water and bring to a boil, covered. Simmer for about 15 minutes, mashing the berries to extract juice.

Purée through a food mill. Strain purée to remove seeds. There should be approximately 3 cups of juice.

Bring juice to a boil. Add lemon juice. Stir low-methoxyl pectin into honey and add all at once to boiling juice, stirring to dissolve pectin completely. Add bonemeal-lemon solution. Pour immediately into hot, sterilized jars. Wipe rims clean and adjust sterilized caps, then leave at room temperature to cool. Check to be sure each jar is sealed before storing.

Yield: approximately 2 pints

Elderberry and Apple Jelly

5 pounds tart apples
5 cups water
2 pounds elderberries
1 cup water
½ cup lemon juice

2½ teaspoons low-methoxyl
  pectin
1¾ cups honey
4¾ teaspoons bonemeal-lemon
  solution (see page 211)

Wash and quarter apples. Add 5 cups water and bring to a boil. Simmer, covered, until apples are tender. Put through dampened jelly bag or several layers of damp cheesecloth and leave to drip for several hours. There should be about 3 cups of juice.

Wash elderberries and cut berries off of main stem. Combine berries with 1 cup water and bring to a boil, covered. Simmer for about 15 minutes, mashing the berries to extract juice. Purée through a food mill and then strain to remove seeds. There should be about 3 cups of juice.

Combine apple and elderberry juices and bring to a boil. Stir in lemon juice.

Soften pectin in honey and add all at once to boiling juice, stirring to dissolve pectin completely. Add bonemeal-lemon solution, mixing in well.

Pour immediately into hot, sterilized jars. Wipe rims clean and adjust sterilized caps. Leave at room temperature to cool. Check to be sure each jar is sealed before storing.

Yield: approximately 4 pints

Ginger and Pear Conserve

3 pounds winter pears (about 21 to 24 Seckel pears)
½ orange, seeded and chopped fine (skin and pulp)

1 tablespoon chopped ginger root or 1 piece about 1½ inches square
½ cup honey
1 tablespoon lemon juice

Wash, peel, core, and slice pears. If desired, to save time, chop half the amount of peeled and cored pears in a blender along with the orange.

Make a spice bag of cheesecloth, and put ginger into it. Combine pears and orange, spice bag, honey, and lemon juice and bring slowly to a simmer in a covered saucepan. Cook until pears are tender. Remove cover and continue to cook, stirring occasionally, until desired consistency is reached.

Pour hot into hot, sterilized jars. Wipe rims clean and adjust sterilized caps. Leave at room temperature to cool and check to be sure each is sealed before storing.

Yield: approximately 3 cups

Ginger and Pear Conserve—2

½ pound green ginger, scraped and chopped
4 oranges for juice and thinly shredded peel
3 lemons, juice and thinly shredded peel
1 pint water

6 pounds honey
8 pounds pears, weighed after paring and coring
2 cups pecans or black walnut meats

Cook the ginger, orange peel, and lemon peel with a pint of water until tender, then add honey, orange juice, and lemon juice. Chop the pears coarsely, add them to the spiced juice, and cook until pears are tender. Add nut meats. Cook 5 minutes longer.

Pour in small, hot jars and seal, boiling hot.

Ginger Quince Jam

2 cups quince purée	½ teaspoon ginger
¾ cup honey	2 tablespoons lemon juice

Combine all ingredients, and cook over low heat until of desired consistency.

Pour into hot, sterilized jars, wipe rims clean, adjust sterilized caps, and leave at room temperature to cool. Check to be sure each is sealed before storing.

Yield: approximately 2¾ cups

Concord Grape Conserve

7 cups grapes, washed and stemmed (about 2½ pounds before preparation)	¼ lemon, seeded and sliced very thin
½ cup honey	½ cup raisins
	½ cup pecans, coarsely chopped

Slip skins off of grapes. Do not discard. Bring pulp to a boil and simmer over low heat until seeds are loosened. Press pulp through colander and/or strainer to remove seeds.

Combine puréed pulp, grape skins, honey, lemon, raisins, and pecans. Simmer over low heat to plump raisins.

Pour hot into hot, sterilized jars. Wipe rims clean, adjust sterilized caps, leave at room temperature to cool, and check to be sure each is sealed before storing.

Yield: 2½ pints

Grape Basil Jelly

2 tablespoons dried basil leaves	2 cups grape juice, unsweetened (see recipe in this book)
½ cup boiling water	
2½ teaspoons low-methoxyl pectin	2½ teaspoons bonemeal-lemon solution (see page 211)
½ cup honey	

Crush basil leaves. Pour boiling water over them and allow to steep in a covered container for 10 minutes. Strain liquid through a dampened cheesecloth-lined strainer.

Dissolve low-methoxyl pectin in honey. Heat grape juice to boiling and add pectin-honey mixture all at once, stirring to dissolve completely. Remove from heat.

Add basil infusion and bonemeal-lemon solution, stirring it in well.

Pour into hot, sterilized jars, wipe rims clean, and adjust sterilized caps. Leave at room temperature to cool and check to be sure each is sealed before storing.

Yield: approximately 3 cups

Waterless Grape Jelly and Jam

8 cups washed and stemmed
blue Concord grapes
(approximately 5 pounds
before separation)
2 tablespoons low-methoxyl
pectin

1½ cups honey
2 tablespoons bonemeal-lemon
solution (see page 211)
½ cup honey

Put grapes in heavy-bottom saucepan and mash well. Bring to a simmer slowly and cook until the grapes are soft, taking care that they don't scorch. Strain through a dampened jelly bag or a strainer which has been lined with several thicknesses of dampened cheesecloth. Leave to drain, or hang bag up until it has stopped dripping. Measure juice. There should be about 5 cups of juice. Save contents of bag for jam.

Bring grape juice to a boil. Dissolve low-methoxyl pectin in honey and stir in all at once as soon as juice is boiling. Stir until pectin is dissolved completely. Remove from heat. Stir in bonemeal-lemon solution.

Pour immediately into hot, sterilized jars. Wipe rims clean and adjust sterilized caps. Leave at room temperature to cool and check to be sure each is sealed before storing.

Yield: approximately 6½ cups jelly

Put contents of jelly bag or strainer through purée utensil. There should be about 2½ cups purée. Add ½ cup honey, cook down to desired consistency, and store as above.

Yield: approximately 3 cups jam

Honey Jelly

3 cups honey   ½ bottle fruit pectin
1 cup water

Measure honey and water into a large kettle and mix. Bring to a boil over hottest heat and at once add pectin, stirring constantly. Then bring to a full rolling boil and immediately remove from heat. Skim; pour quickly into clean, hot, sterilized glasses and paraffin at once.

Yield: 6 glasses

Lemon Honey Jelly

¾ cup lemon juice   ½ cup liquid fruit pectin
2½ cups honey

Combine lemon juice and honey. Bring to a full rolling boil. Add pectin, stir vigorously, and boil about 2 minutes. Pour into hot, sterilized glasses. Cover with paraffin to seal.

Yield: 6 glasses

Raw Marmalade

To make raw marmalade, use fresh fruits, wash and crush, or crush frozen and thawed fruits.

Any of the following fruits can be made up this way, as well as combinations of two or more fruits: peaches, plums, strawberries, raspberries, blackberries, cherries, currants.

For each cupful of fruit, use about ⅓ cup honey. If you have honey which is crystallized, here is a good place to use it up.

Thicken the marmalade with wheat germ flour (blend wheat germ to a fine consistency), using only as much as is needed. Keep refrigerated and covered.

Honey Orange Marmalade

3 medium-sized oranges   6 tablespoons lemon juice
1 cup water   ¼ cup liquid pectin
1¾ cups honey

Run oranges through food chopper, using the fine knife. Measure to make sure that you have at least 1¾ cups of ground orange. Add water.

Bring to a boil and simmer 15 minutes. Add honey and simmer 30 minutes. Add lemon juice, then liquid pectin. Bring to a full rolling boil and boil 30 seconds. Remove from fire. Skim and stir for 5 minutes. Pour quickly into sterilized glasses, then paraffin.

Fresh Orange Marmalade

2 cups fresh squeezed orange juice (about 10 oranges)
1 envelope unflavored gelatin

3 tablespoons honey
1 whole orange

Bring ½ cup of the orange juice to a boil and combine it with the gelatin and honey in a blender. Blend until gelatin is dissolved.

Wash the orange well. If you suspect it has been waxed or sprayed (and if you didn't grow it or buy it from an organic grower, it probably was at least sprayed), let it soak in a weak solution of water and white vinegar (1 tablespoon vinegar to a small bowl of water). Peel the orange, slice it, and remove the seeds. Save about one-fourth of the peel.

Add the sliced orange, one-fourth of the peel, and 1½ cups orange juice to the gelatin mixture in the blender. Blend on low speed just until the peel is coarsely chopped. Pour into containers and refrigerate, or for longer keeping, freeze.

Yield: about 3 half-pints

Peach Jam

4 pounds fully ripened fresh peaches
¼ cup fresh lemon juice

1 package (3½ ounces) powdered fruit pectin
2 cups mild-flavored honey

Wash, peel, and remove pits from fresh peaches. Chop or coarsely grind peaches, blending with lemon juice. Measure prepared fruit, packing down in cup. You should have 4 full cups. Place fruit and lemon juice in large 6- to 8-quart saucepan. Add pectin and mix well. Place over high heat.

Bring to a boil, stirring constantly. (Oil rim of saucepan well and mixture will not boil over.) When fruit is boiling, stir and slowly pour in honey, blending well. Continue stirring and return to a full rolling boil. When boil cannot be stirred out boil exactly 4 minutes. Remove from heat. Alternately stir and skim for 5 minutes to cool slightly and keep fruit from floating.

Pour into prepared glasses, allowing ⅛ inch for paraffin.

Yield: 10 6-ounce glasses

White Peach Preserves

1½ cups honey	3¼ teaspoons low-methoxyl
1 cup water	pectin
¼ cup lemon juice	½ cup honey
2 quarts sliced peaches	6½ teaspoons bonemeal-lemon
	solution (see page 211)

Combine 1½ cups honey, water, and lemon juice and simmer for 10 minutes.

Add peaches and cook until they are transparent. Drain peaches and set aside. Boil syrup hard for 10 minutes, or until it has reduced to approximately 2½ cups. Add peaches.

Mix low-methoxyl pectin into ½ cup honey, then add it all at once to simmering peaches and syrup, stirring well to dissolve pectin completely. Stir in bonemeal-lemon solution. Pour immediately into hot, sterilized jars. Wipe rims clean, adjust sterilized caps, and leave at room temperature to cool. Check to be sure each jar is sealed before storing.

Yield: approximately 4 pints

Peach and Cantaloupe Conserve

4 cups peaches, diced	¼ cup orange juice
4 cups cantaloupe, diced	10 tablespoons honey
rind of 2 lemons	1 cup pecans, coarsely chopped
¼ cup lemon juice	

Combine all ingredients except pecans. Simmer until fruit is soft. If desired, drain syrup off of fruit and cook it down to concentrate and reduce it.

Add pecans to fruit and fill sterilized jars, pouring hot syrup in to cover fruit. Wipe rims clean, adjust sterilized caps, leave at room temperature to cool, and check to be sure each is sealed before storing.

Yield: 2 pints

Raw Pear Conserve

In blender (or through the food grinder) put these ingredients:

2 cups diced pears
½ cup raisins
½ cup honey

½ cup nut meats
¼ cup pineapple

Blend until you have a thick jam. Put in tightly covered glass jars and refrigerate.

Raw Pear and Cranberry Conserve

6 ripe pears, cored
2 cups frozen cranberries

honey to taste
pinch of cinnamon and cloves

Put pears and thawed cranberries in the blender. Whiz them to a fine pulp. Add honey, cinnamon, and cloves. Whiz again. Keep refrigerated in covered dish. Red raspberries may replace the cranberries. Nuts or sunflower seeds may be added.

Red Pepper Marmalade

12 medium-sized red peppers
4 teaspoons whole allspice
½ teaspoon ground ginger
2 cups chopped onions

2 cups white vinegar
3 cups honey
4 teaspoons salt
1 lemon, sliced

Remove stems and seeds from peppers. Cover peppers with boiling water; let stand 5 minutes; drain. Repeat, and drain well. Put through coarse blade of food chopper. The mixture should measure about 4 cups.

Tie spices in a cloth bag. Combine with other ingredients. Boil 30 minutes, stirring occasionally. Let stand overnight.

Next day, bring to boil in large saucepan and simmer 10 minutes. Ladle boiling hot into sterilized ½-pint jars. Seal.

Yield: about 6 half-pints

Raw Pineapple Jam

½ cup cold water
2 cups fresh pineapple
honey to taste

ground nuts
sunflower seeds

Put water and pineapple in the blender and whiz until smooth. Remove from blender and add honey, then ground nuts and sunflower seeds until you have thick jam. Keep refrigerated in a covered dish.

Blue Plum Jam

6 cups blue plums, pitted
   (about 3 pounds)
¾ cup water
12 plum pits
½ teaspoon low-methoxyl
   pectin

½ cup honey
2 teaspoons lemon juice
½ teaspoon bonemeal-lemon
   solution (see page 211)

Combine plums, water, and pits and simmer, covered, until plums are very soft. Remove pits. Push plum pulp through colander.
   Dissolve low-methoxyl pectin in honey. Bring plum purée to a boil. Add pectin-honey mixture all at once, stirring to dissolve completely. Remove from heat.
   Add lemon juice and bonemeal-lemon solution and stir it in well.
   Pour into hot, sterilized jars, wipe rims clean, and adjust sterilized caps. Leave at room temperature to cool, and check to be sure each is sealed before storing.

Yield: approximately 3½ cups

Quince Jelly

8 or 9 quinces (3 pounds after
   coring and quartering)
water to cover
2 teaspoons low-methoxyl
   pectin

⅔ cup honey
2 teaspoons bonemeal-lemon
   solution (see page 211)

Wash, quarter, and core quinces. Cut quarters into large cubes. Cover with water. Bring to a boil and simmer for an hour, until quinces have turned a deeper pink orange color. Mash quinces unless they have cooked to a pulp.

Drain through a colander lined with dampened cheesecloth. Do not discard pulp. Purée pulp and reserve for quince jam. Measure juice. There should be about 5 cups.

Bring 4 cups juice to a boil. (Reserve extra juice for quince jam.) Stir low-methoxyl pectin into honey and add this all at once to the boiling juice, stirring to dissolve pectin completely. Add bonemeal-lemon solution, stirring in well.

Pour into hot, sterilized jars. Wipe rims clean, adjust sterilized caps, leave at room temperature to cool, and check to be sure each is sealed before storing.

Yield: 4⅔ cups

Paradise Jelly

1½ pounds quince (about 4), cored and chopped (not peeled)
2 pounds tart apples, cored and chopped (not peeled)
1 cup cranberries, picked over and washed

water to cover each fruit
2 teaspoons low-methoxyl pectin
1 cup honey
2 teaspoons bonemeal-lemon solution (see page 211)

Bring each fruit to a boil in water to cover and simmer until fruit is soft. Strain each through a colander which has been lined with dampened cheesecloth. Let juice drip through. Do not squeeze cheesecloth or pulp.

Combine 2 cups quince juice, 2 cups apple juice, and 1 cup cranberry juice. Reserve any remaining juice for another use. Do not discard pulp. Purée each one. Quince pulp may be used for jam, and apple and cranberry pulp may be combined to make a delicious sauce.

Bring combined juices to a boil. Dissolve the pectin in honey, and add all at once to boiling juices. Stir until pectin is completely dissolved. Add bonemeal-lemon solution, stirring to blend in well.

Pour into sterilized jars. Wipe rims clean, adjust sterilized caps, and leave at room temperature to cool. Check to be sure each is sealed before storing.

Yield: approximately 6 cups

Strawberry Jam

4½ cups prepared fruit (about 8
    cups fully ripened whole
    strawberries)

1 package powdered fruit
    pectin
3½ cups mild-flavored honey

Gently wash strawberries in ice water. Drain. Hull, slice, and crush 8 cups. Measure 4½ cups crushed fruit in very large saucepan. Add powdered pectin to fruit. Mix well. Place over high heat. Bring to a full boil, stirring constantly. Pour honey in all at once and stir well. Bring to a full rolling boil. Boil hard 2 minutes, stirring constantly. Remove from heat. Alternately stir and skim for 5 minutes to cool slightly and keep fruit from floating. Ladle quickly into prepared glasses. Cover jam at once with ⅛ inch of hot paraffin.

Yield: about 11 6-ounce glasses

Raw Strawberry Jam

Wash the desired amount of strawberries, then hull. Mash and add honey to taste. If the berries are very juicy, as they are some years, stir in either wheat germ flour (blenderized wheat germ) or rice polishings, and add enough to thicken the jam as you want it. Tiny new mint leaves may also be shredded up in the jam for a unique flavor. Keep refrigerated in a covered dish.

Fresh Strawberry Jam

2 tablespoons lemon juice
2 tablespoons water
1 envelope unflavored gelatin
1½ teaspoons arrowroot flour

2 cups washed, hulled, and
    sliced strawberries (about 1
    quart whole berries)
6 tablespoons honey

Combine lemon juice, water, gelatin, and arrowroot in a saucepan. Heat, stirring constantly, until gelatin and arrowroot are dissolved. Add the strawberries and honey to the gelatin mixture and heat to boiling.

Pour into clean containers and store in refrigerator or, for longer keeping, in the freezer.

Yield: 3 half-pints

Fresh Strawberry Preserves

2 cups water
1 envelope unflavored gelatin
2 cups sliced strawberries
(about 1 quart whole
berries)

1 teaspoon lemon juice
⅓ cup honey

Pour 1 cup of water into a saucepan and sprinkle gelatin over it. Let stand 5 minutes.

Purée half the berries with 1 cup of water and the lemon juice and honey in a blender. Add remaining berry slices and the puréed mixture to the saucepan. Heat just to boiling, stirring all the while.

Pour into clean containers and refrigerate, or for longer keeping, freeze.

Yield: about 4 half-pints

Strawberry Preserves

¾ cup honey
2 quarts fresh strawberries,
washed, hulled, and sliced
3½ teaspoons low-methoxyl
pectin

¼ cup honey
3½ teaspoons bonemeal-lemon
solution (see page 211)
¼ cup lemon juice

Drizzle ¾ cup honey over strawberries. Leave at room temperature for 3 to 4 hours.

Bring strawberries and juice to a boil. Stir pectin into ¼ cup honey and add all at once to boiling berries, stirring until pectin is dissolved. Remove from heat. Stir in bonemeal-lemon solution and lemon juice.

Pour immediately into hot, sterilized jars, wipe rims clean, and adjust sterilized caps. Leave at room temperature to cool, and check to be sure each is sealed before storing.

Yield: approximately 3 pints

Strawberry Rhubarb Preserves

6 tablespoons honey
4 cups strawberries, hulled,
washed, and sliced
4 cups rhubarb, washed,
unpeeled, and diced
3½ teaspoons low-methoxyl
pectin

6 tablespoons honey
2 tablespoons lemon juice
3½ teaspoons bonemeal-lemon
solution (see page 211)

Drizzle 6 tablespoons honey over strawberries. Leave at room temperature for 3 to 4 hours.

Combine strawberries, juice, and rhubarb. Simmer over medium heat until rhubarb is tender. Dissolve low-methoxyl pectin in 6 tablespoons honey and add all at once to simmering strawberry-rhubarb mixture. Stir until pectin is completely dissolved. Remove from heat. Add lemon juice and bonemeal-lemon solution, stirring it in well.

Pour into hot, sterilized jars, wipe rims clean, and adjust sterilized caps. Leave at room temperature to cool, and check to be sure each is sealed before storing.

Yield: approximately 2½ pints

Raw Tutti-Frutti Jam

Mash the following fresh fruits:

½ cup red raspberries	1 ripe avocado
½ cup black raspberries	2 very ripe bananas
½ cup blueberries	

Add honey to taste, and thicken if necessary with rice polish. Add a pinch each of cinnamon and allspice. Keep refrigerated in a covered dish.

Fruit Butters

You may find that fruit butters will become your favorite method of preserving fruits. A big advantage of fruit butters is that they can be made in very large quantities. They also aren't as delicate as jellies are: it isn't essential to get them off the stove at exactly the right moment for fear they will not gel. Rather, fruit butters will thicken naturally as they cook down, so you won't have to worry about testing your fruit for pectin or acid, or do a jelly test every other minute to find out if the mixture is thick or not. The test for thickness of a fruit butter is simple, fast, and easy: Put a dab of the butter on a plate, and let it set for a few minutes. If the dab does not separate, that is, if you don't notice liquid at the edges of the drop, the butter is ready to be put up.

However, there are disadvantages to making fruit butters. They have to be cooked for a long time—far longer than jams or jellies—

before they are ready to be put up. If you are making fruit butter on the top of the stove, they must be watched closely and stirred frequently so that they don't burn or scorch. Scorched fruit butter is next to useless, and there are few worse feelings than burning 20 quarts of it. You'll have to make fruit butter when you can spend the entire day in the kitchen so that you can stir it frequently. If, however, you are making the butter in a low oven, your constant attention is not needed. See a recipe for this kind of butter later in this chapter.

Although fruit butters made with sugar have been known to keep for years without spoiling, it is still best to *can* your butters. You can do this by processing both pint and quart jars in a boiling-water bath for 10 minutes. (See page 76 for directions for the boiling-water bath process.)

Don't allow these disadvantages to dissuade you from making at least some fruit butters—they have a long and honorable history. In German, the word *Latwerg* originally meant prune or pear butter which the people of the Rhenish Palatinate made to keep their prunes and pears over the winter months. It often took up to 48 hours to prepare these fruit butters, but the time was spent in celebration as the preparation of the butter became the occasion of a folk festival. The descendants of these Germans, the Pennsylvania Dutch, brought *Latwerg* with them when they came to America. In this country, the term came to be associated with apple butter because the Pennsylvania Dutch made huge quantities of apple butter in the fall, stirring apples and cider in huge kettles over an open fire. As you might imagine, this, too, was a festive event and often used as an "excuse" for courting. Like many other once-regional specialties, apple butter has been made available to most parts of the country by modern transportation and improved marketing facilities; but store-bought spread just can't match the taste of real, honest-to-goodness homemade apple butter.

Equipment

A large kettle or pot with a heavy bottom is best for making fruit butters. Pennsylvania Dutch families used cast-iron pots slung on tripods over outdoor fires. If you're making large quantities of fruit butters, you may wish to invest in a large stainless steel stockpot. (These pots also come in handy when you're making soups, sauerkrauts, and other foods in large quantities.) Stockpots come in various sizes, from small to huge, and you can choose according to your taste and budget. Many restaurant supply stores will give you a good price on a second-hand stockpot. Aluminum stockpots are considerably cheaper than those made from stainless steel, but don't be tempted by the lower price unless you plan to use it only for boiling-water bath processing. The acids in fruits may react with the aluminum and cause both a color and flavor change in your butters. Stainless steel is best because the metals in stainless steel are stable and will not react with acids in fruits and other foods. You may find that the bottom of stainless steel stockpots are thin and do not conduct heat evenly. If this is the case with your pot, buy an asbestos pad for the burner and place it under your pot when you are making fruit butters. This pad will spread the heat evenly over the bottom of the pot and will help to prevent hot spots from developing.

Some recipes recommend "baking" fruit butters in the oven. Again, stainless steel is good for this, and you may be able to buy a stainless steel pan from the same place you got your stockpot. Glass casserole dishes are good, too, if they are deep and wide (it is better to have a shallow layer of butter in a pan rather than a thick one—it cooks up faster). Enamel roasting pans are also good.

Adding Honey to Your Butters

You will notice that in most of the recipes that follow we recommend adding honey to taste when the butter has finished cooking down, rather than adding honey while the fruit is still cooking. If fruits are ripe or slightly overripe, the natural sugars in them are at their peak, and no extra sweetener may be necessary. You may spoil your butter by making it too sweet if you add honey before tasting the finished product.

The ingredients in most of the recipes below are given in parts rather than in actual measurements, like cups or pounds. We have done this so that you can easily adjust the quantities and use just the amount of fruit you have on hand.

Apple Butter

Some people feel that the best apple butter is made from unpeeled, uncored apples. Other people feel that there's little difference in taste between butter made from apples that were peeled and cored before they were cooked down, and butter made from apples that were cooked whole and then put through a food mill or fruit press to remove the skins and seeds. The choice is up to you, but we offer a little advice: If you're not concerned about the slight difference in taste in the finished butter and want to save yourself a little time, peel and core your apples first, because it's more work to remove the skins and seeds later on by using a food mill or fruit press.

The recipe below is given for those who are using unpeeled and uncored apples. If you are peeling and coring yours, leave out the step that tells you to run your fruit through the fruit press.

1 part cider
2 parts apples, unpeeled and uncored, sliced thin

Put the cider in a big pot and bring it to a boil. Add the apples slowly, being careful not to splatter yourself when you do. Allow the apples and cider to come to a boil, then simmer, stirring frequently to prevent sticking (and do remember to stir).

When the apple butter has begun to thicken considerably, the apple slices will start to fall apart as you stir the butter. At some point after the butter has thickened, remove it from the heat and put everything through a fruit press, discarding the peels, seeds, and stems. (If you have peeled and cored your apples, omit this step.) Put the remaining soupy mixture back in the pot, put the pot back on the heat, and simmer until the apple butter is a thick, dark brown mass. (One old cookbook says, ". . . til the liquid becomes concrete—in other words, till the amalgamated cider and apple become as thick as hasty pudding.")

Test to see if the apple butter has thickened. If it has, you may want to sweeten it. Use honey and sweeten to taste. You may also want to add spices to your apple butter. If so, add cinnamon, allspice, and ground cloves to taste. Bring the mixture to a boil, and jar the apple butter in hot, sterilized jars, leaving a ¼-inch headspace. Screw the lids on tightly.

Process pints and quarts for 10 minutes in a boiling-water bath. Remove from heat and seal jars if necessary. Cool them on a wire rack and store cooled jars in a cool, dark place.

Fruit Butter Recipes

Baked Apple Butter

18 pounds apples
water
5 cups honey (approximately)
juice and grated rind of 3
   lemons

4 teaspoons cinnamon
2 teaspoons cloves
¾ teaspoon allspice

Wash and core apples; then quarter them. Place them in a large pot and pour in enough water to almost cover them. Simmer for 1½ hours.

Put the pulp through a strainer, blender, or food mill and pour back into the original pot (if it's oven-proof) or into a roasting pan. To each cup of pulp add ⅓ cup honey. Then add the lemon juice and grated rind and the spices.

Place the apple pulp mixture in the oven and cook slowly at 300°F., stirring occasionally. The butter will thicken as it bakes. It will take several hours—perhaps overnight—to thicken sufficiently.

When it is thick enough for your taste, remove and pack into sterilized jars, leaving a ½-inch headspace. Adjust seals and process pints and quarts for 10 minutes in a boiling-water bath.

Yield: about 7 quarts

Apricot Butter

Cut fruit in halves, remove pits, and place in large kettle with just enough water to prevent scorching. Cook until tender. When tender, put through a fruit press or press through a sieve. Cook until thick, stirring frequently to prevent scorching. When the butter is thick, sweeten with honey if you desire.

Pour into hot, sterilized jars, leaving a ½-inch headspace. Adjust seals and process pints and quarts for 10 minutes in a boiling-water bath.

Pink Crabapple Butter

1 quart (about 20 ounces) small red crabapples	¾ cup cider vinegar
	½ to 1 cup honey
10 ounces tart eating apples (about 2 medium-sized)	8 cloves
	2-inch piece cinnamon stick

Wash crabapples and apples, cut out stems and blossom ends. Do not peel or core. Cut apples into eighths.

Combine vinegar and honey, starting with ½ cup unless apples are very tart. Add crabapples and apples, and cook, covered, until they are very soft. Put through food mill or strainer to purée.

Add cloves and cinnamon stick to purée. Cook uncovered over very low heat until of desired consistency. Add the remaining honey during this process if it is needed. Remove cloves and cinnamon stick and pour into hot, sterilized jar. Wipe rim clean, adjust sterilized cap, and leave at room temperature to cool. Check to be sure jar is sealed before storing. No further processing is necessary.

Yield: 1 generous pint

Grape Butter

1 gallon grapes
4 tablespoons water

Put grapes in kettle with water. Heat and mash the grapes. Continue cooking as the mixture thickens, stirring frequently. When thicker, put through a fruit press and remove skins and seeds. Return to heat and cook until thick.

If you want to sweeten your grape butter, add honey to taste, just before canning.

When thick, pack in hot, sterilized jars, leaving a ½-inch headspace. Adjust seals and process pints and quarts for 10 minutes in a boiling-water bath.

Yield: about 4 quarts

Baked Peach Butter

3 parts sliced peaches
1 part water

Peel peaches and remove pits. (For freestone peaches, scald for 1 minute to remove the skins.) Place peaches in large pot and cook until peaches are soft. Shake pot frequently to prevent sticking (it is better if you divide your peach butter up among several pots for this part of the operation). When the fruit is tender, put it through a mill or a fruit press. Turn the purée into a shallow roasting pan or pans and cook uncovered for 1 hour at 325°F. Continue cooking, stirring every 15 to 20 minutes, until the butter is thick, fine textured, and a rich reddish amber color.

When it has reached this state, take it out of the oven. Ladle it into hot, sterilized jars, leaving a ½-inch headspace. Adjust seals and process pints and quarts 10 minutes in a boiling-water bath.

Yellow Peach Butter

6 cups sliced yellow peaches
¾ cup water
¼ cup lemon juice

10 cloves
2-inch piece cinnamon stick
1¼ cups honey

Combine peaches, water, and lemon juice and simmer until peaches are soft. Put into container of electric blender and blend to a purée.

Return to heavy-bottom saucepan and add cloves and cinnamon stick. Continue to cook, uncovered, over low heat to desired consistency, stirring occasionally.

Remove cloves and cinnamon stick. Add honey. Pour immediately into hot, sterilized jars. Wipe rims clean, adjust sterilized caps, and leave at room temperature to cool. Check to be sure each jar is sealed before storing. No further processing is necessary.

Yield: 1½ pints

Pear Butter

4 cups winter pears, peeled, cored, and sliced
½ cup water
2 teaspoons honey

2 teaspoons lemon juice
2-inch piece cinnamon stick
12 cloves

Combine pears and water and blend to a purée in electric blender. Add honey, lemon juice, and spices.

Strain and pour into hot sterilized jars, wipe rims clean, and adjust sterilized caps. Leave at room temperature to cool and check to be sure each is sealed before storing. No further processing is necessary.

Yield: about 3 cups

Plum Butter

Wash plums and remove all blemishes. Put in kettle and add enough water so that the bottom of the pot is covered. Cook until tender. Put through food mill or fruit press to remove pits and skins. Measure pulp and add ½ cup honey for each cup of plum pulp, if desired. Return to heat and cook until thick. Pack in hot, sterilized jars, leaving a ½-inch headspace. Adjust seals and process pints and quarts 10 minutes in a boiling-water bath.

Prune Butter

1 pound prunes
½ cup white or cider vinegar
honey to taste

Rinse prunes. Cover with water, bring to a boil, and then reduce heat and simmer until tender. Cool slightly. Remove pits. Put the prunes through a sieve or blender to purée. Add the vinegar. You can add spices to the butter now, if you wish. Recommended spices include:

$\frac{1}{4}$ teaspoon grated nutmeg
$\frac{1}{2}$ teaspoon powdered allspice
$\frac{1}{4}$ teaspoon powdered cloves

Cook until thick and then sweeten with honey.

Pour into sterilized, hot jars, leaving a $\frac{1}{2}$-inch headspace. Adjust seals and process in a boiling-water bath for 10 minutes.

Juicing
Your Harvest

Fruits and vegetables too bruised or overmature for canning or freezing need not be thrown away. Even though the skins outside of these foods may be in pretty poor shape, the juice locked inside the plant cells can be salvaged, and then canned or frozen for drinking, or for using in soups, desserts, and casseroles later in the year.

It's true, of course, that whole fruits and vegetables have a proportionately greater food value than their juices; this is because the fiber, pulp, and skin are intact. However, by making juices from less than perfect produce, you're capturing many of the vitamins, minerals, and enzymes that would otherwise be lost if the whole foods were left to spoil. And for the gardener with storage problems, juicing foods might provide a solution. The juice from a pound of spinach will fill a small juice glass half full, so you can store a garden full of excess produce in a small place if you preserve it in the form of juice.

Tomato and Fruit Juices

These juices are relatively easy to make. Specific directions for each juice follow, but the procedure is just about the same for all of them. Basically all you do is simmer tomatoes or fruit in water or its own juice in a stainless steel, glass, or enamelware pot. Don't use aluminum because the metal can react with the food acid. The fruits may be smaller and less perfect than the ones you save for canning and freezing, but they must be very ripe. As the fruit starts to simmer, cut through it with a knife to release some of the juices. Stir occasionally to prevent sticking.

When the fruit is tender, press it through two layers of cheese-cloth, or through a food mill or colander (straining through cheese-cloth will give you the clearest juice). Keep the pulp separate from the liquid you collect, but don't discard it: pulp can be used for fruit leathers, desserts, and sauces.

The juice should not need sweetening if the fruit you started with was ripe. If you do wish to sweeten it, though, add honey to taste; usually ½ cup is sufficient for each gallon of juice. Lemon juice can also be added to peach and apricot nectar, as well as sweet cherry juice; it will add a little zing and help preserve the color. Four teaspoons lemon juice, 2 tablespoons vinegar, or ½ teaspoon citric acid can be added to quarts, and half that amount of any one to pints of tomato juice if you suspect your tomatoes have a low acid content (see page 68).

The juices (and the remaining pulp) can be frozen or canned. To freeze, pour into clean glass jars or freezer containers and leave ½-inch headspace for expansion. Seal and freeze right away. You can also pour the juice into ice cube trays, freeze, and then transfer to plastic bags.

Canning procedures vary from food to food, so check specific directions that follow. The delicate flavor of most fruit juices can be spoiled by the high temperatures of a long boiling-water bath. For this reason, directions state that some juices be poured boiling hot into sterilized jars and sealed without processing, and other juices be processed in a *hot*-water bath, which means that the water is steadily *simmering* all the while and reads at 185° to 190°F. on a candy or hot water thermometer. The flavor of tomato juice is not spoiled by high temperatures, and it can and actually *should* be canned in a *boiling*-water bath where the water is rapidly boiling all the while and reads 212°F. on a thermometer. For more information on boiling-water canning, see page 76.

Apple Juice and Apple Cider

See directions on page 244 for making cider and juice. Freeze, or pour into sterilized jars, leaving ¼-inch headspace. Process pints and quarts in a 185°F. *hot-water* bath for 30 minutes.

Apricot Nectar

See the recipe for Peach Nectar that comes later.

Berry and Currant Juices
(all except cranberries)

Do not add any water; just crush berries in a pot and *simmer* them in their own juice, stirring occasionally, until they are soft. Add a small amount of water only if berries threaten to scorch. Strain by letting the liquid drip through a jelly bag or several thicknesses of cheesecloth for several hours, until the pulp releases no more liquid. If desired, heat to a simmer and add lemon juice and honey to taste. Freeze, or pour into sterilized jars, leaving ½-inch headspace. Process pints and quarts in a 190°F. *hot-water* bath for 30 minutes.

Cherry Juice

Sweet cherries make a rather bland juice, so add some sour cherry juice to it if possible.

Prepare just as berry juice, adding a little water if necessary to prevent scorching. If desired, heat to a simmer and add lemon juice and/or honey to taste. Freeze, or pour into sterilized jars, leaving ¼-inch headspace. Process pints and quarts in a 185°F. *hot-water* bath for 30 minutes.

Cranberry Juice

For every cup of berries add 1 cup of water. Bring to a *boil* in a pot and cook until berries burst. Strain through jelly bag or several layers of cheesecloth. Boil 1 minute, and add honey if desired. To can, pour boiling juice into sterilized pint or quart jars, leaving a ⅛-inch headspace. Adjust caps and seal immediately. No further processing is necessary.

Grape Juice
(white or blue)

Blue grapes will produce about twice as much juice as white grapes.

Place grapes in a pot and cover them with boiling water. Heat slowly to a simmer and continue to *simmer* until fruit is very soft. Strain through jelly bag or several thicknesses of cheesecloth.

If you want clear grape juice, refrigerate for 24 to 48 hours and then strain once more. Heat to a simmer and add honey to taste, if desired. Freeze, or pour hot juice into sterilized jars, leaving a ¼-inch headspace. Process pints and quarts in a 190°F. *hot-water* bath for 30 minutes.

Peach or Apricot Nectar

Although the nectar can be thinned with water to make juice, it is usually canned as is and then thinned at time of serving.

Add 1 cup of boiling water to each quart of ripe, pitted peaches or apricots. Cook until fruit is soft and press through sieve or food mill. Reheat and add honey to taste, if desired. Pour hot into sterilized jars, leaving a ¼-inch headspace. Process half-pints and pints 15 minutes and quarts 20 minutes in a *boiling-water* bath.

Plum Juice

For every cup of cut-up plums, add 1 cup of water and gently simmer together. Continue to simmer until fruit is soft, 10 to 15 minutes. Strain through jelly bag or several thicknesses of cheesecloth. Heat to a simmer and add honey to taste, if desired. Freeze, or pour hot juice into sterilized jars, leaving a ¼-inch headspace. Process pints and quarts in a 190°F. *hot-water* bath for 30 minutes.

Tomato Juice

Chop up clean tomatoes, with ends cut out, and simmer slowly until soft. Press through fine sieve or food mill. Add spices or honey, to taste if desired. You may add 4 teaspoons lemon juice, 2 tablespoons vinegar, or ½ teaspoon citric acid to quarts and half that amount to pints if you wish to raise the acid content (see page 68). Freeze, or reheat juice to just below boiling and pour hot into sterilized jars. Leave a ¼-inch headspace. Process pints 10 minutes and quarts 15 minutes in a *boiling-water* bath.

Here are a few more tomato juices. Because these contain low-acid vegetables in addition to tomatoes, they must be either frozen or processed in a pressure canner. See directions that follow. For specific information on pressure canning, see page 78.

Tomato Juice—2

Varieties considered especially good for juice are: Rutgers, Abraham Lincoln, Roma, Italian Canner, Queen's Jubilee (yellow), and such others as Earlibell, Big Giant, and Spring Set.

15 ripe red tomatoes	3 cloves
½ cup water	½ cup honey
1 onion, chopped	

Simmer the tomatoes and other ingredients, except the honey, in an enamel or stainless steel pan for 25 minutes.

Cool slightly and put the mixture through a blender for 2 minutes, or until well blended. Strain through a coarse sieve and add honey.

Freeze, or reheat and pour into sterilized jars, leaving a 1-inch headspace. Then process pints 15 minutes and quarts 20 minutes in a pressure canner at *10 pounds pressure*.

Green Drink with Tomatoes

3 tomatoes	1 cup cold water
1 large cucumber	1 tablespoon kelp
4 large sprigs of watercress,	6 basil leaves
both the leaves and stalks	4 sprigs of dill
4 spinach stalks, and leaves	1 cup fresh yogurt

This green drink is best if you make it just after you pick the ingredients. Wash and cut up the ingredients and put them a few at a time into a blender that already has the cup of water in it. Start with the tomatoes and greens, and last of all the yogurt. Only blend long enough to liquefy.

This is a very cooling and refreshing drink. You can also include sprigs of wild plants such as lamb's-quarters, wintercress, plantain, and sorrel.

If you wish to can or freeze this drink, don't add the yogurt before doing so. Rather, add it at time of serving.

To can, bring juice to a boil and pour into sterilized jars, leaving a 1-inch headspace. Process pints 20 minutes and quarts 30 minutes in a pressure canner at *10 pounds pressure*.

Tomato Juice and Purée

10 to 15 tomatoes	2 carrots
6 scallions	salt
5 stalks celery	

Cut up all the vegetables and put the tomato pieces in a large enamel or stainless steel pot. Bring to a boil and add the other vegetables and salt to taste. Boil rapidly until the tomatoes are tender. Put the mixture through a colander. Push through a little of the pulp along with the juice and set this aside. Now push through as much of the rest of the pulp as you can and use this for tomato puree.

Both can be frozen or canned for later use. To can, see directions for pressure canning in recipe above.

Steam Juicing Fruits

A steam juicer is a combination of pans that enables water to be boiled, with the steam produced moving to a colanderlike basket holding fruit. When exposed to the hot steam, the fruit easily yields its clear juice, which is caught in a pan below the colander that has an outlet for draining off juice into bottles. Thorough washing of the fruits or vegetables is the only preparation necessary except for cutting the larger or firmer foods for quicker extraction of the juice. This is a marvelous time saver when juicing elderberries, cherries, or other small foods—fruits go into the juicer stems, seeds, pits, skins, and all.

Steam juicers have been around for some time, but until recently they all were made of aluminum. If there is one thing we don't care for, it's using aluminum for preparing foods, especially acid foods that are heated for some time, as juice is. Acids react with aluminum to form an aluminum salt that can then be transferred to food by the liquid. Now there are enamel or stainless steel steam juicers available, and we recommend them strongly. All come with direction booklets. Fruit juices can be frozen or canned, following directions starting on page 235.

Easy Fruit Juice

Helen and Scott Nearing started canning fruit juices on their Vermont homestead many years ago. Their method for "putting up" juices sounds so simple, and yet so successful, that we think it's worth more than a mere mention here. So, we quote from their book, LIVING THE GOOD LIFE (Schocken Books, 1954):

> The glass jars were sterilized on the stove. A kettle or two of boiling water was at hand. We poured an inch of water into a jar on which the rubber had already been put, stirred in a cup of sugar until it had dissolved (we used brown or maple sugar, or hot maple syrup), poured in 1½ cups of fruit, filled the jar to brimming with boiling water, screwed on the cap, and that was all. No boiling and no processing. The raspberries, for example, retained their rich, red color. When the jars were opened their flavor and fragrance were like the raw fruit in season. The grape

juice made thus was as delicious and tasty as that produced by the time-honored, laborious method of cooking, hanging in a jelly bag, draining, and boiling the juice before bottling. Our only losses in keeping these juices came from imperfect jars, caps or rubber. We found that two people could put up 15 quart jars in 20 minutes.

Low-Acid Vegetable Juices

These can be made like fruit and tomato juices, the only difference being that they usually need to be chopped up or shredded and cooked longer to release their juices. Instead of being strained through a sieve or food mill they are sometimes made smooth in a blender. These juices are rarely sweetened, but herbs and other seasonings can be added.

Because vegetables are low in acid, they must be processed in a pressure canner. See individual recipes for processing times and check back to page 78 for more directions on pressure canning.

Mixed Vegetable Juice

If you're missing a few of the vegetables, other than tomatoes or cucumbers, don't worry.

18 large, fresh tomatoes, cored and cut in thin wedges (no need to peel)
6 medium-sized cucumbers, diced
3 medium-sized sweet green or red peppers, cored, seeded, and minced
6 celery stalks, diced
6 medium-sized zucchini or yellow squash, coarsely grated
6 medium-sized carrots, peeled and coarsely chopped
3 large onions, peeled and minced

3 cups coarsely shredded cabbage (minced lettuce could substitute for cabbage)
1 garlic clove, peeled and minced
3 cups water
1/3 cup lemon juice
1 tablespoon honey
1 tablespoon salt
1/2 teaspoon freshly ground black pepper
2 bay leaves, crumbled
3 sprigs fresh parsley or watercress, if desired
2 sprigs fresh thyme or basil

Simmer all ingredients slowly in a covered, large heavy enamel or stainless steel kettle over low heat 45 to 50 minutes, stirring now and then, until vegetables are mushy. Strain through a fine sieve, forcing out as much juice as possible.

Freeze, or reheat juice and pour into sterilized jars, leaving 1-inch headspace. Process pints 55 minutes and quarts 1 hour and 25 minutes in a pressure canner at 10 pounds pressure.

Yield: about 6 quarts

Borscht Juice

4 cups shredded fresh beets	2 tablespoons honey
2 large grated onions	½ cup lemon juice
3 teaspoons salt	

Mix beets, onion, and salt. Cover and cook over low heat for about 1 to 1¼ hours. Stir in honey until dissolved. Slowly add lemon juice and stir until thoroughly mixed. Pour into blender and whiz until smooth. Taste and adjust seasoning if desired.

Can or freeze if you wish, but serve well chilled with a spoonful of sour cream or yogurt mixed in.

To can, heat to boiling and pour into sterilized jars, leaving 1-inch headspace. Process pints 30 minutes and quarts 35 minutes in a pressure canner at 10 pounds pressure.

Yield: about 2 quarts

Using Commercial Juicers

Nowadays, a good selection of juicers is available that make juicing even tricky foods like vegetables easy. There are two basic types of juicers. The most common is comprised of a perforated metal basket that sits atop a high-speed rotary plate. The plate has hundreds of small raised cutting edges. When a carrot, for example, is forced into the hole on top of the machine, the teeth on the plate rip open the tissues and destroy the cell walls; juice and pulp are then free to go their separate ways. The high-speed whirling plate hurls the pulp and juice mixture at the walls of the basket. The pulp is retained by the basket, and the juice sprays off from the centrifugal force and runs down the case into a spout, where you collect it. This is the usual type found in the home. It has to be cleaned out after each use, as the basket fills up with fine, slightly moist pulp.

The second type has a screw-type cutter to rupture the cells; the juice falls by gravity, and the pulp is continuously ejected, so that you don't have to clean it during a day's juicing. This type is more suited to restaurants and health bars.

All these juices can be frozen or canned. To can, heat fruit juices to simmering and pour into sterilized jars, leaving ¼-inch headspace. Process pints and quarts 30 minutes in a 185°F. *hot-water* bath. Vegetable, and fruit and vegetable combination juices should be brought to boiling and poured into sterilized jars, leaving 1-inch headspace. Process in a pressure canner at *10 pounds pressure.* Check back on the vegetable canning chart starting on page 86, and process the length of time needed for the vegetable requiring the longest processing time.

The following recipes that you make with a juicer come from Brownie's—the famous health food restaurant in New York City that started in 1936 as a vegetable juice bar. Brownie's still knows the value of these drinks and has perfected some truly delicious ones. In all these recipes, prepare the other juices first from the fruits and vegetables listed, pour them in an 8-ounce glass, and fill with carrot juice.

Honi-Lulu

2-inch wedge pineapple
juice of ½ orange
carrots

Note: Citrus is better juiced in a citrus or hand juice squeezer.
Yield: 1 serving

Orange Blossom

½ McIntosh apple
juice of ½ orange
carrots

Yield: 1 serving

Sunshine

2-inch wedge of pineapple
juice of ½ orange
dash of papaya syrup
   concentrate

squeeze of lime juice
carrots

Yield: 1 serving

Approximate Nutritional Contents of Important Juices

	BEET (Red)	CARROT	CELERY	CUCUMBER	LETTUCE Romaine	PARSLEY	RHUBARB	SPINACH	TOMATO	WATERCRESS	APPLE	COCONUT	GRAPE	GRAPEFRUIT	LEMON	ORANGE	PINEAPPLE	POMEGRANATE
PROTEIN %	1.6	1.1	1.1	0.8	1.2	3.5	0.6	2.1	0.9	1.7	0.1	1.4	1.3	0.4	0.9	0.6	0.4	1.5
FAT %	0.1	0.4	0.1	0.2	0.3	1.0	0.7	0.3	0.4	0.3	0.2	12.5	1.6	0.1	0.6	0.1	0.3	1.6
CARBOHYDRATE %	9.7	9.3	3.3	3.1	3.0	9.0	3.8	3.2	4.0	3.3	12.5	7.0	19.2	9.8	8.7	13.0	9.7	19.5
CALORIES PER PINT	220	217	89	84	94	283	115	115	112	109	250	700	462	200	210	265	207	472
CALCIUM %	0.140	0.225	0.390	0.050	0.345	0.350	0.220	0.390	0.055	0.785	0.035	0.120	0.055	0.105	0.110	0.120	0.040	0.030
MAGNESIUM %	0.130	0.100	0.140	0.045	0.065	0.160	0.085	0.250	0.065	0.170	0.040	0.100	0.045	0.045	0.045	0.055	0.050	0.020
POTASSIUM %	1.770	1.540	1.460	0.700	1.660	1.50	1.625	2.685	1.335	1.435	0.640	1.500	0.530	0.805	0.615	0.905	1.350	1.600
SODIUM %	0.485	0.385	0.645	0.050	0.100	0.200	0.125	0.445	0.060	0.495	0.055	0.180	0.025	0.020	0.030	0.060	0.080	0.250
PHOSPHORUS %	0.210	0.205	0.230	0.105	0.140	0.130	0.090	0.230	0.145	0.230	0.060	0.370	0.050	0.100	0.055	0.090	0.055	0.050
CHLORINE %	0.290	0.195	0.665	0.150	0.395	0.090	0.180	0.330	0.145	0.305	0.025	0.600	0.010	0.025	0.030	0.025	0.255	0.068
SULPHUR %	0.090	0.110	0.140	0.155	0.130	0.120	0.065	0.180	0.070	0.835	0.030	0.140	0.045	0.050	0.045	0.050	0.045	0.040
IRON %	0.004	0.003	0.003	0.002	0.007	0.016	0.003	0.013	0.002	0.015	0.002		0.0015	0.0014	0.003	0.002	0.002	0.004
SILICON %	0.009	0.007	0.008	0.013	0.018		0.006	0.020	0.009		0.006		0.002			0.0007		
MANGANESE %	0.008	0.0005	0.0014	0.0013	0.0064	0.008	0.0013	0.0042	0.0012	0.0036	0.0003	0.0017	0.0001	0.0001	0.0002	0.0003	0.006	
COPPER %	0.001	0.007	0.001	0.016	0.0003	0.015	0.0005	0.001	0.0005	0.005	0.0008	0.0009	0.0005	0.0003	0.0002	0.0008	0.0004	0.0005
IODINE Parts per billion	230	180		500	650			400	350	180							200	120

U.S. Dept. of Agriculture Chart

This chart, prepared by the United States Department of Agriculture, measures percentages of various nutritive elements in vegetable juices. The bulk of juice is, of course, water. Cabbages eaten whole are 97 percent water, for instance, so by juicing them you're retaining more than 97 percent of their nutritive elements.

Vegetable Garden

1 beet	1 scallion
3 leaves escarole	1 radish
2 leaves chicory	carrots

Yield: 1 serving

Making Apple Cider

What Apple Varieties to Use

Apples for cider need not possess the flawless perfection we seek in table fruit. Here you can use the blemished, the bug-marred, the runts and the "drops," and the otherwise unchoice, mixing all apple varieties together or using only what you have. USDA extension horticulturists tell us that highly flavored cider begins with a band of suitable varieties. Apples should be firmripe, but not overripe. Peak ripeness is indicated by characteristic fragrance and spontaneous dropping from trees. Green, undermature apples cause a flat flavor when juiced. Don't use apples in brown decay because their juice will ferment too rapidly for a prematurely "hard" cider.

For pressing, apples are usually separated into four variety groups—sweet subacid, mildly acid to slightly tart, aromatic, and astringent. The best-flavored cider comes from a blend of varieties from all four groups. Sweet subacid types, which usually make up the highest percentage used in a cider blend, include Baldwin, Rome Beauty, Delicious, Grimes Golden, and Cortland. Some mildly acid varieties, which make up the next largest proportion, are Jonathan, Winesap, Stayman, Northern Spy, York Imperial, Wealthy, Rhode Island Greening, and Melrose. Varieties that add aroma to the juice include Golden Delicious, McIntosh, and Franklin.

Juicing Apples by Hand

When all the fruit has ripened on your apple trees, you are ready for cider-making. If you have no fruit press and only a little fruit, ordinary household gadgets will suffice. Apples can be cut and run through a food chopper or blender, or crushed with a rolling pin on a chopping board. Catch all juice runoff in glass or enameled ware, *but not in an aluminum or any other unglazed metal* container because contact with metal gives apple juice a bitter taste and a dark gray color.

Put your crushed pulp into a clean muslin sack—an old but clean pillowcase will do—and squeeze out all the apple juice possible. Now pour this juice into clean glass jugs or bottles, cover with a cotton wool plug, and let stand at room temperature. The bottles should be filled to just below the brim. Be sure the plug is in not tightly, but snugly. This will allow proper fermentation without adding the air-borne "vinegar-bug."

After 3 or 4 days sediment will begin settling on the bottom as fermentation beads rise to the top. If you wish only a mild, sweet cider this is the time to "rack off" the clear liquid from the sediment and store in a cold place for immediate use—unpasteurized cider like this should be drunk within 4 or 5 days for best taste. "Racking off" is done by inserting one end of a rubber pipe (about 3 feet in length) into the liquid, and siphoning at the other end with your mouth, as you would with a soda straw. As soon as you feel liquid in your mouth, pinch off this end with your fingers and insert into empty container which should stand well below the filled one. Naturally all this equipment should be scrupulously clean. Rack off only the clear liquid; do not disturb the sediment at the bottom.

If you prefer a cider with more zip and on the dry side, allow "must" to keep standing in warm room. In about 10 days it will begin frothing and may foam over the top. Replace with new cotton plug, clean off sides, and let frothing continue until fermentation subsides. If you have an airlock—a curlicue glass "cork" sold by homebrew suppliers—use it in place of the cotton plug. If not, sub-stitute three thicknesses of clean muslin tightly stretched over the bottle opening and secured well around the neck. Be sure you use strong, sound glass, like cider jugs, or the bottle may burst with increasing fermentation.

Since this process turns all available sugars in the "must" to alcohol, it stands to reason this brew will no longer be a sweet drink—it will become "dry" cider. And the longer it stands, the "harder" it becomes. Alcohol—which does not freeze—may be ex-tracted by allowing this drink to freeze solidly—remove corks first—and then running a hot poker through the frozen contents till you reach the free alcohol and pour it off.

If you prefer the mild drink with the definite essence of apple unaltered, fermentation can be arrested by pasteurization at 165°F. Pour fresh into sterilized mason jars and process for 30 minutes in a *hot-water* bath (the water in the process should be simmering, not boiling, and about 190°F.). Then completely seal. Strain before serving. To preserve by freezing, pour into glass jars or freezer containers, leaving a ½-inch headspace, and freeze.

Using a Fruit Press

With a bushel or more of fruit to grind, it is best to use a fruit press. Be sure it is a *hard*-fruit press, with a cutting cylinder that minces the toughest apples with no strain at all. The portable presses are hand-operated, though there are power-driven models too.

Gather all your fruit together and pick over for decayed or wormy specimens, especially if picked from the ground. Hose it down and allow to dry in sun for about 3 days, spread out on racks or benches. A bushel of apples, thoroughly squeezed, will yield 3 gallons of juice, so gauge your container needs accordingly. All jugs, bottles, and kegs should be scrupulously clean. You will also need a large tub—not metal—to hold crushed pulp, a smaller one to catch juice runoff, and a clean muslin sack for juice pressing of pulp.

Set up your fruit press outdoors to facilitate easier handling and subsequent cleanup. If it's raining, or is too cold, work in a garage or shed. Apples should not be used when rain-wet. The operation of the press is so simple and self-evident that no instructions will be needed. With two people at the task—one cranking the grinder, the other feeding the cutting hopper—a bushel of apples can be ground up in 15 minutes. No need to peel or cut up fruit when using a hard-fruit model; the cutting cylinder takes the toughest apple whole in stride. But *be careful,* keep hands and fingers away from rotating blades. As the pulp receptacle fills up, transfer the mash into your tub and repeat until all fruit is ground up. Be sure the smaller vessel is placed where it will catch all escaping juice. Now fill the muslin sack with only enough pomace to fit the juicer well, which is usually built under the mash well. Crank the presser handle until no more juice runs and repeat the process with the rest of the pulp. All squeezed juice may be temporarily contained in a wooden tub or a crock.

Some cider experts contend the finest cider comes from pomace which has been exposed to the air for 24 hours and ground again. It is spread about and turned once or twice for fullest possible absorption of oxygen. If you wish to put this theory to a test, do up half of your pomace this way and compare results.

The rest of the work is a pleasure. Pour the apple juice through a straining cloth into your containers. Fill each almost brimful, and let stand at room temperature at least 3 days before drinking. As you siphon off the clear liquid from the sediment into new bottles, be sure to replace the loss of the settled portion—always keeping fermenting cider jugs full.

Turning Your Apple Cider into Vinegar

The hard part of making apple cider vinegar is extracting the juice, and if you've followed the directions above, you've already done that. The rest is simple. It is just a matter of allowing the extracted apple juice to ferment past the stage of sweet cider and dry cider into the vinegar stage.

Pour the strained apple juice into a crock, a watertight wooden container, dark-colored glass jars, or jugs. If you don't want your fermenting juice to run all over the floor, you'd better leave ample headspace—about 25 percent of the container—for the juice to expand during fermentation. Cover your container with something that will keep dust, insects, and animals out, but let air in. A triple layer of cheesecloth, clean sheet material, or a tea towel will do nicely. Stretch this material over your crock, jars, or whatever, and tie it tightly with string. Store your brew in a cool, dark place, like an unheated basement or garage. Now sit back and wait while the juice does all the rest of the work. Fermentation will take from 4 to 6 months. After about 4 months, remove the cover and taste the vinegar. If it is strong enough for your liking, strain it in a triple layer of cheesecloth, pour it into bottles, and seal with caps or corks. If the vinegar is too weak, let it work longer, testing it every week or so, until it is strong enough for you. If, when finished, your vinegar is too weak, add some store-bought kind. If it's too strong, dilute it with a little water.

The "scum" or layer that forms on top of the vinegar during fermentation is called the "mother." This is what you want to strain off when your vinegar is strong enough so that it stops working. You can save this "mother" to use as a starter for your next batch of vinegar. Just pour your apple juice—or any other fruit juice or wine—into a crock, wooden container, or dark glass jars, add the "mother" and let it ferment. In a little while you'll have vinegar.

Pickling with Homemade Vinegar

It is a risky business pickling food with homemade vinegar because, unlike store-bought vinegar which has a controlled acid content, the acidity of homemade vinegar varies. If you wish to pickle with your own vinegar, we suggest that you make small batches at a time so that you won't stand the chance of ruining all your relish or chowchow.

Herb Vinegars

While we don't advise pickling with homemade vinegar, there is nothing wrong with adding a few herbs and spices to your home brew. Herb vinegar is delicious and very easy to make. Just add individual herbs like tarragon or garlic, or a combination of your favorites to your vinegar.

Louise and Cyrus Hyde, owners of Well-Sweep Herb Farm in Port Murray, New Jersey, make herb vinegar for friends, visitors to their organic farm, and of course for themselves. Although almost any herb can be added to vinegar, they have found a few to be the most popular. The favorites are tarragon vinegar, basil and tarragon vinegar, and basil and garlic vinegar. Other popular kinds are dill vinegar (made from seeds and leaves); rosemary vinegar, a purple-tinted vinegar made from opal basil and sweet basil leaves; orange-mint vinegar for fruit salads; and cucumber-flavored salad burnet vinegar.

Because the Hydes like to have herbs floating in their bottled vinegars, they add fresh sprigs of herbs—about a handful or 3 or 4 sprigs—to each quart of vinegar. The mixture is allowed to set at least

You can make herb vinegar simply by adding fresh sprigs of herbs to bottles of vinegar, as shown here. You may also use crushed dried herbs instead of the fresh ones, but because they are in more concentrated form, add smaller amounts of them to each bottle.

2 weeks before it is used. The herbs stay in the bottle and are poured onto a salad along with the vinegar.

If you prefer herb vinegar without the leaves floating in it, crumble about 3 tablespoons of dried leaves in a jar. If you're adding herb seeds, such as dill or anise, crush them well first. Then warm your vinegar and pour it over the herbs. The warm vinegar decomposes the leaves and extracts oil from the herbs more quickly than cool vinegar. Let the vinegar and herbs set for 2 to 4 weeks in a covered bottle or crock. Stainless steel and porcelain containers are also fine, but never use any other metal because the acid in vinegar will react with it and give your herb vinegar an undesirable appearance and taste. After a few weeks, test your vinegar to see if it is flavorful enough for you. When its flavor is right, strain it into sterilized jars and cap until you're ready to use it.

The Hydes like the distinct flavor of apple cider vinegar, but find that when used alone with herbs, its strong flavor masks the more subtle flavor of the herbs. For this reason they prefer to mix apple cider vinegar with the milder-tasting white, distilled vinegar before they add herbs to it.

Here are a few herb and spiced vinegar recipes that you might like to try in addition to the Hyde's suggestions:

Basil and Garlic Wine Vinegar

Place a handful of fresh basil leaves in a 4-quart stainless steel or enamel pot. Crush the leaves with a spoon or potato masher and add 2½ quarts of cider vinegar. Turn heat up high and bring quickly to a boil. Cover the pot or pour the vinegar and basil in a clean crock or bottle, and then cover and allow to steep for 2 weeks.

After the steeping period strain the basil from the vinegar. Place another handful of fresh basil leaves in the pot, crock, or widemouthed bottle, and pour the strained vinegar, cover, and allow to steep for 10 more days. Strain the vinegar again and add 2 cups of dry red wine.

Sterilize three 1-quart bottles, fill with vinegar, add a sprig of fresh basil to each bottle if desired, and cap or cork tightly.

Yield: 3 quarts

Dill and Nasturtium Bud Vinegar

Bring a quart of white vinegar to a boil. Place 1 tablespoon nasturtium buds and a sprig of fresh dill weed in a crock or stainless steel or enamel pot and pour the boiling vinegar over them. Cover and allow to steep for 1 week.

Then boil again steadily for 4 minutes, and filter while hot into a sterilized quart bottle. Cap or cork tightly.

Yield: 1 quart

Celery Vinegar

Bring a quart of white vinegar to a boil in a 4-quart stainless steel or enamel pot and add 12 peppercorns, 2 teaspoons salt, and 4 cups of white tender celery roots and stems that have been chopped finely. Boil 3 minutes.

Cover the pot, or pour into clean crock or wide-mouthed bottles and cover. Let steep for 3 to 4 weeks. Then strain and pour into a sterilized quart bottle. Cap or cork tightly.

Yield: 1 quart

Cucumber Vinegar

Clean and slice, but do not peel 3 medium-sized cucumbers. Place in clean crock, pot, or large wide-mouthed bottle. Add 1 teaspoon salt and 1 teaspoon peppercorns. Bring a quart of white vinegar to a boil and pour over the cucumber, salt, and pepper. Cover and steep for 3 weeks in a cool place.

Then strain into a wide-mouthed jar or other glass container and let the vinegar settle a few hours. Ladle or pour off the clear vinegar into sterilized bottles. Cork or cap tightly.

Yield: 3 half-pints

Spiced Vinegar

3 garlic cloves
1 small onion, sliced
1½ teaspoons powdered ginger
    root
1 teaspoon prepared
    horseradish

1½ teaspoons mustard seed
1 teaspoon peppercorns
2 teaspoons allspice
2 quarts vinegar (cider is
    preferable)

Place all ingredients but vinegar in a 4-quart stainless steel or enamel pot and crush slightly with a spoon or fork. Add the vinegar and bring quickly to a boil, reduce heat, and simmer for 2 hours. Then take off heat and steep for 3 hours more. Strain and pour into 2 sterilized quart bottles. Cap or cork tightly.

Yield: 2 quarts

Tarragon Vinegar

3 cups fresh tarragon leaves
6 whole cloves
peel of 2 lemons

3 quarts white vinegar (white wine vinegar is good)

Wilt the tarragon leaves by spreading them on trays and placing them in a warm, shady place for 2 days. Place them in a large, clean crock or pot made from stainless steel or enamel. Add the cloves, then slice the lemon peels and add them, too. Bring the vinegar to a boil and immediately pour over the herb and spice mixture.

Cover the pot or crock well and steep for about 2 weeks. Then strain the vinegar and pour into sterilized bottles. Cap or cork well. Don't use for at least 3 weeks.

Yield: 3 quarts

Vegetable and Fruit Recipes

Vegetables

Baked Beans
(to can or freeze)

3 pounds (about 2 quarts) dried
    beans (navy, pinto, or
    white)
3 medium onions
2½ cups dark molasses

3 teaspoons salt
3 teaspoons dry mustard
1½ pounds salt pork, cut in
    2-inch pieces

Wash and pick over beans. Soak overnight in water to cover. Discard soaking water the next day, and then cover again with water. Bring to a boil, then lower the heat and simmer gently until beans are tender. Drain, reserving liquid.

Into each of 3 2-quart bean pots (with covers) place one whole onion. Divide the beans among the 3 pots. Combine the molasses and seasonings with the reserved liquid, and divide the mixture among the 3 pots of beans. If there is not enough liquid to fill the pots to within 1 inch of the top, add more water. Tuck the pieces of pork into the beans near the top.

Cover and bake in a slow oven (300°F.) for as long as possible (up to 10 hours), adding water as necessary. Don't allow the beans to dry out.

To can: Pack into hot jars, leaving 1-inch headspace. Adjust caps. Process at 10 pounds pressure, 1 hour and 20 minutes for pints, 1 hour and 35 minutes for quarts.

To freeze: Allow to cool. Transfer the beans to freezer containers.

When ready to serve: Bake in a 350°F. oven until heated through.

Yield: 6 quarts

Bean and Corn Soup

1 cup dried white beans
4 cups water or milk
1 onion, chopped
2 garlic cloves

1 teaspoon turmeric
1 teaspoon cumin
2 cups corn
salt to taste

Cook beans with onion, garlic, and spices in simmering liquid. When tender, purée, then add corn and salt to taste. Reheat and serve.

Yield: 4 servings

Black Bean Soup

1 cup dried black beans
4 cups water
3 bay leaves
2 garlic cloves
¼ teaspoon dry mustard

1½ teaspoons chili powder
4 cloves
2 onions, chopped
salt to taste

Add ingredients except salt to boiling water and simmer until tender. Purée. Salt to taste and serve.

Yield: 4 servings

(East) Indian-Style Black-eyed Peas

1 cup dried black-eyed peas
1 chopped onion
2 tablespoons coconut
¼ teaspoon turmeric

1 teaspoon honey
¼ teaspoon cumin
3 cups water
salt to taste

Add ingredients except salt to boiling water and simmer until liquid is absorbed and beans are tender. Salt to taste.

Yield: 4 servings

Harvard Beets
(to freeze)

1 tablespoon cornstarch
1 teaspoon salt
½ cup cider vinegar
½ cup water

½ cup honey
2 whole cloves
4 cups sliced boiled beets
butter (for serving)

Stir together in saucepan the cornstarch, salt, cider vinegar, and water. When smooth, place over low heat and stir in honey and cloves. Boil five minutes, until sauce is thick and clear. Add beets.

To freeze: Pack Harvard beets into clean freezing containers. Label, date, and freeze.

When ready to serve: Place in saucepan, add 1 tablespoon butter for each cup of beets, and simmer for about 20 minutes.

Yield: 2 pints

Baked Broccoli
(to freeze)

For each casserole:

3 to 4 cups trimmed broccoli	salt and pepper to taste
1 cup pearl onions	¼ cup grated cheddar cheese
2 eggs, beaten	¼ cup bread crumbs
1 cup ricotta cheese	¼ cup grated cheddar cheese
1 tablespoon minced green onions	butter

Parboil broccoli and onions for seven minutes, drain, and arrange in freezer-wrap-lined casserole dish. Mix eggs, ricotta cheese, green onions, salt and pepper, and ¼ cup grated cheddar. Pour over broccoli and onions. Sprinkle bread crumbs and the rest of the grated cheddar over casserole. Dot with butter.

Freeze. When frozen, remove from pan. Label, date, and return to freezer.

When ready to serve: Preheat oven to 350°F. Unwrap food, place in casserole dish, and bake for about 45 minutes.

Yield: 6 to 8 servings

Corn Pudding
(to freeze)

2½ quarts fresh sweet corn (about 20 large ears)	2 teaspoons salt
1 to 2 tablespoons honey	¼ teaspoon freshly ground pepper
2 tablespoons cornstarch	

Scrape the kernels, pulp, and milk from the corn to make about 10 cups. (It's unnecessary to blanch the corn first.)

Mix all ingredients thoroughly, then pack firmly into 4 well-buttered 9-inch foil pie pans, dividing the whole amount evenly and smoothing the surface as flat as possible. Overwrap each pie pan in heavy-duty aluminum foil and quick freeze.

When ready to serve, thaw the corn pudding completely, unwrap, dot the top with butter, and bake uncovered in a hot oven (400°F.) for 40 to 45 minutes until crusty and golden on top.

Yield: 4 puddings of about 4 servings each

Dried Sweet Corn

Combine:
1 cup dried corn
2 cups cold water
Soak 2 hours or so. Do not drain. Then add: ½ teaspoon salt
1 tablespoon honey
pepper to taste

Cover and cook slowly until the kernels are tender (50 to 60 minutes). Correct seasonings. Add butter to taste. Serve in sauce dishes, broth and all. Two tablespoons milk may be added if you like.

This is simply the basic recipe for serving dried sweet corn as a vegetable. It can be used in casseroles or omelettes, mixed with other vegetables or in salads, and even used for desserts such as puddings. Its nutty flavor is pungent and aromatic.

Yield: 4 to 6 servings

Frittata
(to freeze)

For each frittata:
2 cups cooked vegetables (suggested: 1-inch pieces of asparagus, chopped broccoli, sliced mushrooms)	¼ teaspoon dried herbs (suggested: oregano, basil)
	2 tablespoons butter
	¾ teaspoon salt
1 small onion chopped or 4 green onions, chopped	6 eggs beaten
1 garlic clove, minced	½ cup milk
	dash of Tabasco
	dash of Worcestershire sauce
	⅓ cup grated cheese (optional)

Steam or sauté vegetables until barely tender. Sauté onion, garlic, and herbs in butter for 5 minutes.

Mix together remaining ingredients, and stir in vegetables and sautéed onion mixture. Pour into greased pie plate or quiche pan or foil pie plate. Bake 15 to 20 minutes in a preheated 350°F. oven.

To freeze: Quick freeze frittata, and wrap when frozen. Label, date, and return to freezer.

When ready to serve: May be thawed, loosely wrapped, at room temperature if it is to be served at room temperature. Can also be reheated, in a 325°F. oven, for 15 to 20 minutes.

Yield: each pie serves 6 as
an appetizer, 3 or 4
as a main dish

Green Bean Supper Casserole
(to freeze)

12 cups French-cut green beans
¼ cup butter
¼ cup sifted whole-wheat flour
1 teaspoon salt
½ teaspoon pepper

1 teaspoon honey
½ cup minced onion
1 pint sour cream
Swiss cheese, grated (to be used
    as topping before serving)

Parboil the beans for 3 minutes; drain, and arrange in 2 foil-lined casseroles.

Melt butter; stir in flour and cook, stirring, for several minutes. Add seasonings, honey, and onion, and stir over medium heat until blended. Blend in the sour cream. Pour sauce over beans.

To freeze: Freeze in casserole. When food is solidly frozen, freezer-wrap, label, and date it. Return it to freezer immediately.

When ready to serve: Preheat oven to 350°F. Unwrap casserole, place in baking dish, and top with ½ pound grated Swiss cheese. Bake 40 to 45 minutes or until center is heated through and cheese is melted.

Yield: 2 casseroles serving 6
to 8 people each

Wine Red Kidneys

1 cup dried kidney beans
1 onion, chopped
2½ cups water

½ teaspoon salt
¾ cup wine

Cook beans and onions in boiling water until tender. Mash half the beans, add salt and wine, then simmer 20 minutes more.

Yield: 4 servings

Limas in Barbecue Sauce
(to can or freeze)

4½ cups dry limas (about 2 pounds), soaked overnight
8 ounces salt pork, finely diced
1 cup finely chopped onion
2 garlic cloves, minced
3 tablespoons prepared mustard

2 teaspoons Worcestershire sauce
2 teaspoons chili powder
2½ cups canned tomatoes, chopped
1 tablespoon molasses or honey
¼ cup vinegar

Cook beans slowly until tender in water to cover (about 45 minutes). Brown salt pork with onion and garlic. Mix remaining ingredients with the salt pork. Add beans. Stir gently.

To can: Heat mixture to boiling. Pack hot into hot jars, leaving 1-inch headspace. Adjust caps. Process at 10 pounds pressure, 1 hour and 20 minutes for pints, 1 hour and 35 minutes for quarts.

To freeze: Pour into freezing containers. Cool, label, seal, date, and freeze.

When ready to serve: Reheat gently.

Yield: 3 pints

Onion-Cheese Turnovers
(to freeze)

¼ pound butter
2½ pounds onions, sliced
2 tablespoons whole-wheat flour
2 tablespoons water

salt and pepper to taste
1 tablespoon poppy seeds
½ pound grated cheddar cheese
1 recipe whole-wheat pastry

Melt butter in a large, heavy pan, and add onions. Sauté until soft and juicy, and then thicken with a mixture of the flour and water. Season with salt, pepper, and poppy seeds. Fold in grated cheese.

Roll out the pastry to a ⅛-inch thickness. Cut into 10 5-inch squares. Place 2 to 3 tablespoons of the onion mixture on ½ of each square. Moisten the edge and fold over the other half. Seal the edges with a fork.

Wrap and seal for freezing, and freeze immediately.

When ready to serve, place turnovers on a cookie sheet and bake at 425°F. for 40 minutes or until pastry is browned.

Yield: 10 turnovers

Potato Cheese Casserole
(to freeze)

4 pounds new potatoes
¼ pound butter
¼ cup milk
½ pound cream cheese, at room
    temperature
½ cup grated cheddar cheese

½ cup grated Romano or
    Parmesan cheese
1 green pepper, chopped
½ cup snipped chives
¼ cup minced pimiento
salt and pepper to taste

Boil potatoes until tender. Drain and mash. Melt butter; add milk, cheeses, and other ingredients, and combine with mashed potatoes. Beat until fluffy.

To freeze: Pour into lined casserole dishes. Freeze. When casserole is solidly frozen, wrap in freezer paper, label, date, and return to freezer immediately.

When ready to serve: Unwrap, place in casserole dish, and bake, uncovered, in a preheated 350°F. oven for about 1 hour. Top should be browned and mixture bubbly.

Yield: 3 quarts

Potato Pancakes
(to freeze)

6 medium-sized potatoes
1 large onion
2 tablespoons whole-wheat
    flour
2 tablespoons chopped parsley

3 eggs beaten
salt and pepper
5 tablespoons noninstant dry
    milk
oil for frying

Grate potatoes, and then wrap the grated potato in a muslin towel and wring out all moisture. Grate the onion, and combine with potato. Add the remaining ingredients, except for oil, and stir well. Fry in oil like pancakes, browning both sides. Drain on absorbent paper.

To freeze: Place on cookie sheet in freezer, until frozen. Placing double squares of freezer paper between each pancake, stack them and wrap in foil or plastic bags.

When ready to serve: Reheat on a cookie sheet in a 300°F. oven for 20 to 30 minutes. Serve with applesauce and sour cream.

Yield: 4 servings

Stuffed Baked Potatoes
(to freeze)

12 large baking potatoes	2 teaspoons salt
1 to 1½ cups milk	¼ teaspoon black pepper
½ pound cheddar cheese, grated	2 green onions, cut fine, . including tops
½ cup butter	1 egg

Scrub and grease potatoes. Then bake in a 425°F. oven for about an hour or until soft. Scald milk, and add grated cheese. Stir until cheese is melted. Slit each potato lengthwise, and scoop out all potato pulp, leaving shells intact. Mash the pulp until smooth; then gradually add the milk mixture, butter, seasonings, and onion. Be careful not to add so much milk that the mixture becomes soupy. Add egg and whip until fluffy. Fill potato shells with mixture. Set on a cookie sheet and freeze. Then wrap the potatoes individually in foil and return to the freezer.

When ready to serve: Bake in a preheated 350°F. oven for about 1¼ hours.

Yield: 12 stuffed potatoes

Sweet-Sour Red Cabbage
(to freeze)

¼ cup oil	1 large head red cabbage, about 5 pounds, cored and thinly shredded
1 medium onion, chopped	
6 tart apples, cored and thinly sliced	2 teaspoons salt
¼ cup cider vinegar	½ teaspoon caraway seeds (if desired)
1 cup water	
¼ cup honey	

Heat oil in a large heavy skillet. Sauté onion until golden. Add apple slices, vinegar, water, honey, and cabbage. Stir to combine. Season with salt and caraway seeds. Simmer for ½ hour, stirring occasionally. Remove from heat. Adjust the sweet-sour flavoring by adding more vinegar or honey if desired.

To freeze: Spoon prepared red cabbage into clean freezer containers, leaving ½-inch of headspace. Seal, label, date, and freeze.

When ready to serve: Reheat, with additional liquid if necessary, over moderate heat.

Yield: 2 quarts

Dry Roasted Soybeans

1 cup dried soybeans
3 cups water

2 teaspoons instant vegetable salt
1 teaspoon salt

Wash soybeans and remove any foreign particles. Add 3 cups of water to soybeans and soak overnight in refrigerator.

Pour soybeans and liquid used for soaking in a heavy saucepan. Place over medium heat and bring to a boil; lower heat and add instant vegetable salt, cover, and simmer 1 hour. (If desired, ½ teaspoon oil may be added to soybeans to keep mixture from boiling over.)

Remove saucepan from heat and drain soybeans thoroughly; pour into a shallow pan and bake in a preheated 350°F. oven for 45 minutes to 1 hour or until soybeans are brown. Remove from oven and sprinkle with salt while warm.

Yield: 1½ to 2 cups

Swedish Soybean Soup

2 cups dried soybeans
water to cover
1 medium smoked ham hock or
  meaty ham bone
3 quarts cold water
2 teaspoons salt
½ teaspoon paprika
1 cup chopped celery, with
  leaves

1 cup chopped onions
3 medium raw turnips, diced
¼ cup chopped parsley
⅛ teaspoon cayenne pepper
1 cup tomato purée or canned
  tomatoes
chopped parsley for garnish

Wash soybeans and discard beans with imperfections. Cover soybeans with water and place in refrigerator, covered, overnight.

The following day, place soaked soybeans in a large, heavy soup kettle. (Be sure to use a large enough pot and leave partially uncovered, so as to avoid soybeans cooking over. This can happen very easily.) Add ham hock or meaty ham bone and 3 quarts of cold water. Place uncovered over medium heat and bring to a boil, removing any foam from surface as it accumulates. Reduce heat; add 2 teaspoons salt and paprika. Cover partially and allow to simmer for 3 hours, stirring occasionally.

Add the chopped celery, onions, turnips, parsley, cayenne, and tomato purée or canned tomatoes. Cover partially and allow to simmer for another hour or until soybeans are tender. Continue to stir occasionally while cooking. Taste and correct seasoning. Garnish with chopped parsley.

Yield: approximately 3 quarts

Spinach Croquettes
(to freeze)

2 pounds young spinach leaves
½ cup whole-wheat bread
   crumbs
1 egg

1 small onion, chopped
1 teaspoon salt
vegetable oil

Pick over and cut roots and stems from spinach leaves. Wash until free of sand. Place them moist in saucepan, cover, and cook for 6 minutes, until tender.

In a large bowl put cooked spinach, crumbs, egg, onion, and salt. Mix together and shape into 3-inch round patties. Sauté briefly on each side in oil until light brown.

To freeze: Place croquettes on tray in freezer. When frozen, wrap with double sheets of freezer wrap between patties. Label, date, and return to freezer.

When ready to serve: Preheat oven to 325°F. Reheat croquettes on cookie sheet for about 25 minutes.

Yield: 4 to 6 servings

Spinach-Pasta Casserole
(to freeze)

2 large onions, chopped
2 garlic cloves, minced
¼ cup oil
3 pounds ground beef
1 teaspoon salt
1 pint canned tomatoes,
   undrained
½ teaspoon cinnamon
⅓ cup butter
½ cup sifted whole-wheat flour
1 quart milk

dash of pepper
dash of nutmeg
½ cup grated Parmesan cheese
6 cups uncooked elbow
   macaroni
¼ cup butter
3 eggs, slightly beaten
2 pounds spinach, cooked
1 pound cheddar cheese,
   grated

Sauté onion and garlic in oil until limp; then add beef and brown. Pour off excess fat. Add ½ teaspoon salt, tomatoes, and cinnamon.

Melt ⅓ cup butter, stir in flour, and cook, stirring, for several minutes. Add milk, ½ teaspoon salt, pepper, nutmeg, and grated Parmesan cheese, and stir until thickened.

Cook macaroni in boiling, salted water. Drain and add ¼ cup butter and the slightly beaten eggs.

To freeze: In each of 2 lined shallow baking dishes, layer ¼ of the macaroni, ½ of the meat, ½ of the spinach, ¼ of the cheese, ½ of remaining macaroni, ½ of the sauce, and remaining cheese. Freeze. When casseroles are solidly frozen, remove from freezer, wrap, label, date, and seal. Return to freezer.

When ready to serve: Preheat oven to 375°F. Unwrap casserole; place in greased baking dish, and bake for 1½ hours.

Yield: 2 casseroles serving
6 people each

Giant Stuffed Squash
(to freeze)

1 egg
1 large garlic clove, mashed
½ teaspoon salt
1 slice whole-wheat bread, crumbed
wheat germ
1 giant zucchini squash
1 tablespoon oil
1 large onion, chopped
½ red or green pepper, chopped

2 small carrots, diced
any small amount vegetables on hand—mushrooms, kohlrabi, string beans, cucumber
1 very ripe tomato, diced
3 tablespoons tomato paste
oregano, basil, or other Italian herbs
chopped walnuts
sliced cheddar cheese

Beat egg with garlic and salt. Add bread crumbs and enough wheat germ to make a thick, pasty mixture.

Halve squash lengthwise, scrape out seed cavity, and sprinkle with wheat germ. Spread egg mixture over remaining cut surface. Place squash on a rack set over a pan of water and steam in a 350°F. oven about 10 minutes.

Heat oil, and sauté onion and other vegetables, except tomato, until onion is limp and transparent. Add the tomato, the tomato paste, herbs, and salt to taste. Fill zucchini shells with the vegetable mixture, top with chopped walnuts, and cover with cheese slices.

To freeze: Place filled squash halves on tray in freezer. When frozen, wrap , label, date, and return to freezer.

When ready to serve: Preheat oven to 350°F. Bake squash for about 35 minutes or until cheese melts and squash is tender.

Yield: 4 servings

Succotash
(to can or freeze)

12 ears of corn	salt to taste
6 cups lima beans or green snap beans	paprika (optional)
	chopped parsley (optional)
butter (optional)	

Boil ears of corn for 5 minutes. Cut kernels from cobs ⅔ of the depth of the kernels. Prepare fresh limas or snap beans and boil by themselves for 3 minutes. Measure and mix hot corn with an approximately equal amount of beans.

To can: Fill hot, clean pint jars with the corn-bean mixture, leaving 1-inch headspace; adjust lids. Process pints at 10 pounds pressure for 60 minutes.

To freeze: Season mixture with butter, salt, paprika, and parsley, if desired. Pack in freezer containers. Allow to cool, seal, label, date, and freeze.

When ready to serve: Reheat gently.

Yield: 6 pints

Sweet Potato Casserole
(to freeze)

about 2 pounds sweet potatoes	1 cup milk
grated rind from 1 lemon	½ teaspoon salt
grated rind from 1 orange	½ teaspoon cinnamon
4 eggs	½ teaspoon allspice
½ cup honey	chopped pecans or walnuts,
½ cup melted butter	butter (for serving)

Scrub the potatoes, but do not peel. Boil in water to cover until they are tender (about 30 to 40 minutes). Remove the skins and mash the potatoes well. Allow to cool.

Combine other ingredients in a bowl, and beat well. Combine with mashed sweet potatoes.

To freeze: Pour into freezer-wrap-lined casserole dish. Freeze. Remove from casserole dish. Fold and seal wrap. Label, date, and return to freezer.

When ready to serve: Preheat oven to 325°F. Unwrap food and place in buttered casserole dish. Top with chopped pecans or walnuts if desired. Dot with butter. Bake 40 to 45 minutes, until heated through.

Yield: 6 to 8 servings

Tomato Juice
(to can or freeze)

Use perfect, ripe tomatoes. If you must use any with bad spots, taste each before adding to the kettle, and discard any that taste flat. Cut up, and put the tomato pieces in a large stainless steel or enamel pot. Bring to a boil, and cook until juice flows freely. Put through a fine sieve to remove seeds and skin. Measure the juice, adding 1 teaspoon salt for each quart of juice. You may also add 4 teaspoons lemon juice or ½ teaspoon citric acid to raise the acidity for canning (see page 68).

To can: Return to kettle. Reheat to simmer. Pack hot into hot jars, leaving ½-inch headspace. Adjust lids. Process for 10 minutes in a boiling-water bath.

To freeze: Pour into freezer containers, leaving ½-inch headspace. Seal, label, date, and freeze.

Tomato Juice Cocktail
(to can or freeze)

12 large, ripe tomatoes	¼ cup lemon juice
4 medium-sized carrots	2 tablespoons honey (or to taste)
2 large sweet green or red peppers	1 tablespoon salt
4 celery stalks, diced, leaves included	½ teaspoon black pepper
2 onions, diced	2 bay leaves
1 garlic clove, minced	2 sprigs fresh basil or dill or thyme (if desired)

Wash unblemished tomatoes; remove stems and cores; cut in small pieces. Scrub and grate carrots. Core, seed, and mince peppers.

Combine all ingredients in large stainless steel or enamel kettle, and simmer over low heat 45 to 50 minutes, stirring occasionally, until vegetables are soft. Pick out herbs. Strain through a sieve.

To can: Return strained juice to kettle and bring to boil. Pour hot into hot jars, leaving ½-inch headspace. Adjust lids. Process quarts for 30 minutes at 10 pounds pressure.

To freeze: Pour into freezer containers. Seal, label, date, and freeze.

Yield: 4 quarts

Shortcut Tomato Paste

Here's a simple way to make a plain tomato paste for freezing or canning without cooking it for hours on the stove.

Peel whole tomatoes by first plunging them in boiling water to crack skins. Then cool quickly in cold water and, with a knife or your fingers, remove skins.

Cut up tomatoes and purée them in a blender or food mill. Then pour the purée in a jelly bag or cotton muslin sack, hang it over a bowl, and let it drip for several hours. Reserve the drippings for soup or whatever, and freeze the paste remaining in the bag in ice cube trays. Once frozen, transfer to plastic freezer bags.

The paste can also be canned; follow directions for canning tomato sauce below.

Tomato Sauce
(to can or freeze)

5 pounds (about 25) Italian
    plum tomatoes
2 tablespoons olive oil
2 onions, chopped
4 garlic cloves, crushed
5 tablespoons chopped green
    pepper

2 tablespoons parsley
2 teaspoons basil or oregano
1 bay leaf
salt and pepper to taste
water (or red wine)
¼ cup lemon juice

Loosen the tomato skins by plunging tomatoes into boiling water for a minute, then under running water. Remove skins. Cut into chunks.

Heat the oil in a large, heavy enamel or stainless steel kettle, and sauté the onions and garlic. Stir in the pepper and tomatoes, add parsley, spices, and seasonings, and simmer for 1 hour or longer, stirring occasionally. If it boils down add some water or red wine. Then add the lemon juice.

To can: Pour hot into hot jars, leaving ½-inch headspace, and process in a boiling water bath 45 minutes.

To freeze: Pour into freezer containers. Seal, label, date, and freeze.

Yield: 4 to 6 pints

Blender Tomato Sauce
(to can or freeze)

3 tablespoons olive oil or corn
    oil
3 large garlic cloves, chopped
4 onions, chopped
24 tomatoes, peeled
1 teaspoon fennel seed

4 teaspoons dried basil (or 8
    teaspoons fresh)
3 teaspoons oregano
2 teaspoons salt
¼ cup lemon juice (when
    canning only)

In an enamel or stainless kettle, sauté the garlic and onions in oil until soft. Cut tomatoes into chunks, and add along with seasonings to the garlic and onions. Simmer for at least 2 hours, longer if possible. Stir occasionally to prevent sticking. Cool slightly; blend until smooth. Add lemon juice if you plan to can it.

To can: Return to kettle; reheat. Pour hot into hot jars, leaving ½-inch headspace. Adjust lids. Process in a boiling-water bath 45 minutes.

To freeze: Pour into freezer containers. Seal, label, date, and freeze.

Yield: 4 to 6 pints

Cream of Tomato Soup
(to freeze)

⅓ cup butter
2 garlic cloves, minced
4 cups sliced celery
2 onions, sliced
1 peck very ripe tomatoes
1 quart water or stock
¼ cup minced fresh parsley

2 tablespoons salt
1 teaspoon pepper
1 tablespoon cornstarch
½ cup honey
milk or light cream, egg yolk, butter (for serving)

Melt the butter in a large, heavy enamel or stainless pan. Sauté the garlic, celery, and onions lightly. Add the cut-up tomatoes, stock or water, and parsley. Simmer for 30 minutes.

Put the soup through a food mill. Return to the pot and reheat. Add salt and pepper. Make a thin paste of the cornstarch and some cold water. Add to the soup, and boil until slightly thicker. Add the honey.

To freeze: Pack after cooling into freezer cartons, leaving headspace. Seal, label, date, and freeze.

When ready to serve: Pour the canned soup into a saucepan. Thaw the frozen soup. Reheat. For each pint of soup, add an equivalent amount of milk or light cream, 1 beaten egg yolk (if desired), and 1 teaspoon butter. Heat to a simmer. Serve.

Yield: 6 to 8 pints

Summer Vegetable Casserole
(to freeze)

For each casserole:
2 tablespoons butter

½ cup diced green pepper
½ cup diced onion

1 cup corn	salt and pepper to taste
1 cup diced zucchini	¾ cup bread crumbs
1 cup diced fresh tomatoes	grated cheese (if desired)

Melt butter. Add vegetables and sauté until tender (about 10 minutes). Season with salt and pepper. Pour into freezer-wrap-lined casserole dish. Sprinkle with bread crumbs (and grated cheese if desired).

To freeze: Fold and seal freezer paper. Label, date, and freeze. When frozen, remove from casserole dish and return to freezer immediately.

When ready to serve: Preheat oven to 350°F. Unwrap casserole and place in casserole dish; bake for ½ hour or until top turns light golden brown and center is heated through.

Yield: 6 to 8 servings

Zucchini Bread
(to freeze)

3 cups grated zucchini	1½ teaspoons baking powder
1 cup oil	1½ teaspoons cinnamon
¾ cup honey	½ teaspoon ground cloves
2 eggs, beaten	1 teaspoon salt
1 teaspoon vanilla	1 cup chopped nuts
3 cups whole-wheat flour	1 cup raisins

Grate zucchini and wring out excess moisture by twisting the gratings in a muslin towel.

Beat together the oil, honey, eggs, and vanilla. Sift together the flour, baking powder, spices, and salt.

Combine the 2 mixtures, and stir to blend. Add nuts, raisins, and zucchini. Place in greased loaf pan and bake in preheated 350°F. oven for 1 hour.

To freeze: Allow to cool. Wrap in plastic bags. Freeze.

Yield: 1 loaf

Fruits

Apple Cinnamon Crumb
(to freeze)

1 pound cooking apples, peeled, cored, and sliced	1 to 2 tablespoons water
	1 tablespoon honey

¼ cup raisins (if desired)
3 tablespoons butter
1 teaspoon cinnamon
¼ cup finely chopped walnuts

¼ cup whole-wheat bread
crumbs (or wheat germ or
bran)
3 tablespoons honey

Cook apples with water and 1 tablespoon honey until soft. Mash lightly. Add raisins.

Melt butter in a saucepan, and add cinnamon, chopped walnuts, and bread crumbs. Brown gently. Add 3 tablespoons honey.

Layer the crumbs and mashed apples in a foil pie plate, beginning and ending with the crumbs. Press down lightly. Cool quickly. Freeze. Wrap, label, seal, and date. Return to freezer.

When ready to serve: Preheat oven to 350°F. Remove wrappings, and bake for about 1 hour until crisp and golden brown. Serve hot with whipped cream, ice cream, or yogurt.

Yield: 4 to 6 servings

Applesauce
(to can or freeze)

4 pounds apples*
1½ cups water
¼ cup lemon juice (optional)

¾ cup honey
cinnamon, cloves, nutmeg
(optional)

Wash and core apples. Peel if desired. Cut into chunks or slices. In a large, heavy stainless steel or enamel kettle, cook apples with water, lemon juice, honey, and spices until tender. Put through food mill or sieve, if desired.

To can: Pack hot into hot clean jars, leaving ½-inch headspace. Process for 10 minutes in a boiling-water bath.

To freeze: Allow to cool. Pack into freezer containers. Seal, label, date, and freeze.

Yield: 4 pints

* Varieties suggested for sauce include: Rome Beauty, McIntosh, Jonathan, Baldwin, and Rhode Island Greening.

Apricot-Cranberry Gelatin

(This salad must be made ahead to set.)

In blender:
1 cup hot water

2 tablespoons gelatin

Blend well; add:

½ cup honey
1 teaspoon pure vanilla extract
¼ teaspoon pure almond
extract

½ cup oil
1 cup water
1 cup dried apricots
1½ cups fresh cranberries

Add cranberries last, and blend only until chopped well. Add any extras you want, such as nuts, etc. Pour into cake pan and let set. The chopped fruits will rise to the top, and the bottom layer is a creamylike texture.

Yield: 4 servings

Apricot-Nut Bars

⅔ cup dried apricots
water to cover
½ cup soft butter
¼ cup honey
1⅓ cups whole-wheat pastry
flour

½ teaspoon double-acting
baking powder
¼ teaspoon salt
½ cup honey
2 eggs, well beaten
½ teaspoon vanilla extract
½ cup chopped walnuts

Rinse apricots; cover with water; boil 10 minutes. Drain; cool; chop. Preheat oven to 350°F. Oil an 8″ × 8″ × 2″ pan.

Mix butter, ¼ cup honey, and 1 cup pastry flour. Pack into pan. Bake 25 minutes.

Sift ⅓ cup flour, baking powder, and salt. In large bowl, with mixer at low speed, gradually beat honey into eggs; mix in flour mixture and the vanilla. Stir in walnuts and apricots. Spread over baked layer. Bake 30 minutes or until done; cool in pan. Cut into 32 bars.

Apricot-Prune Soufflé

½ pound dried apricots,
cooked
¼ pound dried prunes (pitted
and cooked)

5 tablespoons honey
5 egg whites
pinch salt

Press cooked apricots and prunes through a sieve into a large bowl. Add honey and mix thoroughly. Beat egg whites and salt until stiff, and fold gently into the fruit. Lightly oil a soufflé dish and put the fruit mixture into it. Bake in a slow oven (300°F.) until firm, about 45 minutes. Serve immediately.

Yield: 4 servings

Fruit Mousse

¾ cup cooked dried apricots, peaches, or prunes
1 teaspoon gelatin
4 tablespoons hot fruit juice

1¼ cups heavy cream
½ teaspoon pure vanilla extract
⅛ teaspoon salt

Purée cooked fruit in a blender or through a sieve. In a small bowl, dissolve gelatin in hot fruit juice. Add the gelatin liquid to the puréed fruit. Place the mixture in the refrigerator and chill until it is almost set.

Whip the cream with the vanilla and salt and fold it lightly into the gelatin mixture. Pour into serving dish or individual dessert bowls and chill well before serving.

Yield: 4 to 6 servings

Fruit-Nut Sticks

2 cups nuts
1 cup dates or prunes, pitted, raisins, or dried apricots

2 eggs, beaten
1 cup honey

Grind nuts and fruit. Blend with eggs and honey. Shape into sticks. Place on lightly oiled cookie sheet. Bake at 375°F. for about 10 minutes.

Yield: 2 dozen sticks

Uncooked Fruit Cake

1 cup ground raisins
1 cup chopped dried figs
1 cup chopped dried dates
1 cup other dried fruits, chopped (such as apricots, pears, prunes, peaches, apples)

1 cup honey
1 cup whole-wheat bread crumbs
1 cup concentrated fruit juice or nectar
1 cup sunflower seeds (ground)

Mix in order given. Pack into oiled mold or bread pan. Ripen, covered, in refrigerator several days. Turn out on platter. Serve with whipped sweetened cottage cheese or yogurt.

Yield: 6 to 8 servings

Baked Peach Whip

¾ cup cooked dried peaches, sieved or put through a blender

4 egg whites, beaten stiff
dash salt
3 tablespoons honey

Fold peach purée into egg whites. Add salt and honey. Mix lightly. Pile lightly into 1-quart casserole, lightly oiled. Bake at 350°F. for 30 minutes or until firm.

Yield: 6 servings

Prune Stuffing

3 cups diced whole-wheat bread
½ cup melted butter
1 cup dried prunes
1 cup cold water

2 tablespoons lemon juice
1 tablespoon honey
½ teaspoon salt
½ cup almonds

Mix the bread and the melted butter. Place the prunes in a pan with the water, lemon juice, and honey. Bring to a boil, then simmer for 5 minutes. Save any liquid that may remain and mix with the bread. Pit the prunes, chop, and sprinkle with the salt. Chop the almonds and mix all together. Stuff the bird loosely.

Yield: about 4 cups

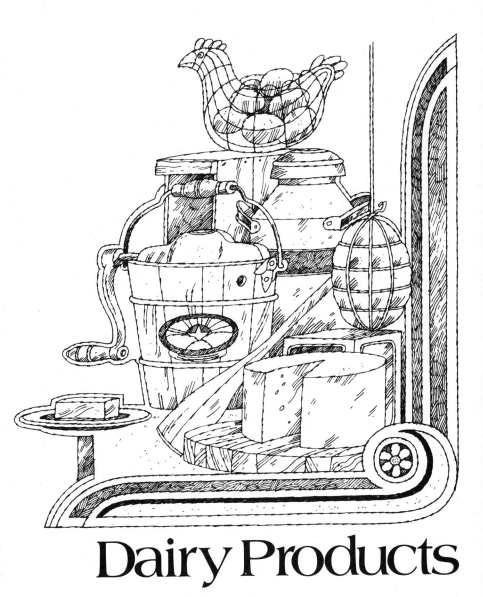

Dairy Products

Freezing Milk and Cream

If you're keeping even one milk cow or goat, there are probably times when you've got more milk on your hands than you and your family can consume daily. There are some very simple things you can do with the excess. You can open a roadside stand (if you can sell raw milk in your state—some don't allow it unless you have certification, and some don't allow it at all), or sell it from door-to-door, you can give it away to friends, or you can store it for times when your animal's milk production is at its lowest. Here, we'd like to spend some time discussing the third alternative: storing milk.

Just how do you preserve this highly perishable food? One of the simplest ways is to freeze it. Milk may be frozen whole or skimmed, pasteurized or raw. Even cream can be frozen. To freeze, pour your milk or cream into glass jars or plastic containers, leaving 2 inches headspace for expansion. Seal tightly and place in the coldest part of your freezer so that it freezes quickly. Whole milk will keep safely in the freezer for 4 to 5 months; cream should not be stored frozen for more than 2 or 3 months. Both milk and cream should be thawed for 2 hours at room temperature before using. The only problem you will encounter when freezing cream is that the butterfat tends to separate out during freezing. For this reason, thawed cream has limited uses. It can't be used successfully as is, poured into coffee or over cereal or fruit. It may not whip properly, but it can be successfully used for frozen desserts, like ice cream. If you want to use it for cooking—to make creamed soups, gravies, custards, and the like—or for baking, beat it a little first just so the butterfat is not floating on top.

Making Butter

A time-honored way of preserving cream is to churn it into butter. Although butter is also a perishable food, it will keep longer than milk or cream under refrigeration if all the buttermilk is worked out of it. Because butter is a concentrated form of cream—1 gallon of cream will yield about 3 pounds of butter—it takes up a lot less storage space than cream.

Separating the Cream

The first step in butter-making is to separate the cream from the milk. This is easy to do if you're using cow's milk. The butterfat in cream is lighter in weight than the other ingredients in whole milk and will rise to the top naturally by gravity in 24 to 36 hours.

There are two simple methods of separating the cream: the shallow-pan method and the deep-setting method. In the shallow-pan method, the milk is drawn from the cow and immediately poured into shallow tubs or pans. These pans or tubs are placed in a cool spot like a refrigerator, basement, or springhouse for at least 24 hours. The cream that has risen to the top is then skimmed off with a flat dipper. Although this is certainly the easiest way to separate cream, some farmers object to this method because during the time it takes for the cream to rise, the surface of the milk is exposed to the air and frequently absorbs or develops objectionable odors and tastes.

A more satisfactory method of separating cream is by the deep-setting method. By this method, the milk is drawn from the cow and immediately poured into cans or buckets which are placed in cold water or, preferably, ice water for 12 hours. The quick cooling of the milk causes the cream to rise more rapidly and more completely. The cream can be skimmed in half the time required by the shallow-pan method, and its freshness and sweet flavor are retained.

Separating Goat's Milk   If it's goat's milk you're working with, the job is a little more difficult. The fat globules in goat's milk are small and well emulsified, which means that the cream will take much longer to separate out than cow's cream. If you let goat's cream take its time to separate out, it may begin to develop a strong "goaty" flavor that most people find unappetizing. To separate the cream properly and quickly, a cream separator, which separates the cream from the skim milk by centrifugal force, is needed. Warm milk (between 80° and 90°F.) is poured into the separator where it is whirled around. The cream and skim milk are released through separate spouts in minutes, while both are still warm and fresh. Some cow owners like to use a cream separator instead of letting their cow's cream separate by gravity, because the separator removes almost every drop of the butterfat from the milk very quickly. The skim milk is fat-free and can be used immediately for drinking or for making cottage cheese.

A cream separator is a delicate piece of equipment and should be cleaned and operated according to the manufacturer's directions. For best results, milk poured into the separator should be warm (not below 90°F.) and fresh. Most separators are left in the milk room or other unheated area. In winter, the cold temperatures can chill the instrument, and if the separator is cold enough, it will cool the first milk that is poured in to below 90°F. To prevent this from happening, warm the separator by running warm water through it before pouring in the milk.

The separator should be cleaned and sterilized immediately after each use. All parts should be rinsed in warm water and then scrubbed with a brush, warm water and soda ash or a cleansing powder made especially for use in dairies. Soap should not be used because it is difficult to wash off completely and may leave a soapy film on the equipment. All the parts of the separator should be sterilized in a farm sterilizer or in boiling water for 5 minutes.

New cream separators aren't cheap. Most run from $300 to $600. Used ones are cheaper, of course, but since the popularity of keeping goats for milk has increased in the last few years, the demand for used separators is growing and the supply is diminishing.

Chilling and Ripening the Cream

It is not very practical to churn a few cups of cream at a time, so many farmers who have just one or two milk-producing animals will collect cream over a few days, waiting until they have enough separated to make churning worthwhile. This is fine, but don't hold cream for more than 4 days before making butter. Butter made from old cream has an acidic, overripe taste, and it spoils quickly. If you are collecting cream over a few days' time, keep its temperature below 50°F. and don't add to it any cream that is not cooled to at least this temperature. The addition of warm cream raises the temperature of the older cream and hastens souring. Mix all the cream together—after it is all chilled—12 to 20 hours before you churn it, and stir it occasionally with a stirring rod (a smooth rod with a 4- to 5-inch diameter disk on one end) or long-handled spoon so that it will have a uniform thickness.

The best way to cool cream rapidly is to cool it in a tub of ice water. Ice water will cool the cream more quickly than will refrigerator temperatures, providing the water is allowed to circulate on all sides as well as under the cream container. If ice water is not available, use plain water, but change it frequently. If you are fortunate enough to have a stream or spring nearby, put your container of cream in it. Its flowing cold water will do a fine job of cooling your cream. Pick a shaded spot—out of direct sunlight—set down your cream container, and cover it with a clean towel or piece of cheesecloth to keep out dirt, insects, or other contaminants.

When the cream has reached a temperature of 50°F. it can be placed in a cool spot until churning. (Get yourself a floating dairy thermometer, available at farm supply and some hardware stores, to measure temperatures accurately.) All cream should be kept at 52° to 60°F. in the summer and 58° to 66°F. in the winter while it is being churned. If the cream is too warm when it is churned, the butter develops too soon and is too soft and greasy. If the cream is too cold when churned, not all of the butterfat will separate out to form butter. This results in creamy buttermilk and less butter.

Cow's cream which is allowed to ripen before it is churned will produce more flavorful butter than that which is made from sweet cream. You can ripen raw cream by allowing it to set at room temperature (65° to 75°F.) until it is thick and slightly sour. To ripen milk quickly, add about ½ cup cultured buttermilk or yogurt. Fresh cream should not be added after ripening begins. Once ripened, the cream should be cooled quickly in a container of ice water until it reaches churning temperature. It should be kept at churning temperature for at least 2 hours before it is churned. Don't try to ripen goat's cream before churning. The goaty or cheesy flavor it will acquire will produce an unpalatable butter.

Churning

The old wooden upright churn with its long dasher is rarely used for butter-making anymore; most of these relics have found their way into antique shops. The wooden churn more commonly found in operation today is the barrel churn. Electrically powered and hand-operated glass churns are also popularly sold. If you want to make just a small amount of butter you don't even need a real churn. You can improvise with a blender, cake mixer, or even a hand rotary beater. Early American settlers made butter by shaking cream in a deep wooden lidded bowl. You can use a glass jar. Pour cream into a jar until it is one-third full and start shaking. This is a rather tedious way of making butter and it calls for a strong arm, but it does work.

Whatever device you use to churn butter, make sure it is thoroughly clean before any cream is poured into it. If you have a wooden churn that is used only occasionally, it is advisable to fill it with water 24 hours before you plan to use it so that the wood will swell and be watertight. Scald the wooden churn with boiling water and then chill it down to churning temperature by filling it with ice water or placing it in a refrigerator, springhouse, or cool basement before using. Glass churns, mixers, blenders, rotary beaters, and glass jars should be sterilized in boiling water and cooled before using.

Pour the cream into your churn, blender, or whatever equipment you are using through a strainer to make sure that there are no lumps in the cream before you begin. Fill your churn only one-third full. Butter made from goat's cream is white. If you wish to color it, now is the time. Add a few drops of vegetable coloring to attain the desired shade of yellow. (Colonial housewives colored their butter with carrot juice.)

If you are using a hand-operated or electric churn, churn about 10 times and then lift up the lid or remove the plug to permit gas to escape. Churn 20 times more and allow gas to escape again. Then resume churning, at about 60 revolutions per minute, until beads of butter about the size of corn kernels form. The churning process should take about 30 to 40 minutes. Approximately 20 minutes will pass before you will hear the splash of the beads forming and feel the thickness of the butter.

Taking Off the Buttermilk

When churning is finished, strain off the liquid. Don't throw it away. This is buttermilk. It won't be as thick as the commercial kinds because it is not cultured. It is lighter than regular milk and has a natural effervescence. It's the real, old-fashioned buttermilk that

makes delicious pancakes, biscuits, and breads. What's left in the churn is mostly butter, with a little buttermilk mixed in. This remaining buttermilk must be removed to obtain the taste and texture of good butter. If it is not removed, the butter will have a shorter keeping quality and have a slightly acidic or sour taste.

Washing and Working the Butter

Wash the butter with clean water. Water temperature may vary according to the temperature of the butter that has formed, but it should be about 60°F. If your butter is too soft and warm, make your wash water cooler than 60°F.; if it is too hard and cold, have your wash water a little warmer than 60°F.

The washing can be done right in the churn. Pour as much clean water as there is buttermilk into the churn after the buttermilk is poured off. Close the churn and churn it a few times to wash the butter. Pour off the cloudy water and repeat the washing process with fresh water. If this water is cloudy when poured off, wash again until the rinse water stays clear. Be patient—you may have to repeat this

To make small amounts of butter in a blender (1) fill the blender container no more than ⅓ full with cream and whip at lowest speed. (2) When yellow beads about the size of corn kernels form, pour off the buttermilk. (3) Rinse the remaining buttermilk out of the butter by pouring cool, clean water into the blender. Turn the blender on low for a few seconds to wash the butter. (4) (see next page) Pour off the cloudy rinse water and repeat until the rinse water pours off clear. (5) Scoop the washed butter into a shallow dish and work out every drop of liquid by pressing the butter against the dish with a rubber spatula or wooden paddle as shown here.

washing process as many as 10 times before all the buttermilk is out. If the butter gets too hard, wash with water that's a little warmer.

You can also take the butter out of the churn and wash it in a large shallow bowl. Pour in clean water and knead or cut the butter to work out the milk. Don't spread the butter, though.

Now pour your butter into a large, shallow bowl. Work out every drop of liquid by pressing and squeezing the butter against the sides and bottom of the bowl with a wooden paddle until no water can be poured off. Do not spread or thin the butter on the sides and bottom of the bowl; this makes the butter greasy.

Making Butter in a Blender

If you're working with a home food blender, pour in the cream until it fills the container about one-third full. Set your blender at its slowest speed. Once the blades begin, remove the cap and watch the cream. It will first get foamy and then begin to thicken. Yellow beads will start to form in about 4 to 5 minutes. Once they get to be the size of a kernel of corn and the liquid seems a bit watery (like skim milk), your butter has formed. Turn off the blender and pour off the buttermilk. Pour in the wash water, turn on the blender for a second or two, pour off the water, and repeat the washing until the wash water

stays clear. Pour the butter into a large, shallow bowl and work all the liquid out in the manner described above.

Using a Rotary Beater or Cake Mixer

If you're churning with a rotary beater or cake mixer, pour the cream into a deep bowl that has been sterilized with boiling water and chilled in the refrigerator. Use your mixer at its lowest speed. If you are beating with a rotary beater, whip at a constant speed which is comfortable for you. Do not stop beating until the butter has formed. Then pour off the buttermilk and add an equal amount of clean water. Beat for a second or two and pour off the wash water. Repeat until the wash water pours off clear. Then place the butter in a large, shallow bowl and work out the remaining liquid.

Churning in a Glass Jar

If a glass jar with a tight-fitting lid is to be your churn, fill it one-third full with cream and shake about 10 times. Then remove the lid to allow the gas to escape. Screw on the lid and shake about 20 times more. Remove the lid and let the gas escape again. Replace the lid and resume shaking without stopping until lumps of butter form and the liquid takes on a thin and slightly watery appearance. Pour off buttermilk and replace with fresh water. Shake jar about 5 times, pour off the wash water, and wash again, until the wash water pours off clean. Place the lumps of butter into a large, shallow bowl and work out the remaining liquid.

If You Have Trouble Making Butter

If your butter takes an unusually long time to develop or it just never comes at all, even after several hours of churning, obviously something is wrong, either with your equipment or your cream. The U.S. Department of Agriculture, in a farmers' bulletin (which is now out of print) describes some of the reasons why you may have problems forming butter:

1. The churning temperature is too low. Normally, it should be 52° to 60°F. in summer and 58° to 66°F. in the winter, but under exceptional conditions it might be necessary to raise it to 65° to 70°F. This is especially true if your churn is very cold and you are churning in an unheated area on an exceptionally cold day.

2. The cream is too thin or too thick. It should be about 30 percent butterfat for best results.

3. The cream is too sweet. Very sweet cream will need to be churned longer than cream which has been ripened until it is thick and slightly sour. Ripening can be speeded up by adding about ½ cup of a starter, like *cultured* buttermilk or yogurt, to the cream.

4. The churn is too full. The churn should not be more than one-third full, no matter what type of equipment you are using for churning. The extra space allows the butterfat to move about freely.

5. Ropy fermentation of the cream preventing concussion. This may be prevented by sterilizing all the utensils and producing milk and cream under sanitary conditions. If additional measures are needed, pasteurize the cream, being careful to keep it from contamination after pasteurization. Then ripen the cream with a starter before churning.

6. Individuality of the animal. The only remedy is to obtain cream from a dairy animal recently fresh or cream that is known to churn easily. Before ripening it, mix it with cream that is difficult to churn.

7. The goat or cow being far advanced in the period of lactation. The effects may at least be partially overcome by adding, before ripening, some cream from another goat or cow that is not far advanced in the period of lactation.

8. Feeds that produce hard fat. Such feeds are cottonseed meal and timothy hay. Linseed meal, gluten feed, and succulent feeds such as silage and roots tend to overcome the condition and make churning the cream into butter easier.

9. Off-flavor of butter. Influenced by cow's feed. Refrain from giving cows strong-flavored and strong-odored foods like turnips. If off-flavor persists, milk cows before, not after, feeding.

Salting and Storing the Butter

Butter which is worked free of all its buttermilk and wash water may be eaten or stored as it is. This is unsalted, sweet butter. Salt, which enhances the flavor and lengthens butter's keeping quality, may be added at this time, if you wish. If you salt the butter, add ⅔ tablespoon of salt for each pound (2 cups) of butter. Work the salt into the butter by pressing and thinning the butter in the bowl with the wooden paddle, then adding a little salt and folding the butter over. Repeat this process until all the salt is worked in and the butter is firm and waxy. Don't spread the butter to thin it; this causes the butter to become oily and lose its firm texture.

When the butter is "worked," it is ready to be placed in appropriate containers and stored. You can roll your butter into a ball (or balls) and wrap it in aluminum foil or heavy-duty plastic wrap. You can press it into small bread pans and cover the pans with wrap. Or you can put the butter into glass jars with lids. It is not a good idea to store butter in plastic containers. These containers are porous and will allow air and strong odors to penetrate their walls. The taste of butter deteriorates the longer it is stored in the refrigerator. Keep it no longer than 2 weeks at refrigerator temperatures. If you wish to keep your butter longer than 2 weeks, freeze it at temperatures of 0°F. or colder. Do not keep it frozen for more than 6 months. Thaw butter for about 3 hours at refrigerator temperatures before using.

Making Hard and Semihard Cheeses

Cheese is one of those foods which has a certain mystique associated with it—perhaps because there are so many kinds. Incorporated in this mystique is the idea that making cheese is something not to be attempted by amateurs. This is just not so. There are many amateur cheese-makers around who make some pretty good cheeses.

Ingredients

The main ingredient in cheese is of course milk—and you need lots of it. To make 1 pound of hard cheese like cheddar you'll need 5 quarts of milk, and for 1 pound of cream cheese or about 1¼ pounds of cottage cheese you'll need 4 quarts of milk. While a pound or so of cream or cottage cheese is an adequate amount to make at one time, there's not much sense in making less than 3 to 5 pounds of hard or semihard cheese at a time because smaller cheeses dry out too quickly. So figure you'll need at least 15 quarts (or 3¾ gallons) of milk to make one batch of hard or semihard cheese.

You can make cheese out of fresh raw milk, or homogenized, pasteurized milk. Both will work, and most of the recipes that follow give directions for both. However, if you're going to buy homogenized, pasteurized milk at supermarket prices, your homemade cheese can turn out to be mighty expensive. Your cheese will be a lot more economical (and many people claim better tasting) if you can find yourself a farmer or homesteader who will sell you fresh raw milk. It goes without saying that if you have your own cow or goat your milk will be the cheapest, freshest, and, in milking season, most plentiful.

Most cheese recipes also call for a starter or an activator; some call for two. This is a substance that contains beneficial bacteria that will make the milk clabber or, in other words, separate the curds from the whey. Raw milk contains the beneficial bacteria that are responsible for making milk clabber, so technically you don't have to use an activator with raw milk. But in most recipes an activator *is* used with raw milk because it speeds up the clabbering process and guarantees that the milk won't sour before it clabbers sufficiently. Pasteurized milk definitely needs an activator to clabber since the beneficial bacteria have been killed off in the heating process.

Cultured buttermilk contains the activating bacteria. Don't use the kind that you pour off your homemade butter because it doesn't have the bacteria. The kind to use is the thick supermarket kind—make sure it's fresh!

Yogurt also contains the right bacteria. You can use fresh, homemade yogurt or an unflavored commercial yogurt, so long as it's really yogurt and not a sweet yogurtlike custard which some "yogurts" are.

Rennin is an extract made from the lining of unweaned calves' stomachs, and it can be bought in the form of rennet tablets. Rennet tablets used to be found in almost all grocery and drugstores, but that was when home cheese-making was commonplace. You still might be able to find them in some such stores, but if you can't, check in a hobby store that sells cheese-making equipment or in one of the catalogs listed on page xiii.

There are some *herbs,* too, that have the milk-clabbering properties of rennet, yogurt, and buttermilk. Nettle, the sorrels (lemon and common), fermitory, and the sap of the unripe fig will all curdle milk. But according to internationally known herbalist Juliette de Baïracli Levy, the giant purple thistle is the best of the herbal milk clabberers. See her recipe for thistlehead cheese later in this chapter.

Plain Semihard Cheese

Current United States Department of Agriculture publications offer no practical information on making cheese, and few cookbooks give explicit enough directions. The best way to get solid information on cheese is to go to someone who makes a lot of it and ask him or her how it's done. This is just what we did.

David Page is an organic gardener and an assistant professor of biochemistry at Bates College in Lewiston, Maine. He is also an amateur cheese-maker who turns out some great-tasting cheeses. Here are his directions for making what he calls a "nondescript, semihard, mild-to-sharp, pleasant-tasting, unadulterated cheese."

After you understand the process, you might like to try the other cheese recipes that follow these directions.

Ripening the Milk

Your cheese is doomed to failure if your milk is not ripened properly. The idea is to inoculate the milk to be used with a lactobacillus culture in order that other strains of bacteria don't get a chance to grow during aging and thus ruin the cheese. Obtaining a culture is simple—just use cultured buttermilk from the store. Take a quart of fresh cultured buttermilk and add about ¼ cup to each of four clean, scalded quart mason jars. Fill up the jars with fresh pasteurized milk. Seal with clean caps, shake, and allow to stand for 24 hours at room temperature. This gives you 4 quarts of fresh buttermilk, which you can also use for other things. You can put your culture in the refrigerator and use as needed. The culture can be perpetuated merely by repeating the foregoing process with your last jar of buttermilk at any given time. This is known as better living through bacteriology.

Get the biggest canning kettle you can find. David uses a 36-quart kettle which is pretty heavy when it is full. A 24-quart canning kettle is all right, but your cheese will be smaller. The kettle must be enameled. Stainless steel is fine but aluminum is a "no-no." Add the contents of 1 quart of buttermilk to whichever kettle you are using. Then pour in sufficient milk to fill your kettle to within 2 inches of the top, and mix with a clean spoon. David uses raw milk for making cheese, but pasteurized milk is OK, too. The milk should not be more than 2 days old (having been refrigerated of course) and may be skimmed if desired. Warm the milk up to 86° to 90°F. over low heat and allow to ripen for 1 to 2 hours. That is, set it aside; go and do something; when you're done—it's done.

Forming the Curd

Dissolve ¼ of a rennet tablet for each 2 gallons of milk in about ½ cup of cool water. (A bit too much rennet is better than a bit too little.) Make sure the ripened milk is at about 86° to 90°F. (Check its temperature with a dairy thermometer.) Add the rennet solution to the milk with plenty of stirring. Cover and let stand for about 1 hour. The way to tell if it's done is to stick your finger (washed of course) into the curd at an angle and lift up slowly. If the curd makes a clean break over your finger, it's done. If it's still the consistency of tired yogurt—be patient.

Cutting the Curd

When the milk has curdled get a long knife and slice the curd up into cubes about ½ inch square—don't worry about getting nice looking cubes. Holding the knife vertically, slice in a parallel fashion in lines about ½ inch apart in one direction and then slice in the same manner in a direction perpendicular to the original slicings. Then slice at a sharp angle across one way following the original lines as best you can in order to undercut the curd and make into cubes. Do this undercutting in several directions until you think you have the curd pretty well cut up.

Heating the Curd

Place the whole works on a very low fire. (David uses the low setting on an electric stove for 8 gallons; smaller amounts require a lower setting.) Stir with a clean spoon constantly in order not to burn the curd at the bottom. As you stir with one hand, have your knife ready in the other ready to hack up any large pieces of curd that escaped the blade previously. Besides, hacking the curd helps to relieve the tedium of standing over the pot and stirring. As you stir, the contents will warm up—slowly. It should take about 45 minutes to 1 hour to get from 86°F. to 105°F. You will note that the big lumps will get smaller and a yellowish cloudy fluid called "whey" separates. Keep the curds in motion during heating.

When you've reached 105°F., take the pot off the stove and let stand for 1 hour. That should allow sufficient time for the curd to harden. What you will have is something resembling a white rubber bath mat at the bottom of a pot full of whey. Pour or scoop off the whey. (Save it to feed to chickens or pigs if you have them, or dump it on the garden or compost pile.) Leave the curds in the pot—it is easier to salt them if you do.

For 8 gallons worth of curds add about 3 tablespoons of salt. The amount you add is purely a matter of taste. Using a spatula (pancake turner) slice into the matted curd and mix in the salt. At this point, the curd is quite well held together. A commercial cheese-making establishment would have a machine to chop up the curd—something like a compost shredder; but save yours for better things. Some whey will be sloshing around with the curds which will help to disperse the salt throughout the curds.

Pressing the Curd

Go to the hardware store and buy 10 or 20 yards of cheesecloth. The stuff you seem to be able to get these days is pretty cheesy so

(1) Begin your cheese-making project by preparing the activator. Pour ¼ cup of cultured buttermilk into each quart jar. Fill the jars with milk and let them sit at room temperature for 24 hours.

(2) Take one of these quart jars, whose contents has now turned into buttermilk, and pour it into a large pot of milk. Heat this milk to 86° to 90°F. and let it ripen for 1 to 2 hours.

(3) In the meantime, dissolve ¼ of a rennet tablet in ½ cup of cool water for every 2 gallons of milk you have in your pot. Add this rennet solution to the warm, ripened milk. Cover the milk and let it stand for 1 hour.

(5) Remove the curds from the heat and allow them to harden for 1 hour. Then scoop off the whey and salt the curds.

(4) When curds have formed, cut them into cubes and heat them over low heat until they reach 105°F. Stir constantly and have a knife at hand to cut up any large curds that should surface. After about 45 minutes the curds should resemble cottage cheese, and the cloudy whey should have separated to the top.

(6) Strain the curds through cheesecloth, and when most of the whey has already drained off, pick up the four corners of the cheesecloth and squeeze the curds into a ball.

(7) **After the curds have been pressed and then aged for about 6 days,** rewrap the cheese in fresh cheesecloth and coat it with melted paraffin. Store the cheese for at least 60 days before eating.

you must use a double thickness of cloth about a yard square. Place the double layer of cloth over a clean pail or bowl which is at least the same volume as the curd you have. Dump the curd (the spatula really comes in handy here) into the cheesecloth and strain out the remaining whey. After all the curd is in the cheesecloth, pick up the corners and form the curd into a ball by twisting the cheesecloth and squeezing the curd in the appropriate places. Hang up and let drain for about 15 minutes.

Now you should have a ball of curd of varying size depending on how much milk you started with. Fold a clean dish towel which is about 2 feet long into a multilayered band about 4 inches wide by the length of the towel. Wrap this band around the side of the ball of curd as tightly as you can and fasten the end of the band with safety pins. What you have now is a ball of curd in a cloth girdle. Place the ball on a towel which is laid in the bottom of an 8-inch shallow bowl or deep plate. This will form the shape of the bottom of your wheel of cheese. Where the twisty part of the original cheesecloth is on the top of the ball, rearrange the cloth so that when the top is compressed, the folds of the cheesecloth do not unduly indent the final wheel of cheese. Place a similar bowl over the top of the ball of curd and pile about 40 to 60 pounds worth of books on top of everything and let stand overnight. Smaller cheeses need less weight. The idea is to press the individual curd granules into a solid wheel of cheese. There are various appliances one can make to do the pressing job, but you may not have the time to make one.

Using the erudite method given above, you have to watch out for several things. First, if your bowls are too deep, they will just come together under pressure and the curd won't get pressed. Also, if the books slip off at an angle, your final wheel of cheese will be lopsided—so be sure to distribute your weights above the curd evenly. A bit of intelligence and ingenuity could come up with a much better way of pressing the curd.

After 12 hours, you will have a nice compact wheel of cheese. Strip off the cheesecloth and wind a fresh band of cloth around where the old girdle was. Fasten again with safety pins. The reason for doing this is that the cheese at this point is quite plastic, and if the sides of the wheel are not supported, the cheese will tend to flatten as it forms a rind.

Turn the cheese several times a day for 5 or 6 days, or until a good even rind has formed over the surface. Some directions say to dry for 1 to 2 days—don't believe them! Some mold may form on the outside, which is of no consequence.

Aging

By this time you are probably wondering if you will ever get to taste your fledgling cheese. Well, you must wait for at least 60 days. Wipe off the outside of the wheel of cheese and wrap tightly with 1 or 2 layers of cheesecloth. Heat 1 to 2 pounds of paraffin in a pot until it is good and hot, and brush the hot wax over the cheese or dunk portions of the wheel into the wax. Be careful of the hot wax. It is important to paraffin the entire cheese or it will dry out. The cheesecloth helps to keep the paraffin coating from cracking. Write the date on a piece of paper and glue it to the outside of the cheese with hot paraffin so you will know when the cheese is ready to eat. Place the finished cheese in a cool place on a clean surface and turn once every few days.

If there is mold growth under the paraffin don't worry, it won't invade the cheese. If the cheese starts to swell, you've got troubles. Such behavior indicates that your milk was not properly ripened and that a "bad" microorganism is enjoying your cheese. Other than bad bacteria, the other big enemy of ripening cheese is most notably small animals. If you have no cats then worry about mice and rats eating your cheeses. If your wheels survive, they will taste pretty good after 60 to 90 days. The cheese will become sharper the older it gets.

This completes the basics of making a good homemade cheese. One can add various herbs and natural coloring agents to the milk after it has ripened in order to obtain a "different" cheese.

Now that you've got the basics, here are a few more semihard and hard cheese recipes sent into ORGANIC GARDENING AND FARMING and tested in our experimental kitchen:

Cheddar Cheese

"Cheddaring" is a step in cheese-making that involves removing whey with dry heat from previously heated and drained curd. In commercial production, cheddaring is accomplished by placing drained curd in a heated chamber. When the steady, gentle heat causes the curd to mat together, it is cut into slabs that are turned for even drainage. Before pressing, the slabs of curd are shredded in a special machine.

The moisture content of cheeses made in this way is around 40 percent on a scale where cottage cheese stands at 80 percent and Parmesan cheese at 30 percent. Many cheeses are cheddared besides the ones called "Cheddar" cheese. Monterey Jack and longhorn are examples.

Because cheddared cheeses have a lower moisture content than the fresh or semisoft types, they can usually be aged for a longer time. It is largely the bacterial action during the aging of the cheese that leads to a fuller taste, and generally, the older the cheese, the sharper the flavor. Cheddared cheeses usually have a higher butterfat content than fresh cheeses, and this creaminess prevents them from developing a dry or crumbly texture as they age.

To add cheddaring to your repertoire no specialized equipment is needed. Think of the heat chamber for drying the curd as a kind of "sauna" for your curd. Then you will be able to devise any number of arrangements depending upon what colanders, strainers, racks, and kettles are on hand. As for the shredding, you could take out your grinder, but you will not have too much curd to break up by hand if that seems easier.

Here is a six-step method for cheddaring cheese with equipment normally available in a well-stocked kitchen. The first two steps are the same as those followed in making the semihard cheese just described.

The secret in cheddaring is removing the whey with dry heat from previously heated and drained curd. Do this by making a "sauna" by placing the colander with the curd on a rack inside a larger kettle. Put a small amount of water in the bottom of the pot and allow the whey to drain in a warm place for 2 hours.

1. Prepare about 4 gallons of milk with rennet as explained in the directions on page 287. Cut the curd when it forms and heat it slowly until a firm "popcorn" curd is reached. (It looks uniformly like popcorn, and not like cut-up Jello cubes.) It is all right if the curd is "squeaky." Let the curd sit for at least 1 hour before draining.

2. Drain curd into the colander of an enamel blancher and let stand over a bucket for ½ hour.

3. Make a "sauna" for the curd by placing the colander with the curd on a rack inside a larger kettle with a small amount of hot water in the bottom. Leave kettle in a warm enough place so that whey will continue to drain for 2 hours or so—the back of the wood stove is excellent. The temperature of the curd during this time should ideally be between 100° and 110°F. During this time the curds should mat together into a single mass. Turn the whole thing once or twice to insure even heating. Pour off whey if necessary to prevent the curds from soaking in it. At the end of the first hour, slice the mass into 5 or 6 slabs an inch or so wide

and continue heating, turning the slabs occasionally. The whole process need not be rushed.

4. Shred up the curd by hand or by putting it through a grinder of some kind. In a large bowl mix the curd with 2 to 3 tablespoons salt and whatever other seasonings are desired. The curd should have a pleasantly salty taste. Because the curds will not press properly unless warm, and because they cool off quickly during the shredding step, it is a good idea to put the bowl of salted curd, covered, in a slow oven for ½ hour or so before pressing.

5. Pack the curd into a press cylinder lined with cheesecloth and press with considerable weight overnight.

6. After pressing, remove cheesecloth, rub the cheese with coarse salt, and allow a natural rind to form over 3 or 4 days. Wax, date, and store the cheese in a cool place, turning once a week or so. If the storage place is between 45° and 50°F. the cheese should keep from 3 to 5 months.

Thistlehead Cheese

The giant purple thistle, used in the Balearic Islands but also found throughout the United States, is the best of all the herbs for clabbering milk. It is speedy in action (usually curdles milk over-night) and it is sure.

The giant purple thistle is a very tall-growing species of thistle, possessing all over very cruel prickles, so that no animals can eat it, and only the bees visit it and the goldfinches carry off the down-topped seeds for their nests and for food. The stems and foliage are grayish, and the flowerheads a rich purple, and of typical thistle shape. The part used are the flowerets when the thistlehead and the flowerets of which it consists have turned brown. When the thistledown begins to appear it is getting too late for the gathering, and the flowerets are less strong and soon will be carried away over the countryside by the wind.

The flowerets should be air dried, either in shallow baskets or perforated brown paper bags, and then stored in jars to last until the next summer.

The herb, in carefully controlled quantity, has to be prepared for adding to the milk. It should be pulverized with a mortar and pestle. The herb is well pounded, then a little warm water (or whey), merely enough to cover it, is added, then left to soak for 5 minutes, pounded again for 5 minutes, soaked again, repounded, usually 3

A simple cheese press can be made from a can with both ends removed, 2 wooden disks that just fit inside the can, each with several holes drilled in them for the whey to drain off, a rod at least as long as the can, 2 washers, and 2 screw bolts. (1) Put one disk in the bottom of the can, line the can with cheesecloth, and spoon in the curds. (2) Fold the excess cheesecloth over the top of the cheese. Run the rod through the cheese and bottom and top disks. (3) Slip the washers on the rod and screw the bolts over them, pressing the disks as close together as possible to squeeze the curds.

times, at least until dark brown-colored liquid forms. The herb is then strained and then a heaping teaspoon of the herbal liquid is added to every quart of warm milk. If too much herb is used, it tastes strongly in the cheese, and it will cause indigestion, being a very potent herb. Therefore, be careful not to add too much.

(Instead of making a liquid from the herb, you can also dip the whole herb in the milk to make it adhesive, press the flowerets together, and bind them with coarse white cotton thread into a sort of rough plait, leaving the long end of the cotton hanging out of the crock, so you can pull out the herb when the milk curdles, which is approximately overnight. Do not add the herb loose into the milk, as it is then like having hairs in one's mouth when consuming the delicious soft curds and whey.)

When really thick curds have formed in the crock, usually after a few hours to a day or night, depending on the warmth of the weather, finish the rest of the cheese-making process by following the directions for semihard cheese, starting with Pressing the Curd.

Greek Headcheese

Although this is called headcheese, it contains no meat like other headcheeses do. To make a small, round head approximately 6 inches across and 3 inches thick, start with 6 quarts of raw milk in a large pot. Although cow's milk may be used, to make this cheese the traditional way use goat's or ewe's milk.

Begin now to soak a straw basket in warm water until it becomes pliable. Make sure it hasn't been colored in any way or it will bleed onto the cheese. Keep the basket pliable until ready to use.

Heat milk until it's lukewarm to the touch. Add 1 crushed rennet tablet dissolved in ¼ cup of water and stir briskly for 1 minute. Remove milk from heat and let set. Do not disturb during this time. Leave until milk has thickened like yogurt, about 25 to 30 minutes. Your finger, when poked into the mixture, should come out clean.

Put thickened milk back on the stove and heat until just warm. Stir briskly 5 to 10 minutes until the mixture turns back to milk. Remove from heat. Using a perforated spoon, stir slowly in one direction only, around and around the pot. You can start to feel the lumps of cheese forming after about 3 revolutions. Gather the cheese to one side of the pot with your spoon and scoop it into your hands. Roll it over and over as you would dough.

Pack cheese into wet basket, draining excess liquid back into the milk pot. Turn cheese out into your hands and then repack it into the basket upside down. Squeeze out more liquid with the backs of your fingers. Sprinkle top with 2 teaspoons of salt and leave the basket draining in a colander or on a cookie rack for 7 or 8 hours.

When thoroughly drained, remove the cheese from the basket and salt the exterior thoroughly. Store uncovered in a dry, airy place to ripen. You can eat the cheese at this point or freeze it if you wish, but it will have a much different taste after being stored in a cool (but not freezing) place in a barrel of brine.

Before brining, turn cheese every other day and salt again. In 2 to 3 weeks it will form a hard rind and can then be stored in a wooden barrel of salt brine where it will keep indefinitely. To make the brine, dissolve 2 cups of pickling salt in 1 gallon of water. Weight down cheese head with a plate and clean rock so it will stay immersed in the brine. When you want to use your cheese remove it from the brine and rinse well.

Storing Cheese

Semihard cheese, like the ones David Page and Juliette de Baïracli Levy make, and others, like cheddar, American, Swiss, and provolone, should be kept under refrigeration after aging. To prevent these cheeses from drying out, they may be wrapped in a thin piece of cloth or cheesecloth which has been moistened with water, a weak salt water solution, or a mild vinegar. Check the cheese occasionally and dampen the cloth when it becomes dry. To prevent the cut edges of a round of cheese from drying out, spread butter or a very light-flavored oil on the cut ends of the round before wrapping and refrigerating it.

If you plan to keep the cheese for more than a few weeks, it may be frozen. Semihard cheeses do not freeze very well—they become a bit crumbly and begin to lose their flavor after a few months at freezer temperatures—but some people do not mind this slight loss in quality. If you wish to freeze your cheese, cut it into small pieces of about ½ pound each, and wrap each piece well in moisture-proof paper. In small pieces, the cheese will freeze quickly, and there will be little damage done to the cheese's flavor and texture. Wrapped properly, semihard cheeses will keep for 6 months in the freezer.

In order to enjoy the full flavor of the cheese, it should be brought to room temperature before serving. Avoid exposing the cheese to high temperatures which cause the cheese to sweat and lose some of the fat captured in the curd.

Smoked Cheese

When you fire up your smokehouse for hams and bacons hang up a few rounds or bricks of homemade cheese as well. The hickory or

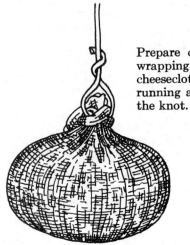

Prepare cheese for smoking by wrapping it in several layers of cheesecloth, knotting it well and running a cord or hook through the knot.

other hardwood smoke will make something very special out of even bland cheeses. Almost any semihard cheese is good for smoking. Mild cheddar, colby, Muenster, and plain brick cheese are some suggestions.

To find out more about smokehouses and cool smokehouse fires, see page 395. As you'll read there, fires for smoking meats should be made with hardwoods only. This is equally important for smoking cheeses, as softwoods emit a resinous smoke that will give your cheese an off-flavor. Wrap your rounds or bricks of cheese in a few layers of cheesecloth and knot the cloth well. Then run a wire hook or good strong cord through the knot and hang the cheese as far from the fire as possible; it should be in the coolest spot in the smokehouse. Cheese hung in an area that gets above 90°F. might melt on you.

Check the cheese regularly to make sure that it's not melting and not getting too dried out. Ten to 15 hours in a cool smokehouse is about right.

Making Soft Cheeses

Cottage Cheese

Cottage cheese got its name from the fact that it is a cheese that can easily be made in the home—or in the cottage, as the term once was. Traditionally, this soft, perishable cheese was made from the skim milk left after butter-making. After the cream had been taken off the top, the remaining milk was poured into a good-sized crock and set in a warm place (usually on the back of the wood stove in the kitchen) for 2 days or so, until the milk had clabbered and much of the whey had separated from the curd. Then the curd was cut into cubes and heated gently over a low fire to firm it up a little. The warm curd and whey were poured into a cheesecloth sack and hung up over a tub until all the whey had drained off and only the soft curds were left. The cheese was then chilled for a few hours and mixed with a little salt and fresh cream and stored in the springhouse, cellar, or ice chest where it could be kept as long as 5 days without spoiling.

Cottage cheese made at home today is made in much the same way as it was one hundred or so years ago. Most of it is made with fresh, raw skim milk, but fresh pasteurized or raw whole milk will do, too.

Raw milk contains the beneficial bacteria that are responsible for making milk clabber. Pasteurized milk has been heated to temperatures that kill this bacteria, and it will not clabber by itself, no matter how long it sits in a warm place. If you're using pasteurized milk, you'll need to add something to activate the curd and start the milk clabbering. You can add a milk product that contains the necessary beneficial bacteria, like cultured buttermilk or plain, unsweetened yogurt, or you can use rennet tablets.

In addition to the milk and activator, you'll need an earthenware crock, glass casserole dish, or stainless steel or enamel pot. Also have on hand a dairy thermometer, a long spoon (preferably made from glass, wood, stainless steel, or enamel), a spatula or wide knife, a large pan or shallow pot that is larger than your crock, cheesecloth, and a colander. All your equipment should be scrupulously clean. Wash it with soap and water and rinse well with very hot water before bringing it in contact with the milk.

As we said before, cottage cheese is simple to make:

Clabbering the Milk

Pour a gallon of fresh raw or pasteurized, skimmed, or whole milk into a crock or pot. If the milk is cold, bring it to room temperature over very low heat. If you are using pasteurized milk or if you want to speed up the clabbering process of raw milk, add one of the following:

4 tablespoons unflavored fresh yogurt	¼ tablet rennet, dissolved in ½ cup warm water (use the kind of rennet you'd use for junket)
½ cup fresh cultured buttermilk	

Mix the milk and the activator together and cover the crock loosely with cheesecloth or a thin towel. Be careful not to smother the milk by covering it with a heavy towel or plate. All you want to do is to cover the milk to keep dust and insects out of it. If air isn't permitted to pass over the milk, the milk can acquire a musty odor and taste which will linger in the finished cheese. Let the covered crock sit in a warm (75° to 85°F.) place until clabbered. If you are using an activator, the milk should be clabbered in 12 to 18 hours. Raw, inactivated milk will need about 48 hours to clabber.

Cutting the Curd

The milk has clabbered when much of the whey, a thin watery liquid, has risen to the top; and the curd, a white substance with a consistency similar to soft cream cheese, has settled to the bottom. Now it is necessary to cut the curd to allow more whey to separate out. If you are using skim milk, cut the curd into 2-inch cubes. Do not make them much smaller than this, because if you do, too much whey may separate out, making the resulting cheese dry and leathery. If you are working with whole milk, the cubes may be cut smaller. To cube the curd, cut into 2-inch strips with a spatula or wide knife. Then slice the curd again crosswise so that you have

(1) The first step in making cottage cheese is clabbering the milk. Pasteurized milk must be activated with buttermilk, yogurt, or rennet before it will clabber.

(2) Once the milk has clabbered, cut the curd into 2-inch cubes to allow more whey to separate out.

(3) Heat the curds over boiling water until they reach 115 °F. If they are heated higher than this, the cheese will be tough and dry.

(4) After heating, check the curds for firmness. They should hold their shape and be neither too soft nor too dry.

(5) Pour the curds through a colander lined with cheesecloth to drain off all the whey, which otherwise would impart a bitter taste to the finished cheese.

(6) You may rinse curds to wash off any remaining whey. To do this, place the colander with the curds in a bowl of water and stir the curds gently. Let the curds drain again before storing in the refrigerator or freezer.

2-inch squares. If the curd is more than 2 inches deep, bring your spatula under the curd and cut across horizontally. It is important to cut the curd carefully so that just enough—but not too much—of the whey can separate out.

Heating the Curd

Pour a few inches of water into another large pan or pot, set the crock with the cubed curd in it, and place it over low heat. Insert the dairy thermometer in the curd and heat until the thermometer reads 115°F. Hold it at this temperature for about ½ hour, stirring occasionally so that the heat will be distributed evenly throughout the curd. Stir gently so as not to break the curd.

Straining the Curd

After about 30 minutes, the curds will have settled to the bottom of the crock. Line a colander with cheesecloth and place it in a bowl to catch the whey. (Whey is an important source of minerals and B vitamins. Instead of pouring it down the drain, add it to soups, beverages, or casseroles, cook rice in it, or feed it to your pets or livestock.) Gently pour the curds and whey into the colander. Allow most of the whey to drain off and then take up the four corners of the cheesecloth and hang the curds in the cheesecloth over the bowl so as to catch the remaining whey that drips out. Let the curds hang this way until no more whey drips out. The curds may be rinsed at this time with clean, cold water if you wish to minimize the acid flavor, although rinsing is not necessary. If you want to freeze your cottage cheese, do not rinse the curds.

Storing Cottage Cheese

If you are freezing your cheese, take it from the cheesecloth without rinsing out the remaining whey, pack it in containers and freeze. If the cheese is to be eaten within a few days, it may be rinsed and should be chilled for a few hours. Then mix it with salt and fresh cream to taste if you wish. The cheese can be seasoned with herbs, like dill, chives, or parsley. If the cheese will be used in cheesecakes, blintzes, or pastries, do not add any seasoning or cream unless the recipe says to do so.

One gallon of milk should make about 1½ pounds of cottage cheese.

Ricotta

Ricotta means "cooked once again" and is literally just that. The economical Italian country wife looked for a use for the whey left over from making her famous hard cheeses, and found that with high heat and acid she could precipitate the albumin (protein) in the whey and thereby obtain a delicious fresh cheese.

Other cheese-producing countries have created similar types of whey cheese, such as the Swiss Hudelziger and Mascarpone, the latter from goat's milk whey; and the French Gran de Montagne, made from whey and enriched with cream. Other exotic names for ricotta-type cheese include Recuit, Broccio, Brocotte, Serac, Ceracee, and Mejette.

Ricotta is most often sold fresh. It is rather sweet as opposed to the slightly tangy flavor of cottage cheese; it is creamy and melts beautifully without separating in baked dishes like Italy's famous lasagne. A less well-known product is dried ricotta, available only in specialized cheese shops, or your own home dairy. This is a piquant grating cheese and does not require the refrigeration mandatory for fresh ricotta. So if you make too much to consume within a few days, you can always press and dry it.

To prepare authentic old-world ricotta, you need whey. This is the nutritious liquid left over from curdled milk when you have removed the curd; 94 percent of the liquid of whole milk. It contains the water-soluble proteins, vitamins, and minerals in the milk, such as the soluble calcium. Most people do not realize that one-third of the calcium in milk is lost in the whey in the cheese-making process, even more when the cheese is made by the acid-coagulation method such as in tangy, small-curd cottage cheese, rather than the renneted method. Liquid whey also contains most of the milk sugar or lactose. However, in the finished whole-milk ricotta cheese only 3 percent lactose remains, so those on a low-carbohydrate diet can enjoy it also.

You will need at least 2 to 3 gallons of whey plus a few cups of whole milk to make only 1 pound of cheese. This is a lot of whey and not very practical for home cheese-making, especially if you must collect whey from the milk of one or two goats. If you do have the whey, however, here is how to prepare it:

Old World Ricotta
(Ricotone)

2½ gallons liquid whey   ⅓ cup apple cider vinegar
1 pint whole fresh milk

Heat the whey until the cream rises to the surface. Add the fresh milk and continue heating to just below the boiling point (about 200°F.). Stir in the vinegar and remove immediately from the heat. Dip out the curds (milk protein) and drain. Salt if desired.

Yield: about 1 pound

Modern ricotta, developed in the early nineteenth century, is made from whole milk and is simply and deliciously prepared at home. This is also the method used by most big commercial American dairies. The whole or part-skim milk is acidified to a carefully controlled level, then subjected to high temperatures. For most of us who do not have equipment for measuring or controlling pH and heat, the following method works very well.

New World Ricotta

These easy-to-handle proportions are adaptable to small-quantity cheese-making. Dried whey is available in natural food stores and is well known for its beneficial action on friendly intestinal flora. This is one of the reasons ricotta is such a highly digestible food. Whey contains about 13 percent protein, has five times the calcium of liquid milk, and is a good source of riboflavin and iron.

1 quart fresh whole or partially skimmed milk
¼ cup dried whey powder
2 tablespoons liquid buttermilk

Stir whey powder into milk with whip and dissolve well. Stir in buttermilk. Incubate at room temperature 24 hours.

Very slowly, bring to scalding temperature. It will separate into curds and whey. Gently drain through cheesecloth; hang to drip a few hours. Salt to taste if desired.

Yield: about 1 cup

Store fresh ricotta in moisture-proof containers, well closed, in the refrigerator. It will keep about 4 days maximum, especially unsalted.

If you wish to dry your cheese for grating, press it heavily in perforated forms, salt it on the surface, and dry in a clean, low-humidity room where the temperature is 100°F. or a bit higher. Otherwise, enjoy your fresh ricotta as is, unadorned, by the spoonful, or try the recipes that start on page 331.

Soft Greek Cheese
(Mizithra)

Mizithra is a very soft cheese resembling ricotta and cottage cheese. It can be stored in the refrigerator for a few days, but will not keep for long.

To make, use the whey left over from the Greek headcheese described on page 298. Or you can use 3 to 4 quarts of whey left over from any hard cheese recipe. To make a traditional Mizithra ewe's or goat's milk should be used. Pour the whey into a pot; stir in 1 cup of whole milk and 1½ cups of water. Heat until liquid is steaming hot, but not boiling. Slowly stir liquid with a wooden spoon all during this stage, trying to disturb the surface as little as possible. Continue stirring for ½ hour. The liquid will start to get slightly foamy on top, then the top will thicken.

Using a large spoon with small holes (if holes are too large, cover spoon with cheesecloth), lift out the thickened mass. Put into a basket or cheesecloth-lined colander and allow to drain. At this stage it's ready to eat, warm or cold, salted or honeyed.

Cream Cheese

This simple, soft cheese can be made with either cream or milk. Be sure that the milk or cream is fresh and at room temperature. It may be either raw or pasteurized.

Measure your cream or milk, and to each gallon add ½ cup cultured buttermilk, mixing well. Dissolve ½ tablet rennet in ¼ cup cold water for each gallon of milk or cream and stir it well into the cream or milk.

Stir for about 10 minutes, or until the milk just begins to clabber. Stop stirring when you feel a slight thickening. Then cover and keep it at about 85°F. until it has completely clabbered. One way to keep the milk above room temperature is to place the bowl in which it is in in a larger bowl of warm water. Or wrap the milk bowl in a bath towel and place it in an unheated oven to insulate it better and help it retain its own heat. This clabbering may take 12 hours or a little more.

When the whey has separated out and the curds have formed one soft but solid mass on the bottom of the bowl, the milk has clabbered. To make sure, tip the bowl and see if the curds break clean from the sides when you do. Now take several layers of new, clean cheesecloth and wet it in clean water. Wring it out and line a colander with it and place the colander in a large bowl to catch the whey. With a clean knife slice through the clabbered milk so that the curds are cut into about 1-inch cubes. Then pour the clabbered milk cubes and its whey into the colander and let it drip for a minute or so. Pick up the cheesecloth by its four corners, being careful not to pour out any of the curds, and tie the corners together to form a bag.

Suspend this bag over a bowl and let it drip until all the whey has drained off and a gelatinous solid mass remains. You should let it drip at least overnight. You can speed up the process by gently squeezing the bag every so often and by changing the cheesecloth once or twice when it gets plugged up with the cream cheese.

When the consistency of the cheese is to your liking, take it out of the cheesecloth and place it in a clean bowl. Salt it if you wish. Start with a small amount of salt, 1 teaspoon, and add more to taste. The salt is not necessary, and some people will prefer not to add any. It will, however, help the cheese to keep a little longer.

Then pack the cheese into rigid containers and keep in the refrigerator. Unsalted cream cheese will keep 3 to 4 days, and if salted, about 4 to 5. Cream cheese can be frozen for longer keeping, although it may become crumbly once thawed. Keep frozen no longer than 4 months.

Making Yogurt

Basically, yogurt is milk that has been fermented by special strains of beneficial bacteria. Yogurt can be made out of any kind of milk, be it raw or pasteurized, cow's or goat's milk, which has been skimmed or left whole. Even dry powdered milk or soy milk (made from soybeans) can be used. The starter bacteria that you introduce into the milk may be commercially prepared yogurt, your own homemade yogurt, or a pure culture, such as *Lactobacillus acidophilus*, from a natural food store.

If you buy yogurt, there are several things you'll have to consider. First of all, buy fresh yogurt. That means you should buy a carton with an expiration date as far in the future as possible, a week and a half or two weeks distant, or more. Also, be sure that you are buying real yogurt. Some yogurtlike substances come in similar cartons and masquerade as yogurt, but don't be fooled—read the ingredients and see what you're buying. It is best to buy plain yogurt or a kind that has the preserves on the bottom of the carton. You will be able to skim enough yogurt off the top of the carton to use as a starter. The kinds of yogurt that have the preserves and flavoring blended through the product can also be used, but even in a small amount the flavoring can permeate the final product to a surprising degree. About ½ of a half-pint-sized carton of yogurt is enough to start 1 quart of homemade yogurt.

Since there are several ways to encourage fermentation, you might like to experiment with a few methods first and then choose the way that works best for you. Your choice will depend upon how much equipment you want to use and how much time you want to spend in yogurt-making. Whatever equipment you do use, make sure that it is thoroughly clean. Remember, you want to encourage the growth of beneficial and not harmful bacteria. An unclean bowl or spoon might give an off-taste to your finished yogurt.

Preparing the Milk and Starter Mixture

To make yogurt with raw or homogenized, pasteurized milk, begin by heating it to the boiling point; this destroys any bacteria already present in the milk and provides a sterile medium for your yogurt culture. If you like thicker yogurt, you can stir in some powdered milk while the whole milk is still warm. Allow the milk to cool to lukewarm. To test, drop a little on your wrist; if the milk feels hot, allow it to cool some more. When it no longer feels hot, add about 3 tablespoons to ½ cup of yogurt or the amount of *Lactobacillus acidophilus* culture suggested on the envelope. Stir well with a whisk or wooden spoon, making sure the culture is thoroughly mixed into the milk. Pour the mixture through a strainer into whatever you're incubating your yogurt in.

After 4 hours, lift the lid of one of the containers and tilt it *gently*. If the yogurt is thick, about the consistency of heavy cream, it is ready. Refrigerate the containers. If the yogurt is still rather watery, allow it to incubate for another hour, then check again. The longer the mixture incubates, the stronger and sourer the final result will be.

To use powdered milk in the above directions, simply mix up a quart of powdered milk, adding extra if you like the end product thicker. Stir in yogurt culture, strain into containers, and incubate. There is no need to sterilize powdered milk.

You can also use soy milk in the above recipe if you are a lacto-vegetarian or simply allergic to other types of milk. Substitute a quart of soy milk per quart of whole milk, stir in the yogurt culture, and incubate.

Using a Yogurt-Maker

A yogurt-maker consists of a constant-temperature, electrically heated base and a set of plastic or glass containers with lids. Most yogurt-makers make four or five individual pints at a time. They are foolproof, cost from $10 to $20, and are available in most natural food and department stores.

Prepare your yogurt mixture and pour it into the containers in the yogurt-maker. Cover the yogurt-maker with the cover that came with it or with a towel. Leave undisturbed for about 4 hours. At the end of this time, remove the lid from one container and gently tilt the glass. The yogurt should be about the consistency of heavy cream. If it's still thin, let it incubate longer and check again. When the yogurt thickens, remove and refrigerate. Don't serve until thoroughly cooled.

Making Yogurt in the Oven

One way to get around using a yogurt-maker is to incubate the milk and yogurt mixture in an earthenware bowl (because it retains heat well) which is kept warm in an oven. Pour the lukewarm mixture into your bowl, cover, and set the bowl in the oven; heat slowly to 120°F. Turn off oven and let it cool gradually to 90°F. Try to maintain temperature between 90° and 105°F. by reheating oven if needed (after about 2 to 3 hours) until milk becomes the right consistency. Check frequently during the last 30 minutes. Chill immediately after milk thickens.

A Covered Casserole and a Warm Place

Prepare the milk and starter mixture. Then warm a casserole dish by running hot water over it or heating it for a few seconds in a very low oven. Pour the mixture in the warm dish and cover. Wrap the casserole dish in a large towel and set in a quiet, warm spot in the kitchen, as on a warm radiator or the warming shelf of a cookstove. Leave undisturbed for at least 6 hours, or let it set overnight. At the end of this time, check the consistency of the yogurt by unwrapping the towel carefully and tilting gently. If it is solid enough for your liking, refrigerate immediately. Serve only when thoroughly chilled.

These last methods of yogurt-making are simple, but not as foolproof as when using a yogurt-maker, since it is difficult to maintain a constant temperature in the milk mixture. The warmer the spot where the mixture sets, the quicker it will thicken and be ready, providing the area is not over 115°F. If you're not successful using one of these last two methods the first time, try it again until you find just the right spot in your kitchen and until you know how long to let it set.

Yogurt in a Thermos

An extremely simple way of making yogurt is to use a thermos bottle. A thermos is an excellent heat retainer. Once the starter has been stirred into the lukewarm milk, pour it into a wide-mouthed thermos, put on the lid and let it set 4 to 6 hours before refrigerating. This is practically foolproof, since the temperature is controlled for you.

Yogurt-making is a simple procedure. Once you become familiar with the method you have chosen you'll be able to make perfect yogurt every time.

Properly made yogurt should be rich and custardlike and have a creamy, slightly tart taste. Homemade yogurt will usually be sweeter than any unflavored store-bought variety. If, after refrigeration, there is a little water (whey) on top of the yogurt, don't worry, you haven't done anything wrong. This is natural, especially after it has set in the refrigerator for a few days. Open a commercial yogurt and you'll find water on top, too. Either mix it in or pour it off, but save it for using instead of water in other recipes, since the whey is high in vitamin B_{12} and minerals.

What Can Go Wrong

If you have trouble making yogurt the first time, check for the following problems:

1. Perhaps the milk mixture was disturbed while incubating. Even a few tilts or knocks can cause the whey to separate from the curd (as it does in cottage cheese). Instead of being a thick and smooth yogurt, it may be watery and lumpy and resemble cottage cheese.
2. Perhaps your mixture was too hot or too cool. If the mixture is too cool, the growth of bacteria will be retarded. If it's too hot, the bacteria may be killed. Add more starter, incubate longer, and adjust temperature to correct.
3. Perhaps the milk or yogurt starter was not too fresh. The older either is, the longer it will take to incubate and the more starter you should use. For best results, neither should be more than 5 days old.
4. Perhaps you used a pure yogurt culture which takes longer to thicken than prepared yogurt.

Storage

Yogurt will keep well for about 8 days under refrigeration if it is kept in an airtight container. It can also be frozen for several months, although it may separate and lose its smooth consistency upon thawing. Thawed yogurt is best used in cooking, rather than eating fresh. Make sure that you save some to start your next batch. The new starter should be used before it is 5 days old.

Yogurt Cream Cheese

This simple-to-make yogurt product can be used just like regular cream cheese, on crackers for hors d'oeuvres, for sandwiches, and in cookies, pies, and other pastries.

To make yogurt cream cheese, make your yogurt by following one set of instructions discussed in this section. Instead of refrigerating the yogurt once it has formed, pour it into a colander lined with a triple thickness of cheesecloth. Catch the whey by placing a bowl under the colander. Allow the whey to drip for 1 minute, then lift up the four corners of the cheesecloth and tie them together. Hang the cheesecloth bag over the sink by suspending the bag from the faucet. Let the cream cheese drip for 6 to 8 hours, then remove it from the bag and store in the refrigerator.

Storing Eggs

Because each egg is intended by nature to house an unborn chicken, nature packages each one in its own protective shell. The shell is porous enough to permit oxygen and other gases to flow in and out through its walls, but its outer coating or membrane prevents bacteria and molds from entering which would otherwise contaminate the egg.

Alone, the shell will protect the eggs for a short time, providing it is kept cool. Brush, don't wash, dirt off eggs before you store them. People who vigorously wash off the dirt are also washing off the egg's protective membrane. If possible, store your eggs in a covered container to keep out objectionable odors that travel with gases through the shell's pores. An egg carton and the egg section in your refrigerator are both fine places to keep them.

Eggs will keep at refrigerator temperatures for a week or two, but after that time their freshness fades. Both the white and the yolk begin to lose their firmness and become watery and runny. The yolk of an old egg will usually break into the white when the shell is cracked open, making separating the yolk from the white of old eggs a difficult, if not impossible, task.

Old-Fashioned Methods

Before farmers had access to freezers, they devised some simple (but not always successful) means of preserving their excess eggs. Some farmers relied solely on the use of salt to keep their eggs from rotting. After gathering their eggs, they packed them in a large barrel or crock with plenty of salt and stored them in a cellar or springhouse to keep them cool.

The majority, however, found some way to clog up the pores of the egg shells so that moisture would not escape and air could not enter. Eggs were rubbed with grease, zinc, or boric ointment, or submerged in a solution of lime, salt, cream of tartar, and water.

Probably the most popular way to seal egg shells was to waterglass them. By this method a chemical, sodium silicate, was mixed with water and poured in a crock which was filled with eggs that were about 12 hours old. The sodium silicate (which is used today to seal concrete floors and as an adhesive in the paper industry) would clog the pores in the shells and make them airtight.

Some people, even today, use waterglassing as a means of preserving eggs, but this storage method has its drawbacks. Eggs preserved this way are not good for boiling because their shells become very soft in the waterglass solution. The whites will not become stiff and form peaks, no matter how long they are beaten. No soufflés, eggnogs, or meringues can be made with waterglassed eggs. There is also a very good possibility that by consuming eggs stored in waterglass you would be consuming some of the undesirable chemical, sodium silicate. If you keep roosters with your hens, waterglassing may not be a successful means of preservation for you. The life factor in fertilized eggs makes these eggs deteriorate more quickly than sterile, unfertilized eggs, and waterglassing may not be enough of a preventive against spoilage.

Freezing

Freezing is the only way to keep eggs safely at home for more than 2 weeks. Eggs, both fertile and unfertile, will keep as long as 6 months in the freezer, if you prepare and pack them properly. The rule for selecting the right food for freezing applies for eggs just as it does for fruits and vegetables: choose only the very freshest. Eggs even a day or two old should be stored in the refrigerator and used within a relatively short time, as recipes call for them. Freeze only just-gathered eggs.

Eggs in their shell expand under freezing temperatures and split open. For this reason, they must be shelled and stored in appropriate containers. If you are storing eggs in rigid containers, leave a little headspace for expansion. You can separate the white from the yolk and freeze each separately, or you can store the eggs whole.

If you are freezing egg whites alone, they can be frozen as is, in airtight containers. For convenience, pack as many eggs together as you will need for your favorite recipes. You can then thaw and use a whole container of egg whites at one time.

If you are packing yolks separately or are packing whole eggs, you will need to stabilize the yolks so that they won't become hard and pasty after thawing. To do this, add 1 teaspoon of salt or 1 teaspoon of honey to each cup of yolks. Twelve yolks make up 1 cup. Break up the yolks and stir in the salt or honey. Of course, it is necessary to mark on the container whether salt or honey was used as the stabilizer so that you won't ruin recipes by adding more salt or honey than you had intended.

If you are packing your eggs whole, you will also need to stabilize them with salt or honey. Add 1 teaspoon of salt or honey to each cup of whole eggs. There are about 5 whole eggs in 1 cup. Scramble the eggs with the salt or honey before packing and freezing. Whole eggs can be packed together in one container, or they can be packed individually by using a plastic ice cube tray. To pack eggs separately, measure 3 tablespoons of whole scrambled eggs (which equals 1 whole egg) into each separate compartment of the ice cube tray. Place the filled tray in the freezer, and when the eggs have frozen, pop them out and store all the egg cubes in a plastic bag. By so doing you will be able to take from the bag and thaw just as many eggs as you need at one time.

Eggs should be thawed completely before using. They thaw at refrigerator temperatures in about 9 hours and at room temperatures in about 4 hours. If frozen properly, thawed eggs have the taste, texture, and nutritional value of fresh eggs and can be used successfully in all recipes calling for eggs. To make up 1 egg from separately frozen whites and yolks, measure out 1 tablespoon of yolk and 2 tablespoons of white. Eggs should be used soon after they thaw, as they deteriorate rapidly.

Homemade Ice Cream

Ice cream has been enjoyed in the United States since the mid-eighteenth century. At first, only the well-to-do could afford ice cream (many of the Founding Fathers were ice cream fans) but as the nineteenth century passed, America became known as a nation of ice cream lovers, and small shops serving ice cream to the masses blossomed throughout the country. The first ice cream factory in the United States was opened in Baltimore in the early 1850s and was so successful that its owner opened another in Washington, D.C., in 1856. At the same time, advances in domestic technology made it easier for people to make their ice cream at home. Ice cream-making became an event reserved for special occasions in many families. American ice cream fanciers began to argue about which recipe produced the best ice cream and whether Philadelphia ice cream (made from sugar, cream, and scraped vanilla beans) was superior to French ice cream (based on an egg custard). Cookbooks included many ice cream recipes.

As good as homemade ice cream is, however, it passed the way of many other homemade treats when cheap commercial brands became available. Based on synthetic products, low in butterfat and with a high overrun (the air whipped into the finished product), the cheap ice creams proliferated during the 1950s. In recent years, however, there has been an increase in the number of good brands of ice cream on the commercial market, and the interest in good ice cream has overflowed into the home, especially for fans awed by the prices of good commercial ice creams.

One good reason, then, for making your own ice cream is because it can be cheaper than buying a good commercial brand. But this is generally only true if you've got your own cow or goats or have a good neighborhood source of raw milk or cream. Making ice cream is also one way that you can use up surpluses of milk. In fact, the commercial producers do just this, reducing the summer dairy surplus by converting it into ice cream, and using about 7.6 percent of the milk produced in the United States in the process. The problem with this for the home is that ice cream really shouldn't be stored in the freezer for more than a month—although you can probably extend that time limit if you have to.

But there are other reasons for making ice cream, too. Making ice cream is nostalgic fun. Rural ice cream festivals and family members taking turns at the hand-cranked freezer are vignettes securely enshrined in the mythology of American life, and even today's kids thrill at the thought of ice cream-making, almost as much as they enjoy eating the products of their labors.

And finally, despite whatever other rationales one can muster for making ice cream at home, the best reason is because homemade ice cream is usually just plain better than any other kind—if you like ice cream, you'll love homemade ice cream. It can be eggier, fruitier, richer, or less sweet than commercial products. You can use honey instead of sugar. You can let your imagination go wild with the flavors. And since you're putting everything together, you also know what *isn't* there—a lot of artificial flavors, colors, stabilizers, emulsifiers, and air.

Hand churning, in a manually cranked maker like this one, is supposed to make the best homemade ice cream.

Manually Cranked Ice Cream

There are several methods for making ice cream. Perhaps the most familiar is the freezer method, using a mechanical ice cream freezer fitted with a metal can and revolving dasher. Modern freezers come in both electrically powered and hand-cranked models. Sears and other mail-order firms sell house brand freezers in a variety of sizes, using either method for power. (See page xiii for equipment sources.)

Manually cranked freezers are a lot more appealing to most people than electrically powered models, especially if they intend making ice cream only occasionally. For one thing, the machines are cheaper initially and don't require electricity—a big plus if you want to make ice cream at your favorite isolated picnic spot (somewhere where the salt runoff won't kill any vegetation). Hand churning also produces a better ice cream, many aficionados believe, and some of the better methods for making commercial ice creams even call for hand finishing. In addition, manual freezers are available in many sizes—White Mountain, one of the most popular makers of manual freezers, even makes one that turns out 20 quarts of ice cream at a time.

Electrically Cranked Ice Cream

Others prefer electrically turned freezers. These come in several models. One kind resembles a hand-cranked machine and uses rock

The only real difference between this ice cream maker and the one in the last photo is the electric motor mounted on the top of this one.

salt and ice to freeze the mix, but has a motor mounted on top to power the dasher and can. A second kind of electric freezer is made to fit inside the freezing compartment of your refrigerator or in your freezer. These machines have heavy woven wire cords so that you can close the freezer door without damaging the cord. A motor turns the dashers while cold air from the freezer freezes the mix. Many models feature a shut-off switch so that the machine stops churning when the mixture reaches the proper consistency.

One problem with these machines is that they have small capacities. Another drawback for many is that they consume electricity. However, if you plan on making ice cream frequently as a means of preserving extra milk, you might decide that using an electric machine will be worth the higher initial cash outlay and the higher operating costs. If you opt for one of these machines, look for a well-made machine with a heavy duty motor and a fairly large capacity.

Freezer Ice Cream

Another method of making ice cream at home calls for little equipment—and produces a product that is passable, if only barely. In this method, the mix is frozen in ice cube trays in the freezer compartment of a refrigerator or the home freezer. Better recipes for this type of ice cream usually call for whipping the product at some point. Sometimes the mix is whipped when partially frozen and re-

Although this type of maker produces only a small amount of ice cream at a time, no ice or muscle power is needed. The unit fits in the freezer compartment and runs on electricity.

turned to the freezer for finishing. Often the mix is frozen completely and, just before serving, is thawed slightly in the refrigerator and then whipped before being dished up. If you're planning on using this method of freezing ice cream, it's best to buy several ice cube trays just for the ice cream, since the dairy products and flavorings tend to impart a taste to the trays which may come out in the ice cubes if the trays are also used for making ice. Directions for making these ice creams are usually given in recipes for them, since the methods vary somewhat.

Cranked Ice Cream How-To's

However, the method for making turned ice cream is standard. You should be able to use these directions for any ice cream recipe which calls for hand cranking, whether you're using a manually cranked freezer or an electrically powered freezer with a bucket. (If you've purchased one of the freezers that fits inside your refrigerator freezer, follow the directions that came with your machine.)

1. Prepare the ice cream mix according to directions and cool in the refrigerator for several hours. (If the mix you're using does not call for cooking, you can mix it in the freezer can to save bowls, but be sure all ingredients are cold when you mix them. Also be sure that the can is clean, which leads us to Step 2.)

2. Wash the can, cover, and dasher well and scald with boiling water. Drain and cool. (Since dairy products and eggs are both ideal mediums for bacterial growth and since many recipes are entirely uncooked, it's best not to take chances.)

3. Pour the cooled mix into the can. Never fill the can more than two-thirds full. For example, never put more than 4 quarts (1 gallon) of mix in a 6-quart can.

4. Put the dasher in the can (be sure it's seated properly), cover, and place in the freezer bucket. Place the cranking mechanism on top; make sure it fits tightly and is securely wedged in place.

5. Pack the freezer bucket with ice and salt. The ice and salt both enable the ice to melt quicker, releasing more cold which chills the ice cream mixture. Use only *crushed* ice. If you don't want to buy crushed ice, freeze ice cubes, place them in a burlap bag, and smash them with a hammer. Don't hesitate to crush them as much as possible, since pieces of ice that are too large are a common cause of ice cream failures (as is too little salt). Layer the ice and salt in the bucket, beginning by filling

the bucket one-third full with ice. Cover with a thin layer of rock salt or ice cream salt, then another thinner layer of ice. Continue until the bucket is full and the can is covered by the ice/salt mixture. Use in all 1 pound of salt to 6 to 8 pounds of ice. Add some water—1 cup for small freezers and 2 for larger models—to hasten freezing.

6. Begin cranking the mixture (or plug in the motor on your electric machine). If you're using a manually cranked model, start cranking slowly, then faster as the mix begins to freeze. Throughout the whole cranking process, water should be flowing from the drain hole in the side of the bucket (keep the hole free from ice). Add more ice and salt and keep the can covered with the mixture. When the cranking becomes difficult—or the motor begins to labor—the mixture is done.

7. Remove the crank (or motor), and take out the dasher. Be careful not to get salty ice or ice water in the can. Scrape what you can from the dasher (and give the dasher to the kids to clean off). Cover the mix with foil or wax paper, plug the hole in the cover (a cork works well), and put the cover back on the can.

8. Freeze the ice cream for 2 or 3 hours wrapped in a towel in your freezer compartment, or empty the bucket, place the can in the bucket, and repack with ice and salt. This process is called packing and is essential to achieving a solid, nicely textured product. However, ice cream does not *have* to be packed—it can be eaten right after churning, although it's very soft at this stage and will melt pretty quickly.

Ice Cream Recipes (Turned Ice Cream)

French Vanilla

This is a rich, eggy ice cream. It takes longer to prepare than most homemade ice creams because the ingredients must be cooked and then cooled. However, it's well worth the effort.

6 eggs
½ cup light-colored honey*
8 cups light cream

½ teaspoon salt
4 teaspoons vanilla

* Light-colored honey is specified in these recipes because honey with a light color has less of a strong taste; although dark-colored honey produces an excellent product, the taste of the honey sometimes overpowers the other flavors. If you enjoy strong honey, don't hesitate to use it in your ice cream.

Separate the eggs and beat the yolks until smooth. Add the honey to the yolks and beat again until well blended. Beat the whites until stiff and stir into the yolk and honey mixture. Add cream, and cook in a double boiler for 15 minutes or until thick, stirring continuously. Add salt and vanilla, and chill. Pour into freezer can and freeze.

Yield: 4 quarts

Philadelphia Vanilla

This recipe produces a dark, speckly vanilla—which is the way that Philadelphians like their vanilla to look. Interestingly enough, people in other areas of the country like their vanilla ice cream to look different; some people prefer a plain white vanilla and others like a yellow vanilla. This recipe is adapted from a formula used by Bassett's of Philadelphia, one of the city's most famous high-quality ice cream makers.

3 vanilla beans (substitute 3 teaspoons vanilla extract* if you wish)	3 quarts light cream 1½ cups light-colored honey

Combine 1 quart of the cream and the vanilla beans in the top of a double boiler over boiling water. Stir continuously until the cream is scalded (about 10 minutes). Remove from the heat, scrape seeds and pulp from the beans, and add back to the cream; discard the pods. Add honey. Cool. When the cream/vanilla mixture has chilled, add the rest of the cream. Freeze.

Yield: 4 quarts

* If you're using vanilla extract, add it just before freezing; there's no need to cook the cream.

Old-Fashioned Vanilla

4½ cups milk 6 tablespoons flour 1½ cups honey	6 eggs 4½ teaspoons vanilla extract 4½ cups heavy cream, whipped

Scald milk in the top of a double boiler over boiling water. Mix the flour with enough milk to make a smooth paste and stir into the rest of the milk. Cook until thick, stirring continuously, and then cover, cooking for 10 minutes more. Then add the honey. Beat eggs slightly and stir into milk mixture; cook for a minute longer. Cool. Add vanilla and fold in the whipped cream. Freeze.

Yield: 4 quarts

Adding Fruit and Nuts

The preceding recipes are all for vanilla ice cream because vanilla is the basis for all other flavors. By using fruits, nuts, and natural flavorings—and your imagination—and experimenting with small batches, you can produce an astounding number of ice cream flavors. Baskin and Robbins is the most famous of the commercial ice cream firms for the wide variety of flavors they have produced in their laboratories in Burbank, California, using mostly natural flavors. In 1971 when the company celebrated its twenty-fifth anniversary, they had developed 401 flavors.

Fruits and nuts should be added to the mix when it is just beginning to harden in the can—and is just beginning to get hard to turn. To flavor a gallon of ice cream, add 5 cups of fruit to the mix. Use crushed or puréed fruits or cut the fruits into *small* pieces—ice crystals tend to form around large pieces of fruit. A small amount of lemon juice often enhances the flavor of fruits and keeps them from darkening, but don't add too much. Swirled ice creams can be made by swirling a small amount of fruit into the finished ice cream, rather than mixing it into the ice cream mixture.

If you are using fresh fruit, you may want to dribble honey over the fruit and let it set a few hours to bring out the juice. Or, juice or blend some of the fruit to provide a syrup and then mix it with the rest of the whole fruit.

If using nuts in your ice cream, roast them lightly first. A small amount of salt added to the mix intensifies the nut flavor, and nuts are best when used with a stronger background flavor—such as maple or vanilla.

Fruit Sherbet
(except lemon)

6 cups milk
¾ cup honey (sweeten to taste)
4 tablespoons lemon juice

¼ teaspoon salt (optional)
2½ cups fruit juice or pulp

Mix milk and honey. Add lemon juice and salt to fruit juice. Gradually mix fruit mixture with milk by constant stirring. Some curdling may be noticed, but this will have no effect on the sherbet.

Cool and freeze in a manual freezer, using 1 part salt to 5 parts ice.

Yield: 3 quarts

Freezer Ice Creams

The following recipes developed by the Fitness House kitchen at Rodale Press are for ice creams that are made by freezing in ice cube trays in the kitchen freezer. This method yields an ice cream of lower quality than that produced by churning, but it is an easier and less time-consuming substitute.

Blackberry Ice Cream

5 cups fresh blackberries,
    washed
¾ cup honey
4 teaspoons (2 envelopes)
    unflavored gelatin

6 tablespoons water
2 tablespoons lemon juice
3 cups light or heavy cream or a
    mixture of both
2 egg whites

Set aside 2 cups blackberries. Drizzle honey over the remaining 3 cups.

In a small saucepan, sprinkle gelatin over water and set aside for a few minutes to soften. Heat slowly, stirring constantly, to dissolve gelatin.

Blend honeyed blackberries to a purée in an electric blender. Put through strainer to remove most of seeds. Add dissolved gelatin to purée, then the reserved 2 cups blackberries, the lemon juice, and the light and/or heavy cream. Pour into freezer trays and freeze for several hours.

Before serving, let soften at room temperature for about 15 minutes. Add unbeaten egg whites, beat in electric mixer until fluffy and of an ice cream consistency. Serve at once.

NOTE: This ice cream can be refrozen several times. But be sure to beat it again each time before serving.

Yield: 8 to 10 servings

Peach Ice Cream

6 tablespoons honey
1 tablespoon lemon juice
3 cups white peaches, sliced

½ teaspoon gelatin
1 cup heavy cream

Drizzle honey and lemon juice over peaches. Stir to coat them well. Set aside, covered, for a few hours. Strain. There should be about ¾ cup juice.

Soften gelatin in juice and then heat it just enough to dissolve the gelatin. Combine it with the peaches in the container of an electric blender and blend briefly, to chop the peaches coarsely. Refrigerate until mixture begins to set.

Whip cream until stiff and fold into peach mixture. Pour into freezer tray and freeze hard.

Before serving, remove ice cream from freezer and leave at room temperature for 15 minutes or so. Beat with electric beater until ice cream is of correct consistency.

Yield: 4 to 6 servings

Strawberry Sherbet

3 cups strawberries, fresh or
    frozen
½ cup honey
2 teaspoons gelatin

¾ cup strawberry juice
¼ of a lemon
2 egg whites

Wash and halve strawberries. Drizzle honey over them and set aside at least an hour. Drain. There should be ¾ cup juice. If not, add other fruit juice to make this amount. In small saucepan sprinkle gelatin over juice and heat slowly to melt, stirring constantly.

Wash, seed, and cut lemon into quarters. Place one quarter in container of electric blender along with melted gelatin and juice mixture. Blend at high speed until smooth. Add half of the strawberries and blend coarsely. Mash the remaining strawberries and fold them in. Freeze until solid.

Before serving, allow sherbet to soften at room temperature just enough to beat with electric beater. Add unbeaten egg whites and beat until fluffy. Serve immediately.

Refreeze any leftover sherbet and beat again before serving.

Yield: 6 to 8 servings

Storing Ice Cream

Unfortunately, ice cream is not a good keeper. It should be stored at 0°F., and not for more than a month at a time. Store homemade ice cream in plastic freezer containers (not ones you've stored strongly flavored foods in before) or glass jars with tight-fitting lids.

Allow the ice cream to soften slightly before serving.

Dairy Recipes

Frozen Maitre D'Hotel Butter

1 pound lightly salted butter
¼ cup lemon juice
¼ cup very finely minced fresh parsley

Cream butter, lemon juice, and parsley until well blended, chill until firm enough to mold, then shape into 4 logs about 1 inch in diameter. Wrap each log, label, and freeze.

Use on fish, broiled meats, and vegetables. Other fresh, minced herbs may also be incorporated into butters and frozen for later use.

Cooked Cole Slaw Dressing

3 tablespoons honey
1 cup homemade yogurt
1 teaspoon salt

2 eggs, beaten
½ cup tarragon vinegar
1 teaspoon chopped celery

Cook in top of double boiler until it becomes a smooth custard.

Yield: 2 cups

Cooked Fruit Salad Dressing

½ cup apricot juice or any
    desired unsweetened fruit
    juice
½ teaspoon salt

4 tablespoons honey
2 beaten egg yolks
1 cup yogurt
cinnamon or mace to taste

Use double boiler in which you mix juice, salt, and honey. Add egg yolks gradually and beat well. Stir constantly until thickened. Remove from heat and let cool. When cool fold in yogurt and desired spice.

Yield: 1½ cups

Fruit Dressing

1 cup yogurt
2 tablespoons orange juice
1 teaspoon lemon juice
1 teaspoon orange rind
½ teaspoon lemon rind

1 tablespoon honey
¼ teaspoon grated or ground nutmeg
⅛ teaspoon ground mace

Blend all ingredients together in order given. Pour into a container with cover and place in refrigerator.
Serve with fruit salads or gelatin salads.

Yield: 1¼ cups

Honey-Yogurt
Fruit Salad Dressing

½ cup yogurt
1 tablespoon lemon juice

½ cup honey
cinnamon, nutmeg, or mace

Blend all together well, add spices, and use on fruit salad.

Green Onion Dressing

¾ cup thinly sliced green onions (including tops)
1 cup yogurt

2½ cups mayonnaise
2 tablespoons lemon juice
½ teaspoon salt

Combine all ingredients. Store in glass container in refrigerator, covered.

Yield: about 4 cups

Russian Dressing

1 cup yogurt
2 cups mayonnaise

2 cups catsup
4 teaspoons horseradish

Combine yogurt and mayonnaise with wire whisk. Stir in catsup and horseradish.

Store in covered glass jar in refrigerator.

Yield: about 5 cups

Vegetable Lasagne

½ pound lasagne noodles
1 small eggplant, sliced into
   ½-inch rounds
4 tablespoons olive oil
¼ pound zucchini, sliced into
   ¼-inch rounds

3½ cups tomato sauce,
   seasoned with oregano,
   basil, and garlic
1½ cups ricotta cheese
¼ cup grated Parmesan cheese
   (or dried ricotta)

Cook and drain noodles. Spread them out in single layers while preparing remainder of recipe so that they will not stick together.

Broil eggplant slices, brushed with oil, until they are light brown and dry. Simmer zucchini slices in tomato sauce 5 minutes.

Place a layer of noodles in the bottom of an olive-oiled baking dish. Spread with 1 cup of tomato sauce. Add another layer of eggplant, and ¾ cup of ricotta. Pour 1 cup of tomato sauce over all. Repeat layers in the same order, ending with 1½ cups tomato sauce. Dribble 4 tablespoons olive oil over top, and sprinkle with grated cheese.

Bake at 350°F. for ½ hour or until bubbly. Cut into squares while hot.

Yield: 4 generous
servings

Ricotta Icing

1 cup ricotta
¼ cup noninstant dry milk
   powder

2 tablespoons honey
1 teaspoon vanilla extract
orange juice

Blend ricotta, milk powder, honey, and vanilla. Add orange juice, blending in, until icing is of spreading consistency. Spread on cake and refrigerate.

Yield: icing for 1
8- or 9-inch cake

Thin Ricotta Omelets

For each omelet you'll need:

1 egg
2 teaspoons milk

salt and pepper to taste
2 tablespoons ricotta cheese

Beat egg with milk, and add seasonings. Heat an oiled 6- or 7-inch pan, and pour in egg mixture, tilting to cover bottom rather thinly. Cook until just set. Place dabs of ricotta on omelet. Roll up like a cigar, and brown slightly. Remove, and keep warm until ready to serve.

Serve two per person with whole-grain bread and a salad for a light supper.

Easy Ricotta Pudding

½ cup ricotta
1 tablespoon honey (if desired)

1 tablespoon jam

Mix ricotta with honey, if used. Serve in an attractive glass dish with a dollop of jam in the center.

Yield: 1 serving

Wheat Germ and Ricotta Tart

Pastry for a 2-crust pie, made from your favorite recipe and sweetened slightly.

For the filling, you'll need:

2 cups hot milk
2½ cups wheat germ
1½ pounds ricotta cheese
¾ cup honey
6 eggs, separated
½ cup chopped honeyed fruit
rind or chopped, pitted
dates

1 tablespoon fresh grated
orange rind
1 tablespoon vanilla extract
1 egg, beaten with 1 tablespoon
water

Divide pastry into 2 unequal balls. Roll out each ⅛ inch thick, and use the larger to line a deep 10-inch pie pan. Cut the smaller into ¾-inch-wide strips for lattice topping.

Heat milk, stir in wheat germ, and set aside to soften 5 minutes. Mix ricotta, honey, egg yolks, fruit rind or dates, and flavorings. Stir in wheat germ mixture, and fold in the stiffly beaten egg whites. Turn into the prepared crust, lattice pastry strips over top, brush with beaten egg and water, and bake in preheated 350°F. oven for about 1 hour, or until firm in center and top is nicely browned. If crust browns too much before the custard is firm, cover with foil. After removal from oven, the custard will continue to cook and firm.

Yield: 1 10-inch tart

Frozen Pizza

Prepare your favorite whole-wheat yeast bread dough recipe, and let dough rise until double. Pound or roll into a very thin round, turning up the edge about ½ inch. Brush the dough's surface with olive oil.

Drain (thawed) frozen or canned whole tomatoes, or peel and seed fresh tomatoes; cut them into pieces and spread evenly over the dough. Or you could use home-canned tomato sauce. Sprinkle with a little dried oregano, salt and black pepper, and any other seasonings you prefer. Spread grated mozzarella cheese over the entire surface of the pie, using about ½ pound for a 12-inch pie.

Bake in a hot oven (400°F.) for about 5 minutes, enough to stiffen the dough but not long enough to thoroughly melt the cheese.

Let cool, and cover cheese with a layer of waxed paper. Wrap in heavy-duty aluminum foil. Label, date, and freeze.

When ready to serve: Garnish the pie with your choice of green pepper rings, chopped onions, extra cheese, cut-up anchovies, sliced mushrooms, and sausage slices or ground meat (both of which should be browned slightly and drained of excess fat). Bake in a 400°F. oven for 10 minutes, then place the pie under the broiler to melt and brown the cheese. Cut into wedges and serve.

Carrot-Coconut Salad

2 cups shredded raw carrots	¼ cup mayonnaise
½ cup flaked or coarsely shredded coconut	¼ cup yogurt
	2 tablespoons fresh lemon juice
½ cup pitted dates,* coarsely chopped	1 tablespoon honey (optional)
	⅛ teaspoon salt

Scrub carrots and shred. In medium bowl combine carrots, coconut, and chopped dates. Mix mayonnaise, yogurt, lemon juice, honey, and salt in a small bowl; blend well.

* Raisins may be substituted for dates.

Pour dressing over salad and blend ingredients together. Chill before serving.

Line serving dish with greens; spoon salad in center and serve.

Yield: 5 servings

Cucumber Salad

2 to 3 cucumbers
salt
pepper

3 tablespoons yogurt
dill
parsley

Slice young cucumbers very thin, place in bowl and add salt and pepper to taste. Add three tablespoons of plain yogurt and some chopped dill. Garnish with sprigs of parsley after it has set an hour or more, then serve. If you prefer, use sour cream in place of yogurt.

Yield: 4 servings

Minted Cucumber Salad

1 teaspoon dried mint
½ teaspoon garlic powder
1 cup yogurt
1 large cucumber, peeled and
    sliced thin

dark green lettuce leaves
4 red radishes, chopped

Mix mint and garlic powder into yogurt. Distribute cucumber slices on crisp leaves of lettuce and pour yogurt on top. Garnish with chopped radishes. Serve cold.

Yield: 2 to 3 servings

Stroganoff

2 medium onions, peeled and
    sliced
¼ cup oil
1½ pounds round steak, cut
    into 1-inch strips
1½ cups fresh sliced
    mushrooms
3 tablespoons water

½ teaspoon salt
few grains pepper
1 tablespoon soy sauce
2 tablespoons flour
1 cup yogurt
sprinkle of nutmeg

Cook onions in oil until tender, stirring occasionally. Add round steak and brown lightly on both sides. Add mushrooms, water, salt, pepper, and soy sauce. Cover and simmer slowly until tender (about 5 minutes). During that time mix together flour and yogurt. Remove round steak and onions from skillet when done. Take ¼ cup mushroom sauce from skillet and blend smoothly with yogurt and flour. Pour this mixture gradually into skillet, mixing well. Heat over low heat, stirring constantly, until thickened. Add round steak and onions just long enough to heat. Garnish with nutmeg.

Yield: 6 servings

Beets with Yogurt

4 to 6 beets
bay leaf
parsley

basil
¼ cup yogurt
1 teaspoon honey

Slice beets very thin or grate, place in steamer over boiling water, add the beet tops along with a bay leaf, a sprig of parsley, and one of basil. Let steam over boiling water for about 5 minutes. Remove the bay leaf and place beets in dish. Mix yogurt with honey, pour over beets and serve.

Yogurt Cheesecake

Crust:
1 cup bread crumbs (dry)
1 tablespoon honey

4 tablespoons oil (corn or soy)
½ teaspoon cinnamon

Combine ingredients and pat into spring-form pan, covering bottom of pan and bringing up sides about 1½ inches.

Filling:
¼ cup cornstarch
1 cup yogurt
1 pound cottage cheese (low fat variety if possible)
rind of one lemon, grated

1 tablespoon lemon juice
¼ teaspoon salt
½ cup honey
2 teaspoons vanilla
4 eggs, separated

Using a blender, blend cornstarch and yogurt to dissolve cornstarch. Then add cottage cheese and blend until smooth. Add lemon rind, juice, salt, honey, vanilla. Blend to combine.

Beat egg yolks in mixing bowl until thick and lemon colored. Add cheese mixture and mix well.

Beat egg whites until stiff, but not dry. Fold into cheese mixture and pour into prepared pans. Bake at 300°F. for 50 minutes or until center is firm. When finished, turn off heat and with oven door closed, let cake cool in oven for 1 hour or more. When cool, loosen sides of spring pan and remove cake.

NOTE: Cheesecake may be topped with a cornstarch-thickened fruit such as unsweetened canned pineapple, or fresh strawberries, or blueberries.

Yield: 12 servings

Quick and Easy Cheesecake

½ cup water
⅓ cup noninstant dried milk
   powder
⅓ cup honey
4 eggs

1 tablespoon lemon juice
1 teaspoon vanilla
¼ cup flour
1 pound cottage cheese

Combine ingredients in blender. Whirl until smooth. Pour into 8-inch spring-form pan. Bake in a preheated 250°F. oven for 1 hour. Turn off oven, and leave the cheesecake in the oven for 1 hour. Then cool.

Yield: 1 8-inch cake

Yogurt Cream Cheese Pie

½ pound yogurt cream cheese
⅔ cup yogurt
2 tablespoons honey

1 teaspoon pure vanilla extract
1 cup coconut

Blend together the cream cheese, yogurt, honey, and vanilla. Fold in all but 1 tablespoon of the coconut. Pour cream cheese mixture into a coconut or chopped nut pie shell and garnish with remaining coconut.

Yogurt Dip

Add to ½ cup yogurt and beat together slightly:

2 to 4 tablespoons lemon juice
  or apple cider vinegar
  (omit if yogurt is tart)
1 finely chopped green onion
  (or 1 grated onion)

½ teaspoon kelp
½ to 1 teaspoon paprika
1 minced garlic clove

Yogurt-Fruit Freeze

1 cup yogurt
1 tablespoon lemon juice
½ cup pineapple or orange
  juice (unsweetened)

¼ cup dried and stewed
  apricots or peaches
1 tablespoon honey
2 egg whites

Mix the yogurt, lemon juice, orange or pineapple juice, stewed fruit, and honey together until all ingredients are blended. Put the mixture into a freezer tray and freeze until firm. Take it out, put into a bowl, and stir until mixture is smooth and creamy.

Beat the egg whites until stiff and fold them into the fruit mixture. Return to freezer tray and freeze.

Yield: 3 to 4 servings

Meats and Fish

Preparing Beef, Veal, Lamb, and Pork for Storage

Obviously, there's quite a bit of work involved in getting meat on the hoof slaughtered, dressed, and cut before it can be wrapped and frozen, canned, or cured and smoked. A butcher can do all this preliminary work for you, or, if you have the proper equipment and the know-how, you can do the whole job yourself.

Doing Your Own Butchering

If you're going to butcher meat yourself, you have a lot of things to consider. First, do you have a place in which to butcher? It must be clean, cool, dry, and well ventilated, with a ready source of water and a stove for heating water. A garage, basement, or outbuilding can be converted for this purpose, providing there is enough head room to hoist the carcass. Many farmers do their slaughtering and dressing outdoors and then bring the carcass inside to cut. If they do work outdoors, they usually choose a dry, cool fall day on which to do the job. If you are working indoors and the temperature of the room you are using is higher than 38°F., a large cooler will be needed in which to chill the meat, and in the case of beef and lamb, also to age it.

Equipment

The equipment to have on hand for butchering includes something to hoist and suspend the carcass, like strong hooks, a brace extending from an outbuilding or tree (if you're working outdoors), or a heavy rod suspended from the ceiling. A block and tackle, windlass, or chain hoist would be very helpful in hoisting the carcass, especially if you're butchering steer. Rope, buckets, a thermometer to measure the temperature of water, a meat thermometer, a stunning instrument, cleaver, meat saw, hog scraper, hand hooks, whetstone, steel, and a good set of sharp knives are necessary. You may want to have access to a meat grinder, also. If you're butchering hogs, you'll need a scalding vat or watertight barrel. For slaughtering lambs, a low bench or box is necessary.

Slaughtering and dressing animals is an exacting job, and space doesn't allow us to explain the specifics of butchering in this book. For good, detailed information, we refer you to the booklets published by the U.S. Department of Agriculture on the subject. These booklets, which contain lists of equipment, diagrams, and easy-to-follow directions, can be obtained for a small charge from your state or county agricultural station or the Superintendent of Documents at the U.S. Government Printing Office in Washington, D.C. There are also some good books on butchering meat that you might want to consult. You'll find a list of these books on page 417.

Chilling Meat

After the animal has been slaughtered and dressed, the carcass must be chilled promptly to rid it of its animal heat. If not chilled, the meat will spoil more rapidly because destructive bacteria thrive at normal body temperatures. Chilling also makes the meat easier to cut. To hasten chilling, cut off excess fat in the crotch and split the carcass. Hang the carcass to chill in a well-ventilated, cold area with a temperature between 32° and 40°F. A lamb or veal carcass kept in an area with a temperature between 32° and 40°F. should be chilled to 40°F. within 24 hours. A beef carcass may require 40 or more hours to chill to this temperature. A meat thermometer inserted into the thickest part of the carcass will show you when the meat is properly chilled.

Slaughtering on the farm is best done on a fall or early spring afternoon so that the carcass can begin to cool down during the night when temperatures just above freezing prevail. (If you will be curing and smoking your meat it is better to slaughter in the fall. Cold winter temperatures help to retard growth of bacteria responsible for spoilage.) If night temperatures should rise above 40°F., cut the carcass in half and then into a few big pieces and immerse them in clean barrels filled with water, ice, and about 3 pounds of common salt. This solution, which is colder than ice water but warmer than solid ice, will help chill meat, but will not bring it down to freezing temperatures.

Never try to chill a carcass quickly by exposing it to freezing temperatures or by packing it in solid ice. A carcass should never freeze because the thin layer of ice that forms on the meat surface will prevent the proper escape of animal heat from the center of the meat. If a carcass should freeze, thaw it slowly at temperatures no greater than 40°F.

Aging Beef and Lamb

Beef and lamb are aged after chilling. This means holding the meat at a temperature of from 32° to 38°F. for a few days to a few weeks to increase tenderness, and in some meats, to bring out the flavor. Aging is best done in a refrigerated area where the temperature does not vary a great deal. If a cooler is not available, aging may be done in a clean, cool, and well-ventilated place, free from animals and insects.

The length of the time meat should be aged varies. Lamb can be held for 1 to 3 days, while older sheep (mutton) can be aged 5 to 7 days. Beef with little external fat should not be held for more than 5 days after slaughter. Beef with a good amount of external fat can be aged 5 to 18 days. Fat acts as a protection against bacteria in the air that would grow on the meat surface if it were left exposed.

People who plan to store much of their lamb or beef in a freezer for more than 6 months should limit the aging period. Experiments at Pennsylvania State University have shown that the length of the aging period has a direct bearing on the storage life of meat because it permits oxygen absorption by the exposed fat.* As we have mentioned previously, the more oxygen absorbed by the fat, the quicker the rate of rancidity. Aged meat shows higher peroxide values and shorter storage life than 48-hour chilled meat. In addition, the experiments at Penn State showed that aging does not influence the tenderness of meat that is frozen more than a month. Experimenters found that although aged meat is slightly more tender during the first month of storage, this advantage disappears in subsequent months, when the aged and the 48-hour chilled meat are on a par for tenderness. Pork and veal should never be aged, but frozen, cooked, or cured as soon as the animal heat is gone.

Cutting Meat

Whether you are cutting your own meat or having a butcher cut to order, you'll want to get those cuts that best suit your family's

* P. Thomas Ziegler, THE MEAT WE EAT (Danville, Illinois: The Interstate Printers & Publishers, Inc.), 1965.

BEEF CHART

RETAIL CUTS OF BEEF —

WHERE THEY COME FROM AND HOW TO COOK THEM

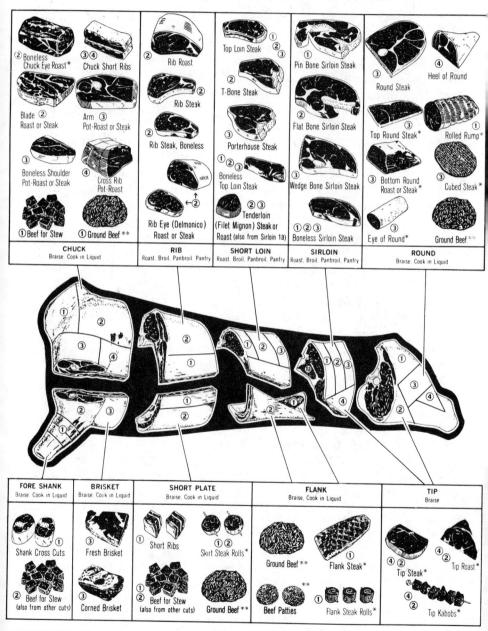

CHUCK
Braise. Cook in Liquid

RIB
Roast. Broil. Panbroil. Panfry

SHORT LOIN
Roast. Broil. Panbroil. Panfry

SIRLOIN
Roast. Broil. Panbroil. Panfry

ROUND
Braise. Cook in Liquid

- ② Boneless Chuck Eye Roast*
- ③④ Chuck Short Ribs
- ② Blade Roast or Steak
- ③ Arm Pot-Roast or Steak
- ③ Boneless Shoulder Pot-Roast or Steak
- ④ Cross Rib Pot-Roast
- ① Beef for Stew
- ① Ground Beef **

- ② Rib Roast
- ② Rib Steak
- ② Rib Steak, Boneless
- ← ②
- Rib Eye (Delmonico) Roast or Steak

- ① Top Loin Steak ②③
- ② T-Bone Steak
- ③ Porterhouse Steak
- ①②③ Boneless Top Loin Steak
- ②③ Tenderloin (Filet Mignon) Steak or Roast (also from Sirloin 1a)

- ① Pin Bone Sirloin Steak ②③
- Flat Bone Sirloin Steak
- Wedge Bone Sirloin Steak
- ①②③ Boneless Sirloin Steak

- ③ Round Steak
- ④ Heel of Round
- ② Top Round Steak*
- ① Rolled Rump*
- ③ Bottom Round Roast or Steak*
- ③ Cubed Steak*
- ③ Eye of Round*
- Ground Beef **

FORE SHANK
Braise. Cook in Liquid

BRISKET
Braise. Cook in Liquid

SHORT PLATE
Braise. Cook in Liquid

FLANK
Braise. Cook in Liquid

TIP
Braise

- Shank Cross Cuts ①
- ① Beef for Stew (also from other cuts)

- ③ Fresh Brisket
- ③ Corned Brisket

- ① Short Ribs
- ①② Skirt Steak Rolls*
- ①② Beef for Stew (also from other cuts)
- Ground Beef **

- Ground Beef **
- ① Flank Steak*
- Beef Patties
- ① Flank Steak Rolls*

- Tip Steak ④②
- ④② Tip Roast
- ④② Tip Kabobs*

*May be Roasted, Broiled, Panbroiled, or Panfried from high quality beef
**May be Roasted (Baked), Broiled, Panbroiled, or Panfried

Courtesy of National Live Stock and Meat Board

VEAL CHART

RETAIL CUTS OF VEAL —

WHERE THEY COME FROM AND HOW TO COOK THEM

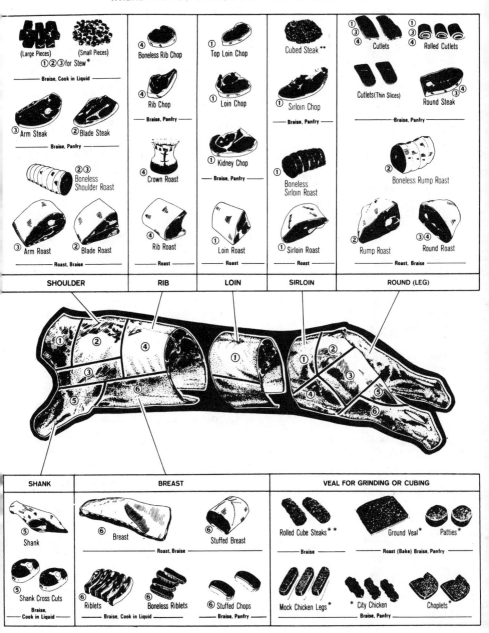

SHOULDER	RIB	LOIN	SIRLOIN	ROUND (LEG)

SHOULDER

(Large Pieces) (Small Pieces)
① ② ③ for Stew *
— Braise, Cook in Liquid —
③ Arm Steak   ② Blade Steak
— Braise, Panfry —
② ③ Boneless Shoulder Roast
③ Arm Roast   ② Blade Roast
— Roast, Braise —

RIB

④ Boneless Rib Chop
④ Rib Chop
— Braise, Panfry —
④ Crown Roast
Rib Roast
— Roast —

LOIN

① Top Loin Chop
① Loin Chop
— Braise, Panfry —
① Kidney Chop
— Braise, Panfry —
① Loin Roast
— Roast —

SIRLOIN

Cubed Steak **
① Sirloin Chop
— Braise, Panfry —
① Boneless Sirloin Roast
① Sirloin Roast
— Roast —

ROUND (LEG)

① ③ ④ Cutlets   ① ③ ④ Rolled Cutlets
Cutlets (Thin Slices)   ③ ④ Round Steak
— Braise, Panfry —
② Boneless Rump Roast
② Rump Roast   ③ ④ Round Roast
— Roast, Braise —

SHANK	BREAST	VEAL FOR GRINDING OR CUBING

SHANK

⑤ Shank
⑤ Shank Cross Cuts
Braise, Cook in Liquid —

BREAST

⑥ Breast
— Roast, Braise —
⑥ Stuffed Breast
⑥ Riblets   ⑥ Boneless Riblets   ⑥ Stuffed Chops
— Braise, Cook in Liquid —   — Braise, Panfry —

VEAL FOR GRINDING OR CUBING

Rolled Cube Steaks * *   Ground Veal *   Patties *
— Braise —   — Roast (Bake) Braise, Panfry —
Mock Chicken Legs *   * City Chicken   Choplets *
— Braise, Panfry —

*Veal for stew or grinding may be made from any cut
**Cube steaks may be made from any thick solid piece of boneless veal

Courtesy of National Live Stock and Meat Board

LAMB CHART

RETAIL CUTS OF LAMB —

WHERE THEY COME FROM AND HOW TO COOK THEM

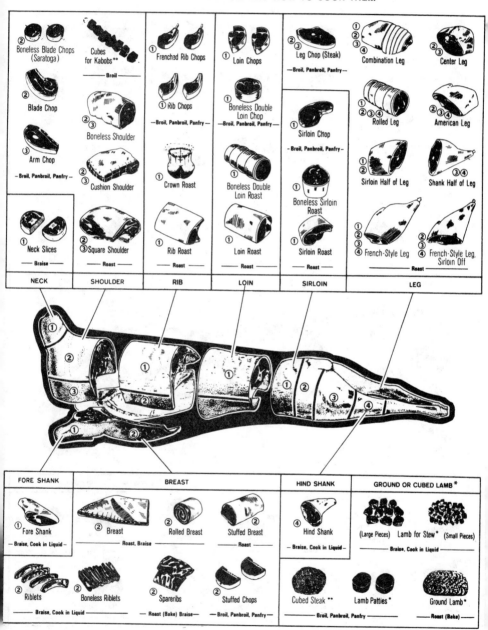

Courtesy of National Live Stock and Meat Board

*Lamb for stew or grinding may be made from any cut
**Kabobs or cube steaks may be made from any thick solid piece of boneless Lamb

PORK CHART

RETAIL CUTS OF PORK —

WHERE THEY COME FROM AND HOW TO COOK THEM

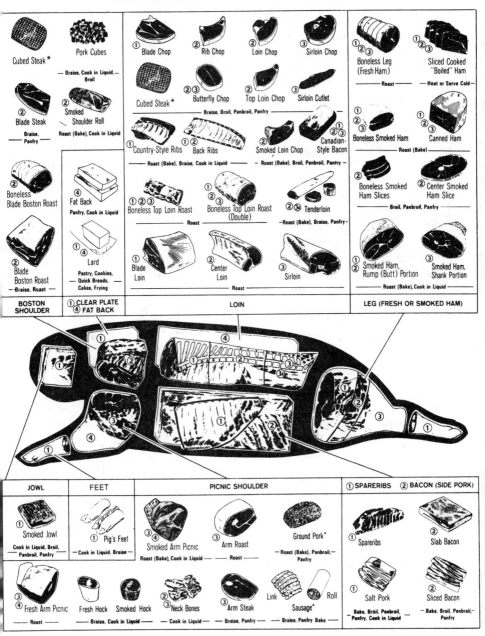

BOSTON SHOULDER	① CLEAR PLATE ④ FAT BACK	LOIN	LEG (FRESH OR SMOKED HAM)

Cubed Steak *

Pork Cubes

— Braise, Cook in Liquid, — Broil

② **Blade Steak** — Braise, Panfry —

② **Smoked Shoulder Roll** — Roast (Bake), Cook in Liquid

② **Boneless Blade Boston Roast**

④ **Fat Back** — Panfry, Cook in Liquid —

② **Blade Boston Roast** — Braise, Roast —

④ **Lard** — Pastry, Cookies, — Quick Breads, — Cakes, Frying

① **Blade Chop**   ② **Rib Chop**   ② **Loin Chop**   ③ **Sirloin Chop**

② **Cubed Steak ***   ②③ **Butterfly Chop**   ② **Top Loin Chop**   ③ **Sirloin Cutlet**

— Braise, Broil, Panbroil, Panfry —

①② **Country-Style Ribs**   ①② **Back Ribs**   ③ **Smoked Loin Chop**   ②③ **Canadian-Style Bacon**

— Roast (Bake), Braise, Cook in Liquid —   — Roast (Bake), Broil, Panbroil, Panfry —

①②③ **Boneless Top Loin Roast**   ①②③ **Boneless Top Loin Roast (Double)**   ②③ **Tenderloin**

— Roast —   — Roast (Bake), Braise, Panfry —

① **Blade Loin**   ② **Center Loin**   ③ **Sirloin**

— Roast —

① **Boneless Leg (Fresh Ham)**   ① **Sliced Cooked "Boiled" Ham**

— Roast —   — Heat or Serve Cold —

① **Boneless Smoked Ham**   ① **Canned Ham**

— Roast (Bake) —

② **Boneless Smoked Ham Slices**   ② **Center Smoked Ham Slice**

— Broil, Panbroil, Panfry —

① **Smoked Ham, Rump (Butt) Portion**   ③ **Smoked Ham, Shank Portion**

— Roast (Bake), Cook in Liquid —

JOWL	FEET	PICNIC SHOULDER	① SPARERIBS ② BACON (SIDE PORK)

① **Smoked Jowl** — Cook in Liquid, Broil, Panbroil, Panfry

① **Pig's Feet** — Cook in Liquid, Braise —

③④ **Smoked Arm Picnic** — Roast (Bake), Cook in Liquid —

③ **Arm Roast** — Roast —

Ground Pork * — Roast (Bake), Panbroil,— Panfry

③④ **Fresh Arm Picnic** — Roast —

Fresh Hock   **Smoked Hock** — Braise, Cook in Liquid —

③ **Neck Bones** — Cook in Liquid —

③ **Arm Steak** — Braise, Panfry —

Link   **Roll**
Sausage * — Braise, Panfry Bake —

① **Spareribs**   ② **Slab Bacon**

Salt Pork — Bake, Broil, Panbroil, — Panfry, Cook in Liquid —

Sliced Bacon — Bake, Broil, Panbroil, — Panfry

---

*May be made from Boston Shoulder, Picnic Shoulder, Loin, or Leg

Courtesy of National Live Stock and Meat Board

needs. Ask yourself these questions before the cutting begins: Do I want more steaks or chops than roasts, or would I prefer it the other way around and get more roasts? Do I want some meat left around the bones for meaty soup bones, or would I prefer it removed and ground? Do I want some meat cut into stewing chunks or have it ground for hamburger meat? You do have these choices because there are several different ways certain parts of the carcass may be divided.

The size of the roasts, the thickness and number of steaks and chops, and the amount of ground beef and stewing beef are determined by the size and preferences of your family as well as your method of storage.

For freezing, you'll want your cuts as regularly square-shaped as possible for easy wrapping and packing. If you're smoking, keep your hams whole and your bacon in large pieces. Canning requires smaller pieces of well-trimmed meat. If you're canning your meat or freezer space is limited, you should consider boning as much meat as is practical. Boned cuts fit into jars and cans easily. They also wrap better for the freezer and can be packed tighter, with less chance of the wrap's tearing when no large bones are protruding. Depending upon the cut, boned meat can be rolled and tied, ground for hamburger meat or sausages, cut for boneless steaks and easy-to-carve roasts, or prepared as stewing meat or bacon.

	Trimmed cuts	Yield	Live weight	Carcass weight
		Pounds	Percent	Percent
Approximate yields of trimmed beef cuts from animal having a live weight of 750 pounds and a carcass weight of 420 pounds	Steaks and oven roasts	172	23	40
	Pot roasts	83	11	20
	Stew and ground meat	83	11	20
	Total	338	45	80

	Trimmed cuts	Yield	Weight of fore-quarters
		Pounds	Percent
Approximate yields of trimmed beef cuts from dressed fore-quarters weighing 218 pounds	Steaks and oven roasts	55	25
	Pot roasts	70	32
	Stew and ground meat	59	37
	Total	184	94

	Trimmed cuts	Yield	Weight of hind-quarters
		Pounds	Percent
Approximate yields of trimmed beef cuts from dressed hind-quarters weighing 202 pounds	Steaks and oven roasts	117	58
	Stew, ground meat, and pot roasts	37	18
	Total	154	76

Courtesy of U.S. Department of Agriculture

Have the larger bones cracked for making stocks. You may wish to make very concentrated soup stock from the bones and freeze this broth instead of wrapping and freezing awkwardly shaped, clumsy bones that take up much space. If you are canning your meat, the only practical way to save the juices and gelatinous extractives from the bones is to make soup stock and can it.

Good information on various ways to cut beef, veal, lamb, and pork can be obtained from your state or local agricultural extension station or the Superintendent of Documents. This information can also be found in most of the books listed on page 417.

	Trimmed cuts	Yield	Live weight	Carcass weight
		Pounds	Percent	Percent
Approximate trimmed pork cuts from a hog having a live weight of 225 pounds and a carcass weight of 176 pounds	Fresh hams, shoulders, bacon, jowls ..	90	40	50
	Loins, ribs, sausage	34	15	20
	Total	124	55	70
	Lard, rendered ..	12	15	27

	Trimmed cuts	Yield	Live weight	Carcass weight
		Pounds	Percent	Percent
Yields of trimmed lamb cuts from a lamb having a live weight of 85 pounds and a carcass weight of 41 pounds	Legs, chops, shoulders	31	37	75
	Breast and stew	7	8	15
	Total	38	45	90

Courtesy of U.S. Department of Agriculture

If You're Having Your Meat Butchered

Not too many people who raise a few meat animals a year for family use do their own butchering any more. Instead, they hire a butcher to do all or part of the work for them.

Judy and Arnold Voehringer, of Kempton, Pennsylvania, raise their own beef cattle organically, and until recently, also raised hogs. They haven't found it economical, on their small-scale operation, to butcher their meat themselves. A local butcher comes for the animals and takes them back to his shop in a truck to slaughter, dress, and cut them. The Voehringers take the cut meat home to wrap and freeze. Although wrapping the meat from a 900-pound dressed steer is an all-day job for one person, the Voehringers feel it is enough of a savings, money-wise, to justify all the work; they save 7 cents per pound by wrapping and quick-freezing the meat themselves. Besides, there are usually enough friends and family members around on wrapping days to make the job easier.

The Voehringers feel that it is very important to take their animals to a butcher with whom they can feel confident. For this reason, they prefer to deal with local butchers whom they can get to know

Beef

Name ....

Address ....

Cost of Hauling.................... Slaughtering ....

Live Weight ....

Cutting Date.......................... Cut by.............................

Whole Carcass...................... Wrapped by ....

Front .... Hind .... Side..................

Weight .... Price ....

Number of People in Family...

You Want:

Round ....

Rib ....   Pickup Date ....

Roasts ....   Phone ....

Steaks ....   We Wrap & Freeze?

Soup Bones ....

Boiling Beef ....

Hamburger ....

Variety ....

personally. They've learned that small operations will give them—and their steer—plenty of attention.

This personal attention was especially important to the Voehringers when their first few steers were slaughtered. The butcher took time to explain the different cuts of meat to them and give them tips for wrapping and freezing at home. Judy still helps the butcher when he is cutting the meat so that she can tell him how she wants the cuts: how thick to make the steaks, what size roasts she wants, how much fat to trim off, and so on. It is against the policy of most larger meat companies to permit customers in the cutting room. Rather, they cut the meat according to order sheets customers fill out when their animals are brought in for slaughter. Typical order sheets might look like the ones here.

Pork

Name ....

Address ....

Date .... Hogs ....

Weight .... Live .... Dressed ....

Back Bone .... Pork Chops...............

Number of People in Family...

You Want:

Shoulders ....

                     Us ....

Hams...................... Cure

                     You ....

Bacon ....

Knuckles....................

Stomachs ....

Sausage.....................

Scrapple ....   Our Dishes ....

                     Their Dishes ....

Pudding ....   Dishes ....

                     Casings ....

Variety ....

Lbs. of Lard................   Our Cans ....

                     Their Cans ....

Lbs. of Sausage ....

Beef for Sausage ....

Pickup Date ....

Phone ....

We Wrap & Freeze?

Although filling out such an order form will allow you to choose pretty much how you want your meat cut, you have more of a say in the matter when you're standing next to the butcher during the whole process.

For practical purposes, most large meat operations collect the beef to be ground from all the steers butchered that day and put it together in a large meat grinder. What comes out is a mixture of meat from all the carcasses. The butcher makes note of the amount of ground meat belonging to each customer so that each gets his fair share, but certainly the ground meat that each customer gets is not all from his own steer. This doesn't matter to most farmers because the beef from one steer is almost the same as that from the next, but it does make a difference to the farmer who raises his steers organically. The Voehringers prefer working with small meat operations because their butchers grind the meat from each steer separately.

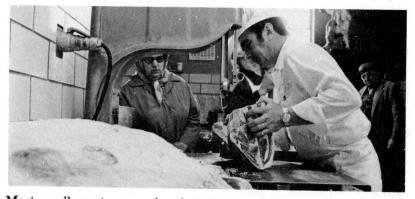

Most small meat companies that do private butchering invite their customers into the cutting room so that the customers can watch the butcher and tell him just how they want their meat cut.

Making Headcheese, Liver Sausage, and Scrapple and Rendering Lard

Early settlers could not afford to waste any part of their butchered hogs; the slaughtered animals were too valuable to them. They ate some of the pork fresh; cured the hams, shoulders, and bacons; and made sausage out of the lean meat scraps. The feet were pickled, the tails were used for stew, the fat was cut off and rendered for lard, and the heads, skins, organs, and bones were boiled to make dishes like headcheese, liver sausage, and scrapple.

Headcheese

Prepare the hogs' heads for boiling by removing any hairs or bristles that remain. Cut out the eyes and quarter the heads. Let the cut-up heads soak in a pot of fresh water at least 7 hours to remove the remaining blood. Then rinse the head thoroughly in running water. Put the heads in a large, heavy pot and add any other scraps you wish, like bones, hearts, and skins. Before putting the skins in the pot, put them in a sack made of a double thickness of cheesecloth so that they can be easily removed when they are tender. Skins are tender when they are easily pierced with your fingers. Cover the head and scraps with water and simmer until meat slips from the bones.

Remove the skins and grind them using a ⅛-inch-hole plate. Prick the meat from the bones and grind it with a ½-inch-hole plate. Mix the ground meat and ground skin together with enough broth to make the mixture the consistency of a soft cake batter (the remaining broth may be saved to make liver sausage or scrapple). Return the mixture to the pot and add your spices. To each hog's head add the following:

1 tablespoon ground mixed spices (this may include garlic, savory, and onion powder, or sweet marjoram and ground cloves)	¼ teaspoon red pepper (optional) 2 tablespoons black pepper 1 crushed bay leaf 2 tablespoons salt

Bring the mixture to a boil and then remove from heat. Now you're ready to form the loaves. Minnesota homesteader Joan Allard shares with us here her tips for making a loaf easily:

Line a container with damp cheesecloth large enough to be tied over the top after it is filled. The bottom layer should be made up of the fattest pieces and skin, with the leaner pieces in the center of the loaf. When the bowl is full, strain boiling broth to the top, gather the edges of the cloth to the center and tie securely. Set this container in a larger one. Place a saucer or lid on the meat and weight it down heavily with cans of food, filled jars, or a cleanly wrapped brick. The excess juices will be pressed out into the larger bowl and can be used in soups or as stock. After unwrapping the pressed meat, rewrap it in aluminum foil or another nonporous material and keep it refrigerated.

Headcheese seems to get firmer with time—but it does not keep well beyond a few weeks. One way you can preserve it longer than usual is to soak it in a salt brine for 48 hours after it is formed. Keep it in the cheesecloth to do this, and weight it down under the liquid at all times.

To best preserve headcheese (or any other sausage or loaf meat), can it at 15 pounds pressure for 90 minutes. Ladle the meat mixture into the jars no more than 1 inch from the top before adjusting seals and pressure canning. (See page 387 for complete canning instructions.) Store it in a cool, dark place as you would for any other canned food. When you are ready to use the canned meat, bring it to a boil and simmer it for at least 20 minutes. Then proceed to serve it, or form a neat slicing chilled loaf as directed above.

Pressed meat dishes similar to headcheese are made using other cuts and types of meat than a hog's head. All are similar in that they are loaves of chopped and cooked meat jellied with their broth and chilled before slicing. The following recipes were also given to us by Joan Allard.

Sylte
(Swedish veal and pork loaf)

4 pounds pork shoulder or similar cut	1 diced onion
3 pounds veal shoulder	1 teaspoon salt
water	1/4 teaspoon pepper
1 teaspoon whole allspice	several whole black
2 bay leaves	peppercorns

Cook meat with spices in boiling water. Use just enough water to barely cover the ingredients. Take the meat out, remove the bones, and chop the meat coarsely. Boil the stock down to 2 or 3 cups to be poured on the loaf when it is formed as directed above.

Potted Hough
(Scottish all beef loaf)

3 pounds beef shank with bones and marrow
1 veal knuckle bone
1 bay leaf and 8 to 10 peppercorns

Follow directions as for making Sylte and form loaves as directed above.

Souse
(Pressed meat with the tang of vinegar)

3 pigs' feet
5 pigs' hocks
1 pound beef
1½ pounds veal
water

1½ tablespoons salt
4 onions
1 teaspoon black pepper
1 teaspoon ground allspice
2 cups vinegar

Barely cover meat with water, add salt, and boil until well done. Pull meat off the bones and grind or chop coarsely. Strain the broth. Replace the meat in it along with the onions and the rest of the seasonings. Let it come to a good rolling boil. Turn it off and then add the vinegar. Let cool and form as directed above.

Zolca
(Jellied pigs' feet)

4 pigs' feet, cut long and
    cleaned
1 veal knuckle
1 small onion, chopped
1 garlic clove, minced
1 tablespoon salt

water
¼ teaspoon sage
1 bay leaf
5 peppercorns
2 tablespoons vinegar

Simmer meat in pot with water to cover with all the other ingredients except the vinegar. Cook until meat begins to fall from the bones. Strain the liquid, chop the meat, and skim off any fat from the broth. Put the meat in a mold and boil the broth down to about one half its original volume. Add the vinegar. Pour this mixture over the meat, weight and chill it.

Rendering Lard

A 225-pound hog will yield about 30 pounds of fat that can be rendered into fine shortening for pastries, biscuits, and frying. The sheet of fat lying just inside the ribs makes the best quality snowy white lard. This "leaf" fat renders most easily, too—and is 90 percent fat. The "back" fat, a thick layer just under the skin, is almost as good, giving about 80 percent of its weight in lard. Far inferior is the visceral fat, which is often dark and off-flavor. Since these various types of fats render at different rates, melt them separately. You can blend them just before storing if you wish.

A slow fire and a heavy pot that conducts heat evenly are most important in making lard. Put ¼ inch of water in the pot to keep the fat from scorching at first. Remove any fibers and bloody spots from the fat and cut it into very small pieces. Put a shallow layer of fat in the pot. Add absolutely no salt or other spices. When the first layer of fat has started to melt, add more. Do not fill the kettle to the top—it can boil over too easily. Stir frequently and keep the fire low.

The temperature of the lard will be 212°F. at first, but as the water evaporates the temperature will rise. 255°F. is the point at which the lard is ready for putting up. Be forewarned that this will take a long time at low heat. As the lard renders, the cracklings (brown bits of crispy fried fat that do not render) will float to the surface. When the lard is almost done and the cracklings have lost the rest of their moisture, they will sink to the bottom.

At this point, turn off the heat and allow the lard to settle and cool slightly. Then carefully dip the liquid off the top into clean containers. Strain the cracklings and residual liquid through cheesecloth or a fine metal sieve. Fill the containers to the top—the lard will contract quite a bit while cooling. Chill as quickly as possible for a fine-grained shortening.

Air, light, and moisture can make lard rancid and sour. So after it has been thoroughly cooled, cover the containers tightly and store them in a dark, cool area. If the water was completely removed in rendering and the lard was chilled thoroughly before capping there will be no souring.

The residual of cracklings are a favorite country treat. Drain them, add salt, and eat the crispy bits as they are. Or make a spread by chopping them finely with onion, salt, pepper, and other seasonings and simmering them in ½ cup of white wine or broth until they are thick and bubbly. Then pack the mixture in a container.

Cracklings can also be used like bacon bits to season eggs and vegetables. You can also add ½ cup of cracklings to your favorite recipe to make "cracklin" biscuits, cornbread, or other quick breads.

Liver Sausage

Liver sausage is a variation of headcheese. It is made by adding cooked pork livers to the cooked heads, tongues, skins, boned meat scraps, and broth. The livers should not constitute more than about 20 percent of the ingredients by weight.

To make liver sausage, cook 3 pounds of liver for 10 to 20 minutes, until done. Do not cook the liver more than 20 minutes because it will become crumbly. Add the cooked liver to 12 pounds of cooked meat scraps and grind the mixture moderately fine. Add 5 pounds of broth to make a soft, but not runny, mixture. Season to taste. The quantities below may be used as guidelines for seasonings:

½ cup salt
3 tablespoons black pepper
2 tablespoons ground sage
  (optional)

1 teaspoon red pepper
  (optional)
1 tablespoon allspice (optional)

The sausage may be poured into loaf pans and stored as headcheese, or it may be stuffed into beef casings and smoked to make the Braunschweiger type of liver sausage.

Scrapple

Scrapple or Pann Haas is a popular Pennsylvania Dutch breakfast dish made famous in Philadelphia. It is made by combining meat scraps with broth (which may be that remaining from making liver sausage or headcheese) and thickening it with cornmeal or other cereal.

To make scrapple, cook heads, bones, or any other meat scraps and water in a heavy pot until meat falls easily from the bones. Remove the meat and grind finely. Strain the broth and return it to the pot with ground meat. Bring to a boil and slowly add the cereal thickener, stirring constantly to avoid lumps.

Usually the cereal added is cornmeal, but some of the cornmeal may be replaced with buckwheat flour in a ratio of 2 parts cornmeal to 1 part buckwheat flour. A small amount of wheat germ may also replace some of the cornmeal.

To make a good-textured scrapple with a rich flavor, we recommend using the following proportions of ingredients:

8 pounds meat
6 pounds broth
2 pounds cereal

Boil the mixture for 30 minutes, stirring constantly to prevent scorching.

A few minutes before removing the mixture from the heat season it with:

1 crushed bay leaf	½ tablespoon ground nutmeg
1 tablespoon sage (if desired)	(if desired)
2 tablespoons salt	1 teaspoon red pepper (if
1 tablespoon sweet marjoram	desired)
2 tablespoons pepper	2 teaspoons onion powder (if
	desired)

When the mixture has thickened and begins to leave the sides of the pot, pour it into loaf pans and chill quickly. Scrapple may be frozen for up to 2 months. To serve, slice and fry quickly, as you would thick bacon.

Sausage

Making sausage is much like making meat loaf; both are a thoroughly mixed combination of meat and seasonings. But sausage also has curing agents and is aged before cooking to bring out its full flavor. Even "fresh" sausage has salt and spices as curing agents and a few days' refrigeration to heighten its taste. You should know that the word "cure" not only refers to variety of spices added to the meat for flavor and preservation, but also to the passage of time that gives those ingredients a chance to do their work. So remember, to make any sausage you need a cure—the type of cure you choose will determine the variety of sausage you will make.

Attention to three other basics will insure high quality in your homemade sausage:

The Meat Itself   For sausage, choose meat scraps with no skin, gristle, blood clots, or pieces of bone remaining. Using two parts of lean meat to each part fat is imperative in getting tender, juicy sausage. Too much fat makes a heavy, greasy sausage that shrinks a great deal during cooking. But do not yield to the temptation of decreasing the amount of fat, thinking you will get a leaner sausage that way; you may be disappointed with the dry, tough, surprisingly stringy texture of the finished product. Happily, the most inexpensive cuts of meat, like pork shoulder or beef chuck, have this ideal lean/fat ratio.

Grinding the Meat   The coarser the grind, the more slowly the flavors of the cure will develop and permeate the meat. So usually highly spiced sausages, like kielbasa, are coarsely ground to keep the flavor from becoming overpowering. Mild sausages, like bockwurst, are usually finely ground to be compatible with the delicate flavor of light spicing. Most manual and electric grinders come with blades that chop the meat into ⅛- to ¾-inch pieces. For a finer grind than ⅛ inch: grind once, add ¼ cup water per pound of meat and free partially, regrinding and repeating the process until you get the texture you desire. For a dice coarser than ¾ inch, cut up partially frozen meat with a sharp knife.

Curing the Meat Meat is cured for preservation and flavor. The oldest, simplest, and safest cure to use is salt, but even it must be used cautiously. Too little will cause spoilage, too much will dry out and harden the meat. Most recipes call for 1 to 2 teaspoons per pound of meat, and you should be careful in varying this amount.

The primary role of peppers, herbs, and spices is to add flavor, but they also improve the keeping qualities of sausage once the cure has been completed. Use them carefully too, since a few days' curing will intensify their flavor. It's a good rule to underseason the mixture, then sauté and taste a bit before adding more.

If you are making a large quantity of sausage, it is a good idea to add seasonings sparingly. More spices can always be added if the meat is too mild, but none can be removed if the meat is too hot. Add your spices, mix them in well, and test the sausage meat by shaping some into a patty and cooking it. Taste the sausage and correct the seasonings. If the sausages are to be smoked, use a little less seasoning, as smoking will slightly dry out the meat and bring out the flavor of the spices. If you plan to freeze your fresh sausage meat, it is a good idea to add the seasoning after the meat has thawed, rather than before you store it in the freezer. Spices shorten the freezer life of meats. Fresh sausage meats keep safely at freezer temperatures for 1 to 2 months.

You will find that the ideal temperature for curing many sausages is that of your refrigerator: 36° to 45°F. Warmer, the meat might spoil, and if cooler, the salt penetration would be slowed. Meat can also be cured at room temperature or by cool or hot smoking (see the section on curing and smoking meats on page 393).

Using Sausage Casings

For professional-looking results and the real fun of sausage-making, you will want to try using natural casings. Alternatives to them are cheesecloth and the new plastic cooking wraps. There is a different technique for using each.

Plastic   Plastic cooking wraps are easy to find and store, and they can be used to make all sizes of sausages. Some people may object to using them, though, because of the chance that with exposure to heat some of the compounds in the plastic may permeate the meat. If you do wish to use them, be sure to get the type of plastic wrap meant for *cooking* and not the type used for wrapping and storing food. These "oven" wraps are easy to use: tear off a length about eight times the diameter of the sausage you are making. Spoon a strip of the sausage mixture along one end and roll it up tightly. Tie off the ends, twist the roll into links of any desired length, and tie with strong string.

Sausages wrapped like this can be baked at 325°F. for about 1 hour or simmered in water for about 25 minutes. Cooking time will vary with the size of the sausage. Don't forget to prick several small holes in the film of each link to let steam escape while cooking. And of course, remove the plastic film before serving. Voilá!—homemade skinless sausage!

Cheesecloth   Cheesecloth or a similar, loosely woven material is a more traditional and ecologically desirable wrapper. It's often used as a substitute for the big 2- and 3-inch natural casings which are costly and hard to find. You will need a length of cloth about twice the diameter of the sausage to be wrapped. There is a trick to tying large sausages so that they hold together well while cooking: use an extra long piece of string to tie off each end and spiral it down the sides to the opposite end and tie it off again. You've probably seen salami or summer sausage wrapped this way. Cheesecloth wrapped sausages can't be fried or roasted, but should be covered with water and simmered for about 30 minutes. Remove the cloth before serving.

Animal Casings   Natural sheep or hog casings are best for ½- to 1-inch sausages. A pound (or cup) of casings will make 30 yards or more of sausage. You can order them in most supermarkets and they are commonly sold around the holidays when many people use them to make traditional recipes. What you usually get is a tangle of long casings packed in salt. Cut off the lengths you need and soak them in water at least 2 or 3 hours to make them pliable. You can repack whatever you don't use in the salt where they keep well for a year or more.

The apparatus you use to get the meat into the casing may be as simple as a funnel or as complex as an electric meat grinder with a sausage-stuffing attachment. A sausage horn is a specially designed funnel found in gourmet cooking shops. They are all used similarly:

ease the casing onto the narrow end of the funnel leaving enough "tail" to tie off securely with a string. Push the meat mixture through the funnel, using your fingers, a spoon, or a wooden mallet. The gathered casing will unravel as it fills. Hold the casing back a little so that there are as few air bubbles as possible in the compact coil. Then twist and tie off links and the other end. Natural casings are most versatile: you can boil, roast, fry, or grill sausages made with them, and eat them casing and all.

Mild Sausage

10 pounds ground pork scraps
6 level tablespoons salt

4 teaspoons sage
4 teaspoons white pepper

Moderately Spicy Sausage

10 pounds ground pork scraps
5 tablespoons salt
2½ teaspoons dry mustard
5 teaspoons black pepper

2½ teaspoons ground cloves
5 teaspoons ground red pepper
6½ tablespoons ground sage

Spicy Sausage

10 pounds ground pork scraps
12½ tablespoons salt
10 teaspoons ground sage
10 teaspoons ground red
   pepper
2½ teaspoons ground allspice

5 teaspoons black pepper
2½ teaspoons honey
2½ teaspoons cayenne pepper
1 teaspoon powdered cloves
¼ teaspoon thyme

Mix the seasonings and then work them thoroughly through the meat. Sausage meat may be packed loose for freezer storage, or it may be stuffed into natural casings, plastic, or cheesecloth and stored in the freezer or smoked. For smoking sausage, see directions for smoking in the section on curing and smoking meats.

Bulk Sausage

You usually find bulk sausage in 1-pound rolls or shaped into patties or links. It's commonly called breakfast or pork sausage, although it's often used for meals other than breakfast and is made from beef as well as pork. Bulk sausage can be packed into plastic or glass containers, empty metal cans, or wrapped in aluminum foil, plastic, or waxed paper—just make sure it is airtight for refrigeration and freezing. You can form links and patties if you lightly oil your hands first.

This is the easiest sausage to make and a slight variation in spices will give you an excellent Italian sausage for pizza or a Mexican chorizo for tacos.

Old-Fashioned Breakfast Sausage

1 pound medium-ground raw
   pork
1 teaspoon salt

¼ teaspoon each ground
   thyme, black pepper, and
   sage
dash of garlic salt

Make sure the lean and fat of the ground meat are evenly mixed and distributed. Combine the other ingredients separately, sprinkle them over the meat mixture, and mix thoroughly. Sauté a bit, taste, and correct the seasonings. Form into one large roll, or several patties or links.

Refrigerate, wrapped airtight for 2 to 3 days to cure and blend flavors. This sausage will keep well for 3 to 4 days in the refrigerator or up to 1 month in the freezer without losing quality.

Italian Sausage

1 pound medium-ground raw
   pork
1 medium onion, chopped fine
1 small garlic clove, mashed
½ tablespoon salt

½ teaspoon each ground black
   pepper and fennel seed
¼ teaspoon paprika
⅛ teaspoon each ground
   thyme and cayenne

(Follow same directions as for breakfast sausage.)

Chorizo

1 pound medium-ground raw
   pork
1 teaspoon each salt and chili
   powder

1 large garlic clove, mashed in
   ½ teaspoon salt
2 teaspoons vinegar
1 tablespoon brandy, tequila, or
   water

(Follow directions as for breakfast sausage.)

Hot Dogs

You can make your own hot dogs much as you'd make sausage. Kids enjoy making these as much as they do eating them, but they should be forewarned that their creations won't be the pink color of store-bought hot dogs because these contain no nitrates. They will be the color of boiled meat. For a variation on this recipe that's sure to please the grown-ups, grind the meat coarsely with lots of onion and garlic to make bratwurst.

1 pound finely ground raw pork
½ pound finely ground raw
   beef
⅛ teaspoon each ground
   marjoram and mustard

1½ teaspoons salt
1 teaspoon black pepper
½ cup dry white wine, flat beer,
   or water

Mix the ground meats thoroughly. Combine the other ingredients separately and then mix them thoroughly into the ground meats. Saute a bit, taste, and correct seasonings. Stuff into casings, forming 4- to 6-inch links. Refrigerate 2 to 3 days to cure and blend flavors.

To cook, simmer in water for 20 to 30 minutes.

Yield: 1½ pounds

Bockwurst

1 small onion, finely chopped
1 tablespoon parsley, finely
   chopped
½ teaspoon ground cloves

1 well-beaten egg
1 cup milk
1 pound finely ground raw veal
   or chicken

Beat the onion, seasonings, and egg into the milk and add it to the meat mixture. Mix thoroughly. Sauté a bit, taste, and correct the seasoning. Stuff into casings, forming 3- to 5-inch links. To cook, simmer in water 20 to 30 minutes. Eat this soon after making since this sausage is very perishable.

Yield: 1½ pounds

Potatiskorv

This recipe, a Swedish favorite, uses potatoes as a filler to stretch a little meat a long, long way.

1 pound medium-ground raw
   pork
2 pounds raw potatoes, peeled
   and grated
1 tablespoon salt
1 teaspoon ground black pepper

½ teaspoon each ground ginger
   and allspice
1 medium onion, finely
   chopped
½ cup water

Mix the ground pork with the grated potato. Mix seasonings, onion, and water and add to the meat mixture. Stuff loosely into casings so that the mixture has room to expand as it cooks. Form 9- to 12-inch links. Eat immediately or refrigerate a few days, covered with water to keep the potatoes from darkening. To cook, simmer 30 minutes in water.

Yield: 3 pounds

Kielbasa

A further refinement in flavor is gained by adding sugar or honey to the cure. It sets up a fermentation that has a softening effect on the meat to counteract the astringency of the salt. Just a touch of sugar or honey will give your homemade sausage a delicate "old country" flavor like the one found in Polish sausage or kielbasa:

2 pounds raw pork, coarsely
   ground or chopped
½ pound finely ground raw
   beef or veal
1 tablespoon plus 1 teaspoon
   salt

¼ teaspoon each ground black
   pepper and marjoram
2 garlic cloves, finely chopped
   or mashed
1 teaspoon sugar or honey

Mix ground meats thoroughly. Combine spices separately, then mix with the meat and honey. Sauté a bit, taste, and correct seasonings. Stuff into casings, forming links 12 inches long. Refrigerate 3 to 4 days to cure and blend flavors. To cook, simmer 30 minutes in water.

Yield: 2½ pounds

Dressing Poultry

Dressing poultry is a far less complicated operation than dressing meat and is commonly done on the farm. No specific area is needed to slaughter and clean birds, unless you have a lot of chickens, turkeys, geese, or ducks to process. You can slaughter your birds as you need them or do many at a time for freezing or canning. With just a little improvisation, you can do the entire job outside, in a basement, garage, or outbuilding. Naturally, when dressing poultry your prime concern should be to end up with a bird that is attractive and free from contamination. This means working carefully and quickly, with equipment that is clean and in good working condition. Your processing area, whatever it may be, should be a place that is clean and free from flies, with a ready source of water and a stove for boiling water. It should also provide enough clean space where birds may be placed between processing steps.

Equipment

Equipment should be clean and ready to use before you choose your birds for slaughter. Like larger animals, birds also require a device to hoist and suspend them for dressing. There are instruments designed just for this purpose. A killing cone is like a funnel with the pointed end removed. The bird is placed inside this device, with its neck through the narrow opening. Killing cones come in various sizes, and it is important that you have one that fits the bird snugly. Shackles are a metal device that suspends birds by their feet. They may be used in place of the killing cone. If neither of these are available, you can suspend poultry by their feet with rope. Tie a short piece of rope to a convenient ceiling beam or support, ceiling hook, or if you're working outside, to a tree limb. Attach a 2-inch square block of wood on the free end of this piece of rope so that you can make an adjustable loop to hold the bird's foot. Suspend the bird by inserting each foot in one of these loops.

For slaughtering and dressing you will need a number of different tools, including at least one boning knife, small rounded knives without cutting edges for scraping pinfeathers, shears for cleaning giblets, and a lung scraper to remove the lungs. Knives and shears should be kept sharp during the operation with a whetstone and steel.

You will also need a few containers. A watertight container that is big enough to hold birds for scalding is necessary. A clean, galvanized 10- or 20-gallon garbage pail is ideal for the job. A container will be needed to hold feathers and inedible viscera until they can be disposed of. A paper-lined cardboard box or plastic-lined bucket, at least 2 feet square, will work well. You will also need a container to hold the giblets, as they should be kept separate from the rest of the bird during processing. Any clean pot, pail, or bucket will be sufficient. A clean pail or bucket big enough to hold dressed birds, ice, and water will be needed for chilling. A work table, big enough to allow you to work freely, is also necessary. Pick one that is of a convenient height for you so that you can work comfortably, without having to stoop over unnecessarily. It should be sturdy, with a clean work surface. Have on hand a durable thermometer that will register temperatures between 120° and 212°F. to measure the temperature of the water for scalding.

For information on the actual slaughtering and dressing of poultry, we refer you to the books and pamphlets listed on page 417.

Preparing Birds for Slaughter

Twenty-four to 34 hours before slaughter, pen those birds to be slaughtered and fast them. The cage should be clean, so that the birds' feathers will not get soiled. It should have a wire bottom so that birds cannot touch the ground to pick up feathers and litter. Fasting reduces the chance of contamination of the carcass because it cleans the digestive tract of feed and ingested matter. Birds should be given water during this fasting period however, so that they will not dehydrate. The skin of dehydrated birds is unattractive when the feathers are removed; it appears dark, dry, and scaly.

Chilling the Bird

As with beef, veal, lamb, and pork, poultry must be chilled after it is dressed to remove normal animal heat. Chilling reduces the temperature of the carcass enough to retard growth of bacteria that would otherwise lead to spoilage of the meat. Chilling also makes the carcass easier to cut and handle. To prechill the carcass, put the bird in a stopped-up sink or in a clean, watertight container filled with clean water that is safe for drinking. Allow this water to run slowly so that there is a constant overflow. If this is not possible, change the water periodically. This prechilling has two purposes: It helps to cool the bird, and it further cleans the animal. If you are dressing more than one bird at a time, add each to this water as it is cleaned.

Once the bird is sufficiently prechilled (it should be cooled down to water temperature), it must be chilled to 40°F. before further processing. This is done by placing the bird in a container filled with ice and water. Large capons will require 3 or more hours to chill to 40°F. Turkeys that are to be frozen should be held in 40°F. chill water for 18 to 24 hours before wrapping and freezing. Do not let birds freeze during chilling by packing them in ice only or exposing them to freezing temperatures. If birds should freeze, thaw them slowly in water no warmer than 40°F. Sudden changes in temperature will lower the quality of the dressed poultry. Once the carcass has reached 40°F., remove it from the ice water, hang it by the wing and let it drain 10 to 30 minutes before wrapping for freezing or refrigeration, or for canning.

Cutting the Bird

If you are canning your poultry, you will want to cut up the bird. The breast should be split and then cut along each side of the

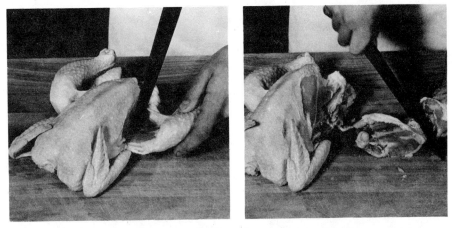

You should cut up your chicken if you are canning it or want to wrap it compactly to save freezer space. Before you begin, chill your bird in ice water and let it drain. Then follow the steps shown here and on the next page.

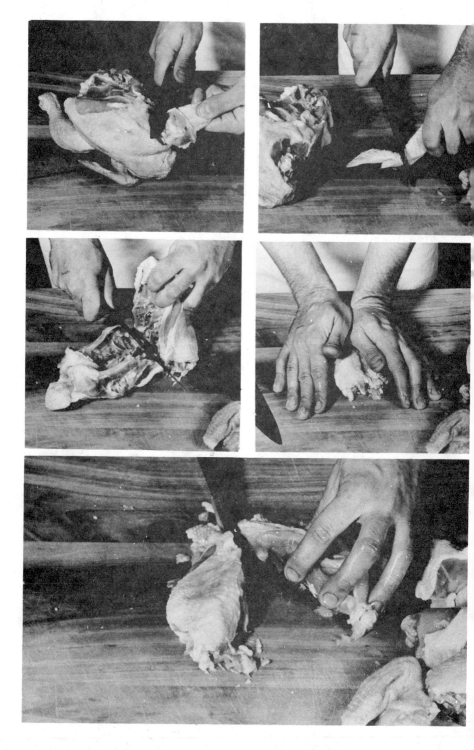

backbone so that this bone can be removed. The breast may then be cut to fit your glass jars or tin cans. Thighs should be separated from the drumsticks. Thighs may be boned, if you wish. Wings should not be canned; there is not enough meat on them to make canning them worthwhile. Rather, add the wings to other bones and simmer in water for broth or soup stock. (For more information on canning poultry, see the section on canning meat.)

It is advisable to cut up poultry for freezing. You'll save much freezer space if you do, because you won't have the wasted space of the body cavity. You may bone or split the breast in half and leave the other pieces whole, or bone the whole bird for compactness. Roasters that are to be frozen whole should be trimmed of excess fat. Oxygen, which causes rancidity, is absorbed by fat, and by cutting off unnecessary fat you are retarding rancidity. To save space when freezing whole roasters, tie the legs and wings tightly around the body of the bird. (For more information on freezing poultry, see the next section on freezing meat.)

Freezing Meat

You may be able to do without a freezer for storing fruits and vegetables, but you'll more than likely find one indispensable when it comes to keeping meat, especially if you've got meat from a steer, calf, lamb, or hog to store for the good part of a year or more. Freezing is unquestionably the easiest and safest way to keep meat for long-time storage.

Most all meats freeze successfully, with the exceptions of processed and spiced meats, canned hams, and cured and smoked products. Luncheon meats, like salami, bologna, and spiced ham should not be frozen, but stored in the refrigerator and used within a week. Cured meats become rancid more quickly than meats that are frozen when fresh because the ingredients used in curing increase meat's ability to absorb oxygen. These products should not be kept frozen for more than 1 to 2 months (see time chart). The period homemade fresh sausage may be safely stored in the freezer can be lengthened if the spices are added after the meat is thawed. Seasonings and fillings for meat loaves, meatballs, and the like should be added after, not before, freezing and thawing ground meat for extended storage life. Seasonings limit freezer life.

If you are freezing, wrap your meat as soon as possible after it has been cut. Wrapping carefully is extremely important in preserving the flavor, texture, and freshness of all frozen food. Wrapping, weighing, and labeling meat from a large animal, like a steer, will require the work of at least two people if it is to be done in less than an 8-hour day.

What Wraps to Use

There are a number of wraps on the market suitable to use for freezer storage. Heavy-duty plastic, aluminum foil, and freezer paper

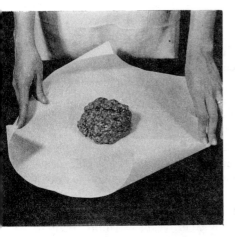

he best way to wrap meats for the
eezer is to make a delicatessen wrap. To
ake such a wrap, grab two ends of the
eezer paper, press them together, and
ld them over several times. Then press
e fold against the meat to squeeze
t all the air and make a compact
ackage. Now fold up the two loose ends
the paper, and seal the paper securely
ith freezer tape. Masking tape may seal
our wrap, but it has a tendency to lose
s adhering ability once it is exposed to
eezing temperatures or moisture. Re-
ember to label your packages.

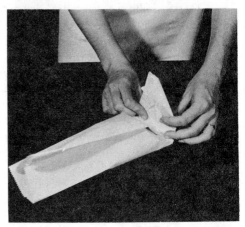

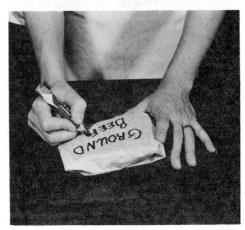

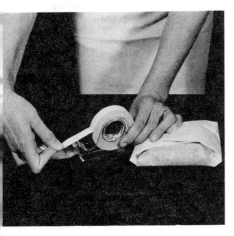

will protect frozen meats. Be certain that the wrap you use is moisture-proof. The air in freezers is relatively dry. If the wrapper is porous or poorly sealed, dry air will get in and draw moisture from the meat, dehydrating the surface and causing freezer burn. Although these burns are not harmful to meat, the dried area will have an unappetizing color and be tough and tasteless when cooked. This is particularly detrimental to foods with high moisture contents, like meats.

Your freezer wrap should be pliable so that it will mold itself to irregularly shaped meats and eliminate as much air as possible. Oxygen from the air which is absorbed by the meat will hasten the rate of rancidity.

The wrap must also be strong. Wraps not made for freezer storage are often too weak to use. These weaker wraps are more likely to tear and allow oxygen and dry air to enter when meat is put in, taken out, or shuffled around inside the freezer. Take special care with cuts of meat that have sharp corners or protruding bones. These may be protected from tearing by placing a plastic bag, stockinette, or old nylon stocking over the freezer wrap.

Meats that are highly perishable, like pork and cured meats, should be wrapped in a double thickness of wrap if they are to be kept longer than 5 months. Beef, veal, and lamb need only be wrapped in a single thickness if they are to be used within 1 year's time. If they are to be kept longer, it is advisable to cover them in wrap of double thickness. It is a good idea to place wax paper or plastic wrap between hamburgers, sausage patties, chops, and steaks so that individual frozen pieces may be separated and cooked separately. Some people wrap their meats in plastic before they wrap them in freezer paper. This insures an airtight, waterproof seal. All bloody cuts, like the organ meats, should always be wrapped first in plastic and then in freezer paper.

After wrapping, every cut should be weighed (a kitchen scale will do nicely), and the weight, type of cut, and date frozen should be marked on the wrapper. With proper labeling, you'll be able to go to your freezer a few weeks or months later and choose the size and cut you want at a glance. As with any frozen food, use it in the order in which it was frozen. Cuts labeled with the earliest freezing dates should be used first, whenever practical.

Freezing Soup Stock

The juices in meat bones may be extracted by simmering the bones in water to make soup stock. Bones should first be cracked or crushed to free juices and gelatinous matter. Marrow bones make for a better stock, but if too many are used the stock will be too gelatinous and have a thick, gluey consistency. Reserve such stock for sauces and gravies. Make a concentrated stock by adding just enough water to cover the bones. Add spices, if desired, and simmer the stock in a heavy pot with the lid tilted at an angle so that the pot is practically covered. Simmer for at least 12 hours. Cool the stock and skim off the fat with a ladle. Fat remaining in the stock will hasten rancidity. Once skimmed, strain the stock through a double layer of cheesecloth and pour stock into plastic containers or heavy glass jars, leaving a 1/2-inch headspace for expansion. Cover tightly and freeze. When you are ready to make soup from your frozen stock, thaw the stock and water it down to desired strength. If you are adding meat, simmer it in the stock until almost tender, and then add your vegetables. Simmer again just until meat is tender and vegetables are chewy, but neither hard nor soft.

Quick Freezing

For best quality, meat should be frozen quickly. Slow freezing gives water within the meat tissues time to separate out and form large ice crystals which stretch and rupture surrounding tissues. Meat frozen rapidly results in little water separation and smaller ice crystals that do little damage to the meat tissues. Butchers have a special freezer that maintains a very low temperature to flash-freeze meat. Many home freezers have quick freeze compartments to freeze fresh meat solid. Once the meat is completely frozen in this quick freeze section, it is then transferred to the regular freezing compartment which will keep it frozen at about zero degrees.

If your freezer doesn't have this special compartment, but it does have a temperature control, turn it to the coldest position and wait 24 hours after the fresh meat has been placed in the freezer before turning the control back to storage position. Don't overburden your freezer by putting a large amount of fresh meat into it at any one time. The amount of fresh meat placed in the freezer at one time should not exceed 2 to 3 pounds per cubic foot of freezer space in a 24-hour period. More than this amount raises the temperature and slows the freezing process.

When adding fresh meat to your freezer, put it in the coldest parts; this is usually along the bottom and walls of the freezer unit. To aid your freezer in freezing meat quickly, pack fresh meat loosely so that the cold air can circulate freely around your cuts and freeze the meat rapidly. After the meat is frozen solid, you can repack it tightly to make the most of your freezer space.

Freezer compartments inside regular refrigerators are seldom cold enough for long-time storage of meats. Use only a chest or upright freezer or the separate freezer compartment of a two-door refrigerator-freezer. Or rent a freezer locker at a freezer plant. Lockers may be rented by the month, quarter, half, and full year at reasonable rates. Temperatures in such plants are kept at 0°F. or below, and some plants employ butchers who will cut and wrap meat to order. Although the number of freezer locker plants has shrunk in recent years due to the popularity of home freezers, there are still thousands in operation throughout the United States.

Should your freezer stop operating and you are unable to keep temperatures in it below freezing, a freezer locker can hold your meat until your freezer resumes operation. What to do in such emergencies is discussed on page 65.

Thawing and Cooking Frozen Meat

Meats need not be thawed before cooking. There is little or no difference between the quality of meat thawed before cooking and meat cooked frozen. If you do cook meat when still frozen, you'll have to be more patient, because frozen meat takes more time to cook than fresh or thawed. While thin steaks take about the same time to broil or panbroil, frozen or thawed, larger cuts like roasts take almost twice as long to cook.

When you do thaw meat, it is best to thaw it in its original freezer wrapper. In order to insure uniform defrosting throughout your piece of meat, defrost it at low temperatures—in the refrigerator if possible. You can also thaw meat in cold water, provided it is in a watertight wrapping. If you must thaw poultry at room temperature, take special precautions to keep the surface of the bird cool during thawing. Place the bird, still in its original wrapper, in a closed double bag until it is pliable. Don't let frozen meats, especially poultry or pork, thaw on surfaces or trays where other foods are kept.

Thawed meats that have not reached temperatures above refrigerator temperatures (35° to 40°F.) may be refrozen, although each time meat is refrozen, there is some deterioration of quality; the ice crystals tend to rupture the fibers, breaking down the texture and letting more juices escape. If meat is above refrigerator temperature, do not refreeze. Thawed meat will keep in the refrigerator as long as fresh meat.

Freezer Storage Time Chart
Freezer Temperature (0°F. or Colder)

Meat (in Freezer Wrapper)	Recommended Maximum Storage Time
Fresh meat	
Beef	6 to 12 months
Veal	6 to 9 months
Pork	3 to 6 months
Lamb	6 to 9 months
Ground beef, veal, and lamb	3 to 4 months
Ground pork	1 to 3 months
Variety meats	3 to 4 months
Sausage and ready-to-serve	
Luncheon meats	not recommended
Sausage, fresh pork	2 months
Frankfurters	1 month
Cured, cured and smoked	
Bacon	1 month
Smoked ham, whole or slices	2 months
Beef, corned	2 weeks
Cooked meat	
Leftover cooked meat	2 to 3 months
Frozen combination foods	
Meat pies	3 months
Swiss steak	3 months
Stews	3 to 4 months
Prepared dinners	2 to 6 months
Poultry	
Chicken (ready-to-cook)	6 to 7 months
Turkey (ready-to-cook)	6 to 7 months
Cooked chicken	2 to 3 months
Cooked turkey	2 to 3 months

Courtesy of Home Economists in Business

Timetable for Thawing Frozen Meat

Meat	In Refrigerator	Low Temperature Oven (155°F.)
Large roast	4 to 7 hr. per lb.	1 to 1½ hr. per lb.
Small roast	3 to 5 hr. per lb.	¾ to 1 hr. per lb.
1-inch steak	12 to 14 hr.	1 hour

Timetable for Thawing Frozen Poultry

Poultry	Ready-to-Cook Weight	In Refrigerator	Low Temperature Oven (155°F.)
Roasting	3 to 8 lb.	1 to 2 days	2 to 3½ hr.
	8 to 12 lb.	1 to 2 days	3½ to 5 hr.
	12 to 20 lb.	2 to 3 days	5 to 7 hr.
Frying chicken, cut up	about 2 lb.	4 hours	1¼ hr.

Courtesy of Home Economists in Business

Canning
Meat

While the popularity of canning meats in the home has diminished with the increased use of home freezers over the last 30 years, this method of food preservation, which was brought into the home nearly a century ago, is still used by many American families to preserve fresh and cured meats. Although more time, work, and equipment is needed to can, canning has some advantages over freezing. There is no freezer to maintain year round at 0°F. and no damage done to the food if there should be a power failure or freezer breakdown.

Because meat is much more vulnerable to destructive enzymes and toxic bacteria than are fruits and vegetables, great care must be taken during the preparation and process of canning meats. Before considering canning meat, please read the section on preparing meat for storage.

Use only good-quality meat. Fresh meat should be chilled to 40°F. after butchering. Lamb and beef should not be aged more than 48 hours. If meat cannot be processed right after the animal heat is gone, store it at temperatures of 0°F. or lower until canning time. Meat may be processed for canning while it is still frozen, but be sure to allow extra time for the meat to cook sufficiently. If you wish to thaw meat before processing, it is best done gradually at low temperatures until most of the ice crystals have disappeared. Poultry should be rinsed and drained before it is processed.

Equipment

Equipment needed for canning includes a cutting board or other smooth, clean surface, sharp knives, a kettle for boiling water, thermometer, tongs, pot holders, glass jars with lids and bands or tin cans, lids, and a sealer, and a pressure canner or pressure saucepan. All equipment should be scrupulously clean. Wash metal utensils in hot, soapy water and rinse with boiling water. Scrub surfaces of wooden equipment with hot, soapy water and a stiff brush, and rinse with boiling water.

For further protection against bacteria, wooden equipment should be disinfected with a home disinfectant. Prepare the disinfectant following the directions on the container, and soak wooden boards and utensils in the solution for 15 minutes. Wash the disinfectant off well with boiling water. If you use this wooden equipment in daily food preparation, it should be scrubbed well and rinsed with boiling water before it is put away so that no particles of food remain to attract bacterial growth.

Like other low-acid foods (vegetables and dairy products) meat may contain toxic bacteria that cause botulism, a severe form of food poisoning. There is no danger of botulism if canned meats are processed at 240°F. for the required length of time. Only a pressure canner or pressure saucepan can reach this temperature practically in the home. No other method of cooking, using an open or covered pot or steamer without pressure, is safe to use with meat and foods containing meat, such as stews, soups, and gravies. Oven canning is impossible with tin cans and is hazardous with glass jars. A temperature of 240°F. inside the container cannot be reached. There is a chance that jars in the oven may burst, blowing out the oven door and causing considerable damage to the kitchen and to the person canning.

Before canning, make sure that your pressure canner is clean and is working properly. Remove the lid and wash the kettle in hot, soapy water. Do not wash the lid, but wipe it with a damp cloth. Clean the petcock and safety valve by running a string through the openings. Gauges should be checked. If you live in a high altitude area, increase the pressure by 1 pound for each 2,000 feet above sea level. If your canner has a dial gauge, it should be examined every year for accuracy by the manufacturer or dealer. Most all processing directions call for 10 pounds pressure. The U.S. Department of Agriculture gives the following adjustments to correct the pressure on dial gauges:

If the gauge reads high:
1 pound high—process at 11 pounds.
2 pounds high—process at 12 pounds.
3 pounds high—process at 13 pounds.
4 pounds high—process at 14 pounds.

If the gauge reads low:
1 pound low—process at 9 pounds.
2 pounds low—process at 8 pounds.
3 pounds low—process at 7 pounds.
4 pounds low—process at 6 pounds.

Do not use a pressure canner with a gauge that registers as much as 5 pounds high or low.

If you are using glass jars, be sure that all jars, lids, and bands are in perfect condition. Discard any with cracks, nicks, or chips. Even slight imperfections may prevent proper sealing. Use new rubber rings each time you can. Jars need not be sterilized, but they should be washed in hot, soapy water and rinsed well with boiling water. Pint and quart jars are good for canning meats, but half-gallon jars take too long to process and are not recommended.

If you're using tin cans instead of jars, use plain tin cans, without enamel lining. Use only perfect, rust-free cans and lids. Cans should be washed, rinsed well, and drained just before packing. Don't wash the lids, but wipe them with a damp cloth if they are dirty. Washing may damage the gaskets. Make sure that your sealer makes an airtight, smooth, finished seam. It is a good idea to test the sealer before canning by sealing a can of water. Submerge the sealed can of water in boiling water for a few seconds and look for air bubbles. If they rise from the can, your sealer needs adjusting.

See page 71 for more information on cans and jars.

Canning Procedure

1. Prepare meat.

Beef: Trim off fat and debone to save space and make packing easier. Cut tender meats, like roasts, steaks, and chops, into pieces the length of the can or jar with the grain of the meat running lengthwise. Tougher pieces of meat should be ground or cut into chunks for stewing meat. Bony pieces may be used to make broth or stock for canning.

Poultry: Cut into container-sized pieces. Remove the bones from the meaty parts, like the breast. Separate the thighs from the drumsticks. Keep the giblets to process separately.

2. Make broth, if desired, to cover meat.

Place meat bones in lidded pot with water and simmer until meat is tender. Skim off fat and save broth for packing.

3. Pack meat.

Meats should always be packed loosely. Containers may overflow if contents are packed too tightly or too full. Hearts may be packed after precooking (hot pack) or they may be packed before cooking (raw pack).

The chart that follows, which was adapted from the U.S. Department of Agriculture Bulletin, "Home Canning of Meat

(1) To can chicken in a hot pack, cut your bird into jar-sized pieces and cook them in broth or water until only a slight pink color remains. (2) Then pack the hot meat into clean jars by placing the thighs and drumsticks on the outside and the boned pieces in the center. Cover the poultry with boiling liquid, leaving a 1-inch headspace. (3) Screw the lids on tightly and set the jars on a rack in the pressure canner, allowing room around each one for steam to circulate freely. Put about 2 inches of water in the bottom of the canner. Fasten the lid on the canner and process for required time.

and Poultry," will give you an idea of the number of glass jars you will need to can your cuts of meat.

Yield of Canned Meat from Fresh

	Pounds of Meat per Jar	
Cut of Meat	Pints	Quarts
Beef:		
Round	1½ to 1¾	3 to 3½
Rump	2½ to 2¾	5 to 5½

Pork loin	2½ to 2¾	5 to 5½
Chicken:		
Canned with bone	1¾ to 2	3½ to 4¼
Canned without bone	2¾ to 3	5½ to 6¼

Hot pack: Meat is cooked to at least 170°F. in a skillet, pot, or in the oven. It is then packed in cans or glass jars, and boiling water or broth is poured over the meat, leaving a 1-inch headspace in the container. Salt may be added to taste, if desired, but it is not necessary. Wipe the rim of the jar carefully and adjust closure.

If you are using a mason jar with two-piece metal cap, put the lid on so that the sealing compound is next to the glass rim. Screw the metal band on tightly by hand. This lid now has enough give to let air escape during processing.

If you are using a mason jar with zinc porcelain-lined cap, fit the wet ring on the jar shoulder, but don't stretch it more than is necessary. Screw the cap down tightly and then turn it back ¼ inch before processing.

[If you are using tin cans, place the lid on each can with the gasket side down. Seal cans with sealer immediately, following manufacturer's instructions.]

Raw pack: Fill jars or cans with raw meat, leaving a 1-inch headspace. Set open, filled containers on a rack in a pan of boiling water. Keep the water level 2 inches from the top of the containers. Heat the meat slowly to 170°F., or for 75 minutes, if a meat thermometer is not available. Remove jars or cans from the pan and add salt to the meat, if desired. Wipe the rim of the container clean and adjust closure, according to directions above, under hot pack.

4. Processing.

a. Put 2 or 3 inches of water in the canner and heat water to boiling.

b. Set filled jars or cans on rack in the canner. Pack them into the canner loosely, allowing room around each one for steam to circulate freely. If there is room for two layers of jars or cans, place a rack between the two levels and stagger the containers so that none are directly over any of those below.

c. Fasten canner cover securely so that all escaping steam exits only through the petcock or weighted gauge opening.

d. Let pressure reach 10 pounds (which is 240°F.). Note the time as soon as the gauge reads 10 pounds and start counting processing time. If you're using a pressure saucepan, add 25 minutes to required processing time. Maintain 10 pounds on the gauge by adjusting heat if necessary. Varying pressure may cause containers to overflow and lose some of their liquid.

e. When processing time is up, remove the canner from the heat immediately.

f. If you are using jars, do not pour cold water over the canner to reduce pressure quickly, but let the canner stand until the gauge reads 0°F. Wait a few minutes after gauge reads 0°F. and then slowly open petcock or take off weighted gauge. Unfasten the cover and tilt it away from you so that steam can escape without rising in your face. When all the steam is gone, remove the jars.

[If you are using cans, remove the canner from the heat as soon as gauge reads 0°F. and open petcock or remove weighted gauge to release steam. Unfasten cover and tilt it away from you so that steam can escape without rising in your face. Remove cans.]

5. Cool containers.

As soon as glass jars are taken from the canner, complete seals, if necessary. Only mason jars with zinc porcelain-lined caps need to be sealed. Do this by quickly screwing the cap down tightly. Mason jars with two-piece metal caps are self-sealing. Cool the jars right side up on a rack or folded cloth. Don't cover them, and don't cool them in a drafty place.

[Put tin cans in cold water as soon as they are removed from the canner. Change the water frequently for rapid cooling. Remove the cans from water while they are still warm and allow them to dry in the air. If the cans are stacked, stagger them so that the air can circulate freely around them.]

6. Check your seals.

When containers are thoroughly cool, examine each carefully for leaks. Press on the center of the jar lid. If this lid does not "give" when you press on it, the jar is sealed. Jars that have lost liquid during canning should not be opened and reprocessed. Although the meat inside such cans may darken during storage, the meat is not spoiled.

[Check your cans by examining all seams and seals. Can ends should not be bulging, but almost flat, and seams should be

smooth with no buckling. If you suspect a can or jar of having a faulty seal, open the container, reheat the meat, and process all over again for the complete time required, to insure safety.]
7. Mark jars and store them.

Write directly on the top of the jar lids, or use adhesive tape, freezer tape, or special labels for cans and jars. Label each container with the type of food canned and date of canning. Canned foods should be used in the order they were canned.

Storage

Jars and cans that contain meat should be stored in a cool, dry place. Do not subject the canned foods to warm temperatures or direct sunlight, as they will lose quality. Freezing does not cause canned meat to spoil, but it may damage the seal so that spoilage begins. To protect against freezing in an unheated area like a cellar or garage, cover the jars and cans with a clean blanket or wrap them separately in newspapers. A damp storage area invites rust which corrodes cans and metal jar lids and causes leakage.

Meat that has been processed, sealed, and stored properly will not spoil. If you suspect meat of being spoiled, don't test it by tasting. Destroy it by burning or dispose of it where it cannot be eaten by animals or humans. It is a good idea to boil all home-canned meat 20 minutes in a covered pot before tasting or serving it. Twenty minutes of rapid boiling will destroy any dangerous toxins that remain in foods that were improperly processed. This precaution should certainly be followed by anyone who is canning for the first time or is not certain that his or her gauge is accurate. Boiling is the best way to find out if meat is safe to eat, because heat intensifies the characteristic odor of spoiled meat. If your meat develops an objectionable odor, dispose of it without tasting.

Jars and cans should be checked during storage and before opening for signs of spoilage. Any abnormality in your can or jar— bulging jar lids or rings, gas bubbles, leaks, bulging can ends—can mean spoilage. If you notice any of these signs, dispose of the container and contents without tasting. If you notice an off-odor, discoloration, or spurting liquid as you open a can or jar you know that the food is not safe to eat. If there is a discoloration on the metal lids or cans, this is most likely caused by sulphur in the meat; this does not mean the meat is spoiled.

DIRECTIONS FOR PROCESSING MEAT

Type and cut	Preparation and processing	Processing Time
Cut-up beef, veal, lamb, or pork	Use tender cuts for canning in strips. Making sure that grain of meat runs lengthwise, cut meat the length of the can or jar. Cut less tender cuts into cubes for stewing or soups.	
	Glass jars: Hot pack—Precook meat in skillet or saucepan in just enough water to keep meat from scorching. Stir occasionally so that all pieces heat evenly. Cook until medium done. Pack hot meat loosely, leaving a 1-inch headspace. Add ½ teaspoon salt to pint jars and 1 teaspoon to quarts, if desired. Cover the meat with boiling water or boiling juice from meat, leaving a 1-inch headspace. Adjust lids and process in pressure canner at 10 pounds pressure for required time.	Quarts 90 min. Pints 75 min.
	Raw pack—Pack cold, raw meat loosely, leaving a 1-inch headspace in jar. Exhaust air by heating meat-filled jars in a pot of hot water. Cook at a slow boil until meat is medium done, or reaches 170°F. Add ½ teaspoon salt to pints and 1 teaspoon salt to quarts, if desired. Adjust lids and process in pressure canner at 10 pounds for required time.	Quarts 90 min. Pints 75 min.
	Tin cans: Hot pack—Precook meat in skillet or	

saucepan in a small amount of water until meat is medium done, stirring occasionally so that meat heats evenly. Pack cooked meat in cans, leaving a ½-inch headspace. Add ½ teaspoon of salt to No. 2 cans or ¾ teaspoon salt to No. 2½ cans, if desired. Fill cans to top with boiling water or boiling juice from meat. Seal. Process in a pressure canner at 10 pounds for required time.

No. 2
65 min.

No. 2½
90 min.

Raw pack—Pack cans with cold, raw meat to top. To exhaust air, cook meat in cans in a pot of hot water. Cook at a slow boil until meat is medium done or reaches 170°F., about 50 minutes. Add ½ teaspoon salt to No. 2 cans and ¾ teaspoon salt to No. 2½ cans, if desired. Press meat down into can so that there is a ½-inch headspace, and add boiling water or broth to fill to top. Seal cans. Process in a pressure canner at 10 pounds pressure for required time.

No. 2
65 min.

No. 2½
90 min.

Ground meat

Glass jars:

Hot pack—If you are making patties, press the meat tightly into patties that will fit into your jars without breaking. Meat may also be left loose although it is harder to remove from jars. Precook patties in a slow oven or over low heat in a skillet until medium done. Pour off all fat; do not use any fat in canning. Pack patties or loose meat in jars, leaving a 1-inch headspace. Cover with boiling water or broth to cover meat, leaving 1-inch headspace. Adjust lids and process in pressure canner at 10 pounds for required time.

Quarts
90 min.

Pints
75 min.

Raw pack—Pack raw ground meat into jars, leaving 1-inch headspace. Cook meat-filled jars in a pot of hot water at a slow boil until medium done or to 170°F., about 75 minutes. Adjust jar lids and process in a pressure canner at 10 pounds pressure for required time.

Quarts
90 min.

Pints
75 min.

Tin cans:

Hot pack—If you are making patties, press the meat tightly into patties that will fit into your cans without breaking. Meat may also be packed loose. Precook patties in slow oven or over low heat in a skillet until medium done. Pour off all fat. Pack loose meat or patties into cans within ½ inch of top. Cover with boiling water or boiling juice from meat to the top of the cans. Seal and process in a pressure canner at 10 pounds pressure for required time.

No. 2
65 min.

No. 2½
90 min.

Raw pack—Pack raw ground meat solidly to the top of the can. Cook meat in pot of hot water at a slow boil until medium done or 170°F., about 75 minutes. Press meat down into cans to allow for a ½-inch headspace and seal. Process in pressure canner at 10 pounds pressure for required time.

No. 2
100 min.

No. 2½
135 min.

Sausage and headcheese

Sausage may be packed and processed just as ground meat. Use your own recipe or one of the ones starting on page 361. Use seasonings sparingly in sausage intended for canning, as spices change flavor in storage. If you normally use sage, omit it; it makes canned sausage bitter.

Poultry, small game, and rabbits

Sort meat into meaty and bony parts and can separately.

Glass jars:
Hot pack, with bone—Bone the breast and cut bony parts into jar-sized pieces. Trim off excess fat. Heat pieces in pan with broth or water. Stir occasionally until meat is medium done, when only slight pink color remains. Pack meat loosely, with thighs and drumsticks on the outside of the jar and boned pieces in the center. Leave a 1-inch headspace. Add ½ teaspoon salt per pint and 1 teaspoon per quart, if desired. Cover poultry with boiling water or broth, leaving a 1-inch headspace. Adjust lids and process in pressure canner at 10 pounds pressure for required time.

Quarts
75 min.

Pints
65 min.

Hot pack, without bone—Remove meat, but leave skin on meaty parts. Cook meat as above. Pack jars loosely, leaving a 1-inch headspace. Add ½ teaspoon salt per pint and 1 teaspoon per quart, if desired. Pour in boiling broth or water, leaving a 1-inch headspace. Adjust lids and process in pressure canner at 10 pounds pressure for required time.

Quarts
90 min.

Pints
75 min.

Raw pack, with bone—Bone the breast and cut other pieces, with bone in, into jar-sized pieces. Trim off excess fat. Pack raw poultry loosely by placing thigh and drumsticks on the outside and breasts in the center of the jar. Leave a 1-inch headspace. Place filled jars in a pan of hot water and cook at a slow boil about 75 minutes, or until meat is medium done. Add salt if desired: ½ teaspoon per pint and 1 teaspoon per quart. Adjust lids and process in pressure canner at 10 pounds pressure for required time.

Quarts 75 min.

Pints 65 min.

Raw pack, without bone—Remove bones, but not skin, from meaty pieces. Pack loosely, leaving a 1-inch headspace. Place filled jars in pan of hot water and cook at a slow boil until meat is medium done, about 75 minutes, or until thermometer reads 170°F. Add salt, if desired: ½ teaspoon per pint and 1 teaspoon per quart. Adjust lids, and process in pressure canner at 10 pounds pressure for required time.

Quarts 90 min.

Pints 75 min.

Tin cans:
Hot pack, with bone—Bone the breast and cut bony pieces into can-sized pieces. Trim off excess fat. Heat pieces in a pan with broth or water. Stir occasionally until meat is medium done, when only slight pink color remains. Pack cans, leaving a ½-inch headspace. Add ½ teaspoon salt to No. 2 cans and ¾ teaspoon salt to No. 2½ cans, if desired. Fill cans to the top with boiling broth. Seal and process in a pressure canner at 10 pounds pressure for required time.

No. 2 55 min.

No. 2½ 75 min.

Hot pack, without bone—Remove bones, but not skin, from meaty pieces. Cook as above. Pack loosely, leaving a ½-inch headspace. Add salt, if desired: ½ teaspoon per No. 2 can and ¾ teaspoon per No. 2½ can. Fill cans to the top with boiling broth

No. 2 65 min.

No. 2½ 90 min.

and seal. Process in a pressure canner at 10 pounds pressure for required time.

Raw pack, with bone—Bone breast and cut thighs and drumsticks into can-sized pieces. Trim off excess fat. Pack poultry loosely, with thighs and drumsticks on the outside and breasts in the center. Pack to the top of the can. Place filled cans in a pan of hot water and boil slowly until meat is medium done, or until thermometer reads 170°F. Add salt if desired: ½ teaspoon per No. 2 can and ¾ teaspoon per No. 2½ can. Seal and process in a pressure canner at 10 pounds for required time.

No. 2
65 min.

No. 2½
90 min.

Raw pack, without bone—Cut poultry and remove bone, but not skin. Pack raw meat to top of the cans. Place filled cans in a pan of hot water and slowly boil until meat is medium done, about 75 minutes, or until thermometer reads 170°F. Add salt, if desired: ½ teaspoon per No. 2 can and ¾ teaspoon per No. 2½ can. Seal and process in a pressure canner at 10 pounds pressure for required time.

No. 2
65 min.

No. 2½
90 min.

Giblets   Can in pint jars or No. 2 cans. Separate hearts and gizzards from livers and cook and can them separately to avoid blending of flavors.

Hot pack—Cook giblets in a pan with water or broth until medium done. Stir occasionally so that meat heats evenly.

Glass jars—Pack meat, leaving a 1-inch headspace. Add boiling water or broth just to cover giblets. Adjust lid and process in a pressure canner at 10 pounds pressure for required time.

Pints
75 min.

Tin cans—Pack giblets, leaving a ½-inch headspace. Fill cans to top with boiling broth or water. Seal and process in a pressure canner at 10 pounds for required time.

No. 2
65 min.

Corned beef Use your own recipe or the one on page 406. Wash and drain corned beef. Cut it into container-sized strips. Cover the meat with cold water and bring to a boil. Taste broth; if it is very salty, drain and boil meat in fresh water. Pack while hot.

Glass jars—Pack meat loosely, leaving a 1-inch headspace. Cover meat with boiling water or broth, leaving a 1-inch headspace. Adjust lids and process in a pressure canner at 10 pounds for required time.

Quarts
90 min.

Pints
75 min.

Tin cans—Pack meat loosely, leaving a ½-inch headspace. Fill cans to top with boiling water or broth. Seal and process in a pressure canner at 10 pounds pressure for required time.

No. 2
65 min.

No. 2½
90 min.

Liver Wash, remove skin and membranes. Slice into container-sized pieces. Organ meats are hot packed. Drop into boiling water for 5 minutes. Follow hot pack directions for cut-up meat.

Heart Wash and remove thick connective tissue. Cut into container-sized strips or 1-inch cubes. Put in a pan and cover with boiling water. Cook at a slow boil until medium done. Follow hot pack directions for cut-up meat.

Tongue Wash, put in pan, and cover with boiling water. Cook for about 45 minutes, until skin can be removed easily. Skin and slice into container-sized pieces or into 1-inch cubes. Reheat to simmering in the same boiling water. Follow hot pack directions for cut-up meat.

Soup stock or broth Crack or saw bones. Simmer them in water until meat is tender. To save work and storage space, make a concentrated stock which can be diluted before reheating. Simmer bones in just enough water to cover them. Strain stock and skim off fat. Cut up meat and return it to stock, if desired. Reheat to boiling and pack into glass jars. Adjust lids and process in a pressure canner at 10 pounds pressure for required time.

Quarts 25 min.

Pints 20 min.

Curing and Smoking Meats

Our chief concern in this book is to explain how best to preserve food for optimal food value. This means preserving as much of the original supply of nutrients in food as possible by giving great attention to the many steps involved in harvesting, processing, and storing foods. It also means being careful not to add anything to our stored foods that is harmful to our bodies, like adding artificial preservatives or any other questionable chemical additives. For this last reason we must begin this chapter on curing and smoking meats with a few words of caution.

There are a number of things that we find objectionable in cured and smoked products. Basically, curing meat means salting it so much that it will resist the bacteria that would otherwise cause the meat to spoil. Anyone who has ever tasted bacon, dried beef, or country ham knows that a lot of salt has been absorbed by the product. And salt, especially in the great quantities that you'll find in cured products, doesn't do our bodies any good. Too much salt can be the cause of hyperacidity, high blood pressure, and excessive water retention. It may interfere with the absorption and utilization of food by overstimulating the digestive tract, and it can aggravate heart disease.

Salt is the basic curing ingredient, but if used by itself it tends to make meat oversalty, dry, and tough. So along with salt, most cures contain sugar. Our objections to sugar, discussed elsewhere (see page 199), hold true for cured and smoked meats also.

Beef and pork cures may contain saltpeter as a color fixative. We object to the consumption of saltpeter because it has been proven to be toxic to man. The nitrate in saltpeter combines easily with amines in food, drugs, alcohol, and tobacco smoke to produce nitrosamines, and these compounds have been found to cause cancer.*

* Beatrice Trum Hunter, "Nitrite Additives in Meat Products," CONSUMERS' RE-SEARCH MAGAZINE 58:5 (May 1975), pp. 14–15. Also Michael Jacobson, DON'T BRING HOME THE BACON: HOW SODIUM NITRITE CAN AFFECT YOUR HEALTH (Washington, D.C.: Center for Science in the Public Interest, 1973).

Smoked meat tastes smoky because the meat has actually absorbed smoke emitted from slow-burning hardwoods. The smoke is a coal-tar derivative, and coal-tar derivatives have long been known as cancer-causing substances.

We recommend freezing, canning, or even drying meat before we advise curing and smoking it. But if your homestead doesn't have the facilities for freezing or canning meat, you'll have to rely on curing, and maybe smoking it too, if you want to keep it for any length of time. The salt in the cure extracts water and checks the growth of most spoilage organisms. The smoke colors and flavors the meat, dehydrates the cured meat further, and slows down the development of rancidity in the fat. However, meat that has been cured properly does not also need to be smoked to insure good preservation.

There aren't too many old-time butchers and smokers left; they've gone out with the introduction of commercially prepared cures and automatic smokehouses. Luckily, here in Pennsylvania Dutch country you can still find a few country butchers who are doing their own smoking—just as it's been done for one hundred and more years. Thomas Meyers, of Fleetwood, Pennsylvania, is one of

Before you decide if you are going to cure and smoke your own meat, ask yourself if you can build, buy, or borrow a smokehouse. It is the most important—and most expensive—piece of equipment you will need.

these men. Some 60 years ago his father taught him the art of smoking, and he's been smoking meats the same way ever since. His smoked bacon, ham, beef, and chicken have an aroma and taste distinctively different from the smoked foods sold by the bigger, modern operations that forced him to give up his business a few years ago. Now he has a small shop and smokehouse which serves customers of a popular country restaurant.

Mr. Meyers is a friendly man and was glad to talk to us about his trade, but when we asked him just how he smokes his products he had a difficult time explaining the procedures to us. "I can't tell you; I just do it," was his reply. "You'd have to see for yourself." He didn't learn how to smoke from a book; he learned by working with his father and practicing on his own until he developed a real eye and feel for his work.

And this is just what you'll have to do if you want to produce a good smoked product. We can give you fine old, traditional recipes for cures, tell you how to build smokehouses, and provide you with a set of general instructions, but the thing that we can't give you here in this book is the practice and know-how that can make the difference between a fair-tasting ham or side of bacon and a great-tasting one. This will have to come from you. If you're serious about curing and smoking your own meats, find yourself a small country butcher—one who takes great pride in his own meats—and work with him at least once when he's curing and smoking.

Building the Smokehouse

The first thing that we're going to discuss is the smokehouse because a smokehouse is the most important and most complicated piece of equipment that you'll need if you intend to preserve meat by both curing and smoking. You can build a smokehouse out of just about anything—an old refrigerator, a packing crate, a barrel—providing it has four essential things: a source of smoke, an area in which to confine the smoke, racks to hold the meat, and a draft.

The smoke is usually produced by burning hardwood, a charcoal fire with hardwood chips or sawdust sprinkled over it, or even by burning corncobs. Never use softwoods; their pitch gives an off-odor and taste to the food.

A draft is absolutely essential to the smoking operation; without it the smoke will stagnate in the smokehouse, and the smoked meat will have an objectionable, sooty taste. In a barrel smokehouse, the draft can be created by a plywood cover with a hole cut in it. In a more permanent arrangement, such as a cement-block smokehouse, movable bricks can be used to vary the draft. If it is possible, construct your smokehouse so that there are draft holes in the bottom of the smokehouse as well as the top. With draft holes at both ends you'll be able to control the circulation of the smoke more efficiently.

A Barrel Smokehouse

A 50-gallon barrel can be made into a temporary smokehouse to accommodate a small amount of meat:

Dig a pit smaller than the diameter of the barrel and about 2 feet deep. Place the barrel, with the top removed and a hole cut in the bottom, over it. A fire pit should be dug 10 to 12 feet away from the barrel in the direction of the prevailing winds. The pit should be deep enough so that it can be covered. The intensity of the fire in the pit and the amount of draft can be controlled by placing a piece of sheet metal over the pit. The pit should be connected to the bottom of the barrel with a trench made from a stovepipe or sewer tile. It is important that this trench be airtight so that little smoke will be lost.

In a cement-block smokehouse movable bricks near the roof may be used to vary the draft.

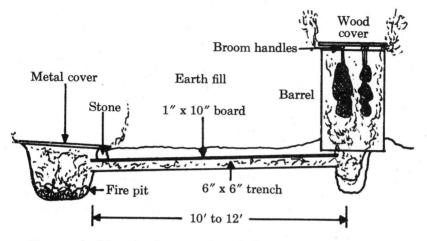

You can build a simple type of smokehouse from a barrel. Dig the fire pit at least 10 inches from the barrel and connect the two with a pipe or tile trench.

A lid for the barrel should be made from plywood. Strips of wood or broom handles to support the hanging meat should be placed across the top of the barrel, as shown in the diagram below. The wooden cover is then placed over the wood strips. The openings made by the wood strips provide draft holes for escaping smoke.

Converting a Refrigerator into a Smokehouse

An old refrigerator can also be converted into a smokehouse. Because it is more durable than a barrel and is insulated, it will provide a more permanent smokehouse that retains heat well. In addition, its removable metal racks and full door make it very convenient for hanging and removing meat.

Although the heat can be placed directly underneath the refrigerator, or even inside it, this arrangement is not advisable because the insulating material inside the refrigerator walls may catch on fire if it is subjected to direct heat. Rather, it is best to dig a fire pit several feet from the refrigerator unit and direct the heat to it by using an underground trench, as we recommended for the barrel smokehouse.

If the motor and compressor are contained in the bottom of the refrigerator, remove these. Place the unit on four cinder blocks, cut a hole in the bottom, and attach a tight-fitting stovepipe or sewer tile leading to the fire pit from this hole to prevent smoke from escaping.

Provide a draft at the top of the unit by cutting a 3-inch hole in the top and attaching a 2-foot length of stovepipe with a butterfly damper.

This type of conversion will allow you to make maximum use of the refrigerator space and prevent any chance of the insulation from catching on fire.

A Permanent Smokehouse

A permanent smokehouse can be built out of wood, cement blocks, or concrete, and can be modified to suit your taste. One such smokehouse is shown here. It is sufficient for smoking meats of 8 to 12 hogs. Pipes or wooden rods are hung on the wooden supports, and from these pipes the meat hangs. The oven behind the smokehouse

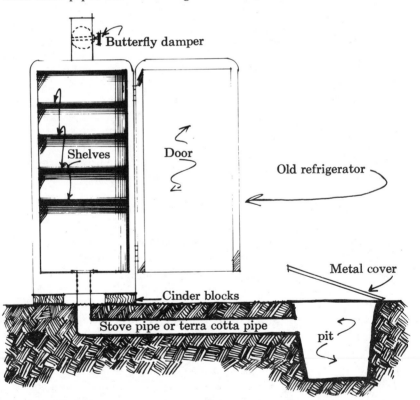

An old refrigerator can be converted into a more permanent smokehouse. Be sure to place the pit at least 10 feet from the refrigerator to prevent insulation in the refrigerator walls from catching on fire.

can be used for rendering lard, or for cooking scrapple, if you're doing your own butchering. When the setup is to be used for smoking, water is placed in the kettle, damper B is closed, and damper A is opened. When the kettle is to be used for cooking, A is shut and B is opened.

It is also possible to modify an existing shed by adding a firebox at one end of the building. If a frame shed is used, it is advisable to construct a firebox outside the shed for safety. Dig an underground trench leading from the firebox to the center of the shed floor for the smoke. The trench can be made from stovepipe or sewer tile.

When to Cure and Smoke

It is vital to hold the temperature of the meat down while it is being cured: the temperature should not go over 38° to 40°F. Tempera-

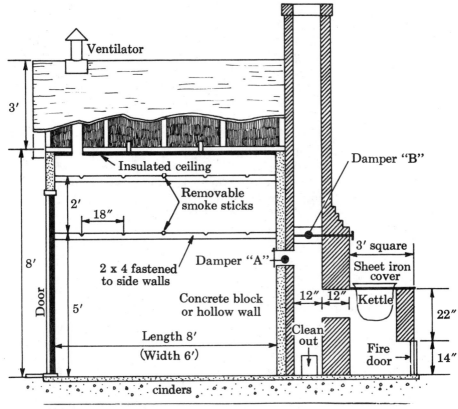

This 6-by-6-by-8-foot combination smoke and storage house and cooker is large enough to smoke the meat from 8 to 12 hogs at one time.

tures greater than this will cause the meat to spoil, and lower temperatures retard the salt penetration. This is where a refrigerator or springhouse comes in handy, especially in warm weather. Most farmers, if given the choice, do their butchering in the fall and curing and smoking in the early winter. Cool weather helps to keep meat chilled when it is most important. By the time warm weather rolls around again, the meat is well aged.

Equipment for Curing

For curing you'll need a sturdy container for holding the meat. Glass or earthenware containers are excellent. Many farmers use large crocks. Never use aluminum containers. Wood containers aren't acceptable unless you are using a dry cure. A hardwood barrel will be all right, so long as it is watertight. If the barrel is new or has not been filled with a liquid for a long time, fill it with water and allow the water to set in it for at least 24 hours before you pour it out and use it for brining. The water will cause the wooden boards to swell and make the barrel watertight. If you are brine curing, the container should be large enough to hold the meat and brine.

Curing

There are two types of cures: the dry cure and the brine cure. Either (or sometimes both) cure is used to mildly preserve, tenderize, and flavor meat intended for smoking or drying. Brine curing, sometimes called pickling or (when used with some cuts of beef) corning, may also be used as a means of preserving meat all by itself, without additional smoking or drying. In such cases the meat is left completely covered by the brine and kept in a cool, dark place until it is eaten. Brining like this is not the best means of keeping meat, and meat should not be stored this way for more than a few weeks, and perhaps less, depending upon the recipe and the weather.

If you wish to make a brine cure, be certain that your water is clean and has a pleasant taste. If it is heavily chlorinated, we suggest that you boil the chlorine out of it first. Don't use bleach or any other chemicals to purify water; use bottled water instead. If you are unable to get good-tasting, clean water, you'd do better to use the dry cure.

Dry curing is usually the more successful of the two methods for the beginner and the person who does not have a refrigerated curing room. Since the dry cure is applied directly to the meat in its full strength, the rate of cure is more rapid than the brine cure. This is especially important in the southern parts of our country where the weather is warm most of the year. With this cure there is less chance of spoilage if the temperature should vary somewhat during the curing period. Curing time is not as exacting as it is with the brine cure, and there is no need for a brine pump or watertight barrel or crock in which to soak the meat.

Commercially prepared cures are available from some salt companies and chemical manufacturers. Although these ready-made mixes are easy to use because they are premeasured and premixed, they generally contain antioxidants. Most companies offer a choice of cures, but you have much more control over taste and degree of saltiness when you mix your own because you can regulate the amount of each ingredient.

Many recipes for cures, even old-time recipes, use both sugar and saltpeter. Some use molasses instead of sugar, and some require both sugar and molasses. We haven't found anyone yet who uses honey, but the butchers that we have questioned see no reason why honey couldn't be used instead of sugar and molasses. The only drawback may be its high price (unless you keep your own bees). Most people agree that one-half the amount of honey may be substituted for sugar. That is, if a formula calls for 2 pounds of sugar, replace it with 1 pound of honey. One pound of honey can be substituted for 1 pound of molasses. Molasses and honey should only be used in the place of or in addition to sugar when the temperature during curing can be kept between 38° and 40°F. Warmer temperatures may make your brine ropy.

Before you decide what cure you are going to use, consider what each ingredient does. Salt preserves the meat; there's no way to get around using it. Sugar is generally added to tenderize and flavor the finished product, but it plays no actual part in preserving the meat. Saltpeter is added primarily to preserve the color of pork and beef; it is not an essential ingredient, if you don't mind your meats a little darker in color. In other words, you can get away with just using salt. As a matter of fact, many old-timers did.

Make sure that all your meat is chilled to 38° to 40°F. before curing. Keep it at this temperature during curing. Don't let it freeze because the cure won't work when the meat is frozen. If the meat does happen to freeze, extend the curing time for as many days as the meat is frozen. For example, if the curing time is 28 days, and the meat freezes for 2 days, cure the meat for a total of 30 days. Follow the suggested times carefully until you develop a feel for curing. Never alter curing times until you know what you are doing. Meat that is underexposed to a cure will have a poor keeping quality, whereas meat that is kept in cure too long will shrink excessively and be too salty.

Weigh the meat and curing ingredients. It's very important to use the right kind of salt. Never use iodized salt. Good salts are marketed under such names as rock salt (the kind you use for freezing homemade ice cream and melting ice on sidewalks), dairy salt, or water softener salt. It might be necessary to crush or grind the crystals, since a medium-grind salt is best for curing meats. (Fish, however, are best cured with a very fine salt, as noted in the section on fish.) A small hand-operated grain mill, like a Corona or a Quaker City mill, will do a good job of grinding the crystals.

The following directions for dry and brine curing are intended for pork. If you are interested in curing (corning) beef, see page 406.

Dry Curing

Dry curing involves rubbing the cure on the outside surfaces of the meat and allowing it time to penetrate. Remember, even though we give the original formulas for cures, it is possible to produce a good cured product without using either the saltpeter or sugar.

The U.S. Department of Agriculture recommends the following dry cure formula for 100 pounds of ham or shoulder:

> 8 pounds pickling salt
> 2 pounds white or light brown sugar or syrup
> 2 ounces saltpeter (we think this is optional)

Use 1 ounce of this mix per pound of ham and ¾ to 1 ounce per pound of bacon. Rub this mixture into the meat to be cured. Rub the required amount of cure into bacons all at once; rub shoulders twice, with one-half the amount each time; and rub hams with one-third the amount three times. Resalt hams and shoulders 6 to 8 days after the last salting. Be sure that the mixture covers the surface; avoid breaking outside membranes or meat will become hard and dry during aging. Pack some of this mixture in the shank ends to make sure that this part gets well cured.

Place the meat in a clean, dry container. If you are curing meat on open shelves, cover them with plastic so that the cure does not drip on meats resting on shelves below. Make sure meats don't rest in the brine that will result as the moisture from the meat is drawn out by the salt.

It takes 7 days for the cure to penetrate an inch of solid meat; therefore, if a piece of bacon is 2 inches thick, cure it 14 days; a ham that is 5 inches thick through the aitchbone should be cured 35 days. The cured meat may then be smoked, if desired.

If you measure the amount of cure, the meat won't get too salty even if it is left in the brine too long because the correct amount of curing mix per each pound of meat was used.

We found another method of dry curing, using salt only, in a book written a century ago: DR. CHASE'S FAMILY PHYSICIAN, FARRIER, BEE-KEEPER AND SECOND RECEIPT BOOK, by A.W. Chase (Ann Arbor: R.A. Beal, 1873):

> After cutting out the Hams, they are looped by cutting through the skin so as to hang in the Smoke-Room, shank downwards; then take any clean cask of proper dimensions, which is not necessarily to be watertight.
>
> Cover the bottom with coarse salt, rub the Hams with fine salt, especially about the bony parts; and pack them in the cask, rind down, shank to the center, covering each tier with fine salt one-half inch thick; then lay others on them letting the shank dip considerably, placing salt in all cases between each Ham as they are put in, and between the Hams and the sides of the cask; and so on, putting salt in each layer as before directed; giving the thick part of the Ham the largest share. As the shank begins, more and more, to incline downward, and if this incline gets too great, put in a piece of pork as a check. I let them lie five weeks, if of ordinary size, if large, six weeks, then smoke them . . .

Brine Curing

Brine curing means soaking the meat in a mixture of salt and water. A brine that also contains sugar is called a sweet pickle. There are several methods for testing the strength of the brine. The old way was to add salt to water until there was enough salt in the water to float an egg or potato. The modern method is to use a salimeter, which will measure the strength of the brine very accurately and allow you to make a brine as strong or as weak as you like. The degree of salinity in a brine or pickle ranges from 60° to 95°. The higher the degree, the greater the concentration of salt.

The standard brine formula for each 100 pounds of meat is:

8 pounds salt	2 ounces saltpeter (optional)
2 pounds sugar	4 to 6 gallons water

Once the brine is made, the chilled meat is packed into a vat or watertight barrel. To retain the regular shape of cuts of bacon, line up the bacons evenly against the sides of the container. Stand them upright with the rinds against the container walls. Add enough brine to cover the meat. Four gallons of brine will cover 100 pounds of closely packed meat. More brine will be needed if the meat is packed loosely. The meat must be totally covered by the brine. If pieces float to the surface, which they probably will, push them down and keep them under the brine by placing over them a plate, piece of wood, or

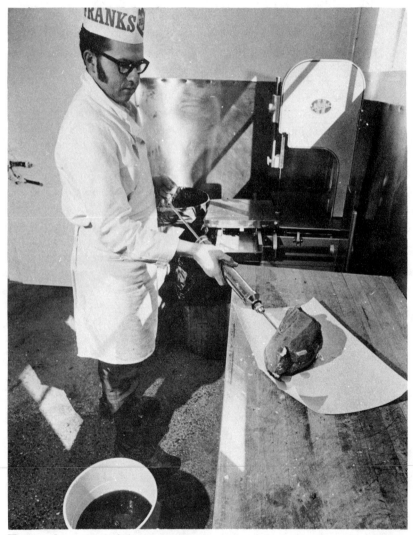

To speed up the brining process, you can pump brine directly into the meat tissues with a brine pump. It is still necessary to soak pumped meat in brine, but the curing time is shortened because brine has already penetrated the meat.

something similar, weighted down with a stone, brick, or other heavy object.

Curing time varies with the degree of salinity. An 85° pickle requires 9 days of curing for each inch of meat thickness, a 75° pickle requires 11 days per inch, and a 60° pickle requires 13 days for each inch. After curing, the meat may be smoked, if desired.

In the brine cure it is extremely important that all parts of the meat be exposed to the brine. Special care should be taken if the brine container is packed tightly. Stir the contents every third day; or remove all the meat once a week, pour off the brine, and repack the meat, pouring the same brine, which has been stirred well, over it. The stirring counteracts the tendency for the brine to separate and become stronger at the bottom as the heavier ingredients settle out. This process is called overhauling.

There is a chance that the brine may become sour (or ropy, as it is sometimes called). This results when the temperature of the room is too high. If the brine has an objectionable odor or adheres to your fingers in long strings when you dip your hand in it, change the brine immediately, wash the meat well, and add new brine. Throw out the old brine.

Some meats, such as bacon, will cure faster than larger pieces, such as hams. Remove the smaller pieces from the brine when they have been cured for the required time, and hold them in a cold place until the rest of the meat is finished curing.

Pumping the Meat

A variation on brining is injecting some of the brine directly into the meat tissues with the aid of a brine pump. The pump, which looks like a giant hypodermic needle, injects the brine into the meat,

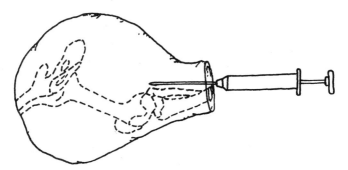

When injecting brine with a pump be sure to insert the needle close to the bones where spoilage is most likely to occur.

especially around the bones where spoilage is most likely to occur. The meat is still allowed to soak in brine, but the pump is used to assure penetration to and around the bone.

If the meat is to be pumped, it should be chilled and given about four guns, or cylinders, of the brine mixture. The brine is generally a little stronger, perhaps 5° to 10° stronger, when it is to be pumped. When pumping, drive the needle into the thickest part of the meat, slowly, but with steady pressure. Insert the needle close to the bone if there is a bone. Release the brine slowly as you plunge in and also as you draw the needle out so as to distribute the brine evenly around the path of the needle. Remove the needle when it is empty and hold your finger over the hole that the needle has made for a few seconds to prevent any brine from oozing out of the meat. Repeat this procedure in the thickest parts with the remaining cylinders. Curing time is shortened almost one-third for pumped meats because the brine has already penetrated the meat.

Combination Dry and Brine Cure

P. Thomas Ziegler, in THE MEAT WE EAT (Interstate Printers and Publishers, Inc.), recommends giving hams cured on the farm, where temperature during curing and curing time may vary, a combination dry cure and brine or sweet pickle. This method hastens salt penetration, and the quicker the salt gets to the center of the ham, the less danger there is of loss from spoilage. Experiments have shown that where 1½ to 2 pounds of salt for each 100 pounds of pork are rubbed into hams 24 to 48 hours before they are placed in a 75° brine, they will cure in 9 days per inch thickness of meat. According to Ziegler, this one rubbing of salt is all absorbed overnight and has more rapid penetrating qualities because it is not mixed with other ingredients or dissolved in water.

Curing Beef and Specialty Meats

Up to this point we have limited our discussion of curing meats to pork because pork is generally the only type of meat that is popularly cured and smoked. However, certain cuts of beef, usually the cheaper, tougher cuts, may be cured for corned beef.

Corned Beef

To prepare meat for corning, choose well-chilled meat. Cuts usually chosen for corned beef include the plate, rump, chuck, and brisket. Remove the bones to make packing easier. Cut the meat into pieces of uniform size. Weigh all the pieces, and allow 8 to 10 pounds of salt for each 100 pounds of beef. Before packing, sprinkle a layer of salt on the bottom of a clean, watertight container like a tight barrel or a crock. Place a layer of meat in the container, then more salt, and then another layer of meat. Continue in this manner until all the meat is packed or the container is full. Cover the top layer with a good amount of salt.

Allow the packed meat to remain in the salt for 24 hours and then cover the meat and salt with a solution made of:

4 pounds sugar	2 ounces baking soda
4 ounces saltpeter (optional)	4 gallons water

For flavor you can prepare a spice bag by tying together in cheesecloth 4 ounces of pickling spices and either 2 bulbs of garlic or 2 pounds of onion and placing it in the brine with the meat.

This solution is sufficient for 100 pounds of meat. Be certain that the meat is completely covered by the brine solution. If the meat is packed loosely, it may require more brine mixture. Readjust the above formula to make the extra amount of brine needed to cover the meat. To keep the meat totally immersed in the brine, push down with a clean board or something similar, and weight this board down with a rock, stone, or other heavy object.

The meat should be allowed to cure for 30 to 40 days at temperatures of between 38° and 40°F. Brine that is permitted to get warmer than 40°F. may become ropy. Check the brine frequently if curing temperature is difficult to control. Dip your fingers into the brine. If the solution sticks to your fingers, forming strands, it has become ropy. Remove the meat from the brine and wash it. Throw out the old brine and wash the container thoroughly with boiling water. Repack the meat and pour newly made brine over it.

At the end of the curing period, remove meat from brine; wash and drain before wrapping or smoking.

Cured or Pickled Tongue

Beef, veal, pork, and lamb tongue can all be cured and then cooked later as they are or lightly smoked. Cure them all just as you would corned beef. Curing times vary for each: beef tongue, 28 days; veal tongue, 13 days; pork tongue, 10 days; and lamb tongue, 8 days.

For every tongue that you plan to pickle, prepare a brine made of:

1 quart clean water
1/3 cup pickling salt

Weigh the tongue and then place it in a stainless steel or crockery container and pour the brine over it. Cover and weight down the lid with a clean rock or plastic bag filled with water so that the tongue is completely submerged. Let it stand in a cool place for 48 hours. Pour off the brine and rinse it in clean water. Then place it in a large pot, cover it with fresh water, and add the following for each tongue used:

2 pounds veal bones   2 bay leaves
3 stalks celery   12 peppercorns
12 cloves   2 cups sugar or 1 cup honey

Cook slowly until tender, about 1 hour for each pound of meat. Drain meat and place it once again in the stainless steel or crockery container. For each tongue cook together:

2 cups vinegar   3 medium onions, sliced
2 cups water

Cook for about 10 minutes, then cool and pour over the tongue. Cover as before so that the meat is completely covered with the brine. Store in a very cool, but not freezing place and use within a month.

Jerky

Jerky is meat that is cut in narrow strips and allowed to dry and become somewhat leathery. Traditionally it is made from beef, venison, or other game meat. In its dry state, it is chewy, flavorful, and convenient for campers and hikers because it is lightweight (about one-third the weight of fresh meat) and doesn't spoil easily. Jerky can also be reconstituted for soups and stews. To reconstitute, add extra water to soups and stews, toss in some pieces of jerky, and let them rehydrate as the soup or stew cooks.

Jerky should be made from lean meat only, as fatty meat will not dry out as thoroughly or keep very well. It may be just sliced and dried as is, but it is most often soaked in a mild brine or rubbed with salt and spices. It may be dried over coals, in a smokehouse, or in a low oven, depending upon your taste and equipment. Jerky may be dried outdoors, too, but this is tricky because you need a few continuous warm breezy days with strong, but not too strong sun. The drying area must be scrupulously clean and protected from insects and animals.

Here is one good recipe:

Prepare lean meat for jerky by cutting away all fat and gristle. Cut it on the bias into strips about ½ inch thick and 2 inches wide. Make them as long as you wish; 4 to 5 inches is convenient. The smaller the strips, the quicker the meat will dry.

To brine the meat mix together:

> 1½ cups pickling salt
> 1 gallon clean cold water

Soak the meat strips in this for 1 to 2 days. Drain.

Two ways to dry jerky outdoors over a low fire. The coals should be just warm so that they dry, not cook, the meat.

To dry salt the meat instead of brining, measure your meat and plan to use 1 teaspoon salt for each pound of meat. If you wish to season the meat, put the salt, some pepper, garlic powder or fresh garlic, and dried herbs in a blender and whiz well. Then rub them into the meat very carefully.

When the meat has been salted, arrange on racks in a warm (150° to 175°F.) oven. Do not let the oven temperature rise above 175°F. Lower is actually better. Turn the meat at least once during the drying process, which should take about 4 to 6 hours. The salted strips may also be laid over dowels or green sticks that are supported by forked branches, or other setup over a coal fire made from hardwood. The coals should not be hot enough to cook the meat, just hot enough to provide a drying warmth. This method should dry the meat in about 24 hours.

You'll know that it's done when it bends before breaking. Let it cool before testing, as it will seem more flexible than it really is while it's still warm. It is better to have jerky a bit overdry than underdry, as too much moisture will encourage bacterial growth.

Once cool and dry, wrap jerky in moisture-proof material and either freeze or refrigerate, or store it in a lidded container in a cool, dry place.

Dried Beef

Dried beef, also called chipped beef, is usually made from the round. If you have a particularly large piece to dry, cut it into 5- or 6-inch square pieces so that it will absorb the brine more readily and dry quicker. Weigh the meat, and for each 100 pounds plan to use 8 pounds of salt. Select a clean crock or tight wooden barrel that will easily accommodate the meat and allow for some headspace. Sprinkle a thin layer of salt over the bottom and pack the meat as tightly as possible, layering it with salt as you do. Save enough salt so that you can generously cover the top layer of meat with it.

Leave the meat as it is overnight or for 8 to 12 hours. Then pour a well-mixed solution of the following over the meat:

5 pounds sugar
2 ounces baking soda

4 ounces saltpeter (optional)
1 gallon lukewarm water

This is sufficient for 100 pounds of meat. After pouring over the meat, pour 3 more gallons of fresh water over it. Weight the meat down by placing a cover that fits just inside the crock or barrel and put a large, clean rock or big water-filled bag over this. All the meat should be submerged in the brine. Allow 2 curing days for each pound of meat.

Then wash off and drain the pieces and hang them up in a protected area to dry for 24 hours before placing them in the smokehouse. Cool smoke them (100° to 120°F.) for 70 to 80 hours or until they're quite dry. They are ready to use as they are, or wrap them up and hang as hams (see page 412) if they're in good-sized pieces. Smaller pieces should be refrigerated. Use within 6 months.

Pickled Pork
(Salt pork)

Although this may be sliced and cooked like bacon, it doesn't have the rich bacon flavor that most of us prefer because it is not smoked and usually doesn't have strips of meat running through the fat. Most often salt pork is used for larding roasts and flavoring bean dishes.

Weigh the fat back or pieces of fat pork. Then rub noniodized pickling salt all over the surface. Pack the pork into a clean crock and let it stand overnight or 8 to 12 hours.

Then prepare the following brine, cool it, and pour it over the pork. This amount is sufficient for 100 pounds of meat:

10 pounds pickling salt   4 gallons boiling water
2 ounces saltpeter (optional)

Cover the meat with a wooden lid that fits snugly inside the crock, and weight it with a clean rock or large plastic bag filled with water. The pork must be completely covered with the brine. Store in a cold, but not freezing place; the temperature should stay between 34° and 40°F. If you can, keep it in the refrigerator. Use within 3 weeks.

Pickled Tripe
(Stomach)

Thoroughly clean and rinse the tripe in cold water. Then place it in a pot, cover with water, and cook it just until the water shows signs of beginning to boil (in other words, scald it). When scalded, remove the inside lining of the stomachs by scraping them, which will leave a clean, white surface.

Boil the tripe until tender (usually about 3 hours) and then place in cold water so that you may scrape the fat from the outside. When you have done this, peel off the membrane from the outside of the stomach. Then the tripe is ready for pickling.

Weigh the tripe and then place in a clean, tight barrel or crock. Prepare a brine by mixing together for every 10 pounds of tripe:

> 1 pound pickling salt
> 1⅔ quarts water

Pour this brine over the tripe, cover, and weight down the lid with a rock or plastic bag filled with water so that the tripe is completely submerged in the brine. Let stand for 3 or 4 days.

Then rinse the tripe in cold water, clean out the barrel or crock, and put the tripe back in it. Cover with pure cider vinegar, cover, and weight it down as before so that no tripe is above the vinegar. Keep in a very cool, but not freezing place, and use within a matter of weeks.

Smoking

Hanging the Meat

Cuts of meat that are to be smoked should have a strong cord strung through them so that they will be easy to hang. Hams and shoulders should be strung through the shanks. If you don't have a regular stringing needle, make a hole in the shank with a boning knife and run a strong cord through the opening. In order to have your bacon hang straight and smooth without wrinkles (wrinkled bacon is unattractive and difficult to cut), it is best to insert a skewer through the flank end of the bacon strip and insert two cords, one on each side of the center of the bacon, just below the skewer.

Once the meat is strung, it should be scrubbed with water and a stiff brush to remove excess surface salt and grease. Rinse the meat and allow it to dry overnight before smoking. If not dry before smoking, the meat will come out of the smokehouse streaky.

Hang the meat in the smokehouse so that no two pieces are touching one another. If you're hanging your meat in two levels, hang it so that the top pieces will not drip on the ones below and make them streaky. If the fire is inside your smokehouse, make sure that the meat is at least 6 feet from the fire. Hang bacon and sausage further from the source of smoke and watch them carefully to prevent them from being oversmoked.

Building the Fire

Build a fire using green hardwoods, corncobs, or hardwood sawdust and shavings. Don't start your fire with paper, sawdust, or shavings. The ashes from these are very light and may rise and adhere to the hanging meat. However, sawdust or wood shavings may be added to cool the fire once it has started. As we said before, never use softwoods, like pine. The U.S. Department of Agriculture tells us that the Indians made an ideal smokehouse fire by radiating branches and sticks from a central point as if making spokes in a

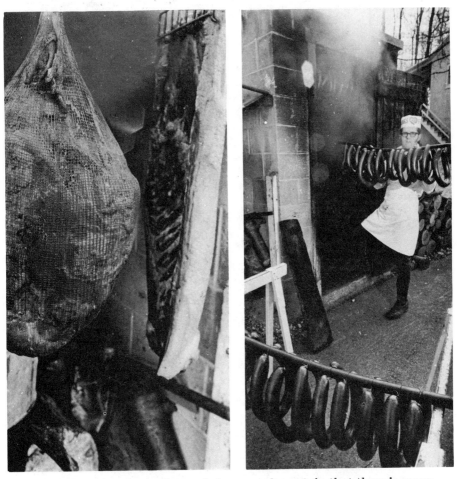

When hanging meat in the smokehouse make certain that there is space on all sides of each piece for smoke and hot air to circulate freely. No two pieces should be touching one another.

wheel. This arrangement produces a low, cool fire. If the fire gets too hot, or flames rather than smokes, smother it with damp leaves or sawdust. Do not allow the fire to get too hot. The ideal smokehouse

The ideal smokehouse fire is a cool, smoky one. Such a fire will dry meat slowly and give it a mellow, smoked flavor.

temperature is between 80° and 90°F. This is called a cool smoke and is used to smoke meat that is to be aged or held for many months. This smoking and gradual drying process helps to preserve the meat and give it a mellow smoked flavor.

Hot Smoking There is another method of smoking meats—with a hot fire, one that brings the smokehouse up to a temperature of 130°F. or more. Hot smoking is often done to meats that have not been cured, and if this is the case, the meats should be eaten immediately or soon after they have been smoked because this method does not preserve them. This method cooks the meat and gives it a smoky flavor at the same time. Poultry and fish are often smoked at high temperatures this way. Meats that are first cured and then hot smoked usually can be wrapped and hung. Dried beef (see page 410) is one such meat.

For hot smoking, your fire has to be hot and you may want to build your fire right in the smokehouse itself. Check it often to make sure that the fire doesn't get too hot and that the meat does not catch on fire.

How Long to Smoke Meats

Old-timers held their meat in the smokehouse for long periods of time. This long smoking considerably reduced moisture in the meat and provided for extended periods of keeping. We prefer our meats much moister now. A good average smoking time for hams is 2 around-the-clock days, or 5 to 6 hours each day on 6 consecutive days. Good-sized pieces of meat will generally need 48 to 72 hours of smoking. Small pieces of meat, like sausages and bacons, must be watched carefully to avoid oversmoking. Color is one guide to length of smoking time. Meat should be amber to chestnut brown in color. You'll have to be the judge here because a lot depends on your tastes, and also on the length of time you intend to keep your meat.

The agricultural extension service at the University of Minnesota makes a suggestion to those people who are using smoking as a means of preserving their meat. They suggest smoking meats lightly at first and leaving them to hang in the smokehouse. Pieces are pulled out, cooked, and eaten as they are needed. As the season progresses, smoking is repeated, a little at a time, until all the meat is used. This method dries off mold and moisture collecting on stored meat. In addition, a permanent smokehouse is a good place to store smoked meats, providing it is fly-, vermin-, and thief-proof.

Storing the Meat

After smoking, cool the meat, and if you wish, sprinkle it with ground black pepper mixed with a little red pepper. Pack it in parchment paper that is thick enough to absorb any grease that may run from the meat, and store it in tightly woven cotton bags. Don't use airtight bags; without some air circulation the meat will spoil. Crumpled newspaper packed around the meat will keep the meat away from the sides of the bags and help to prevent insect attack. Tie the bags securely to keep out insects, making sure that all cords, paper, etc., inside the bags are not sticking out and providing a convenient pathway into the bags for insects. Hang the bags in a cool (45° to 55°F. is ideal), well-ventilated, dry room. Examine the meat frequently for insects. If fly larvae infest the stored meat, cut out and burn all the infested parts. The uninfested part, which is safe to eat, should be put into the refrigerator and eaten as soon as possible.

The U.S. Department of Agriculture offers a suggestion for protecting your cured meats from insect attacks. They recommend burying the meat in a box or barrel filled with cottonseed hulls or in a bin of oats or other grains. If you are using a box or barrel, make sure that it is large enough to allow 3 or 4 inches on all sides of the hams you want to bury. Fill the container 3 or 4 inches deep with cottonseed hulls, place the ham in the middle on the hulls, and cover with hulls to a depth of at least 3 inches. Inspect the ham about once a month for infestation. If the hulls do become infested with grain beetles, replace the old hulls with clean, new ones. If you are burying the meat in a grain bin, attach a string to one end of the meat and run this string to the surface of the grain and attach a tag to it, so that you can easily locate the meat when need be.

If you notice a sour smell when the meat is removed from the smokehouse, don't be alarmed. This may last for a week or two after smoking, but does not mean that the meat is spoiled. Test the soundness of the meat with a ham trier, or make a similar instrument by taking a 10-inch piece of stiff wire and sharpening one end of it. For hams, run the trier along the bone to the center of the ham from both hock and loin ends. Try shoulders in the shank, at the shoulder point and under the blade-bone. If the trier brings out a sweet odor, the meat is sound. If there is an unpleasant odor, cut open the piece and examine it for spoilage. If there is a definite odor of putrefaction, destroy the entire piece.

Mold on the surface of the meat may give it a moldy flavor, but this does not necessarily indicate spoilage. Most mold and mold flavor can be scrubbed or trimmed off. Smoked meat may be brushed with vegetable oil or lard every month to delay mold growth.

Ham may be kept a year or more, but after a year, the hams tend to get hard, the fat may turn rancid, and the flavor gets more pronounced. The exceptions are Virginia and Westphalia hams which are best eaten after 1 to 2 years' storage. Bacons do not keep as well as hams and shoulders and should be cooked and eaten soon after they are cured and smoked. Cured meat may be frozen, but for best flavor, do not keep it frozen for more than a month or two. For freezing cured meats, we refer you to the previous section in this book on freezing meats.

All meats that are cured and/or smoked in the manners described in this chapter must be thoroughly cooked before they are consumed. Curing and smoking are not substitutes for cooking and do not destroy any trichinal larvae (which cause trichinosis) that might be present.

Suggested Readings for Preparing and Processing Meat for Storage

A COMPLETE GUIDE FOR HOME MEAT CURING. Chicago: Morton Salt Company.

Ashbrook, Frank G. BUTCHERING, PROCESSING AND PRESERVATION OF MEAT. New York: D. Van Nostrand Co., Inc., 1955.

Eastman, Wilbur F., Jr. THE CANNING, FREEZING, CURING & SMOKING OF MEAT, FISH & GAME. Charlotte, Vermont: Garden Way Publishing Co., 1975.

HOME PROCESSING OF POULTRY. University Park, Pennsylvania: The Pennsylvania State University, College of Agriculture Extension Service.

Levie, Albert, THE MEAT HANDBOOK. Westport, Connecticut: The Avi Publishing Co., Inc., 1970.

Reynolds, Phyllis C. THE COMPLETE BOOK OF MEAT. New York: M. Barrows and Co., 1963.

Wanderstock, J.J. and Wellington, G.H. LET'S CUT MEAT. Ithaca, New York: New York State College of Agriculture at Cornell University.

Preparing and Storing Fish

We've stressed elsewhere that freshness is the key to a good stored product, and this quality is especially important for fish. Fish is the most perishable of all foods, and it must be frozen, canned, or dried, or eaten very soon after it is caught to prevent rapid deterioration.

You can tell fresh fish by its bulging eyes and the reddish tint of its gills. The scales are shiny and are firmly attached to the skin, and the flesh is taut. One of the easiest ways to tell a fresh fish is by its odor—actually, it's lack of a "fishy" smell, especially around the gills and belly. The final test of a fresh fish is to place it in a tub or sink of clean water. If it's very fresh it will float.

Kill the fish (except shellfish) as soon as it is taken from the water. If the fish is allowed to flop around until it suffocates, the flesh may become bruised, inviting quicker deterioration. Keep shellfish alive by putting it in clean salted water if you can (see page 422).

If you can't eat or process your fish right away, pack it in ice and keep it well packed until you can dress and store it properly. Fish not held at temperatures below 70°F. will spoil in 1 or 2 hours. Drain off water from melting ice often so that the fish doesn't become "waterlogged" and lose some of its flavor to the melting ice.

Better yet, clean the fish immediately after killing it and wash the body cavity clean with fresh potable water. Spoilage will not set in as quickly if the entrails and body wastes are removed. Then pack the body cavity with ice and cover the fish with more ice.

If for some reason ice is not available, wrap the fish in damp moss, ferns, wet newspapers, or burlap and keep it out of the sun. This will not keep fish as well as ice, but it will do for a short period of time in an emergency.

418

Fish may also be kept in the coldest part of a refrigerator until you can clean and properly store it. Keep it wrapped or in a covered container to prevent any fish odor from permeating other foods.

Cleaning and Dressing

Fin Fish

Fish to be filleted need not be scaled and cleaned—skip this first section and jump to the section about filleting.

Begin with all fin fish not to be filleted by washing it in clear water to remove dirt and to wet the scales so that they are easier to remove. Then remove the scales by holding the fish by its tail on a flat surface and scraping the fish gently against the direction of the scales, or from the tail to the head, with the dull edge of a knife or with a fish scaler. Hold the knife or scale slightly diagonally so that you raise the scales as you scrape against them.

To clean a fin fish begin by (1) cutting off the head right behind the pectoral fins. (2) Then make a cut the length of the belly and remove the entrails.

Cut off the head by inserting a sharp knife right behind the pectoral fins through the backbone and down through the neck.

Now make a cut the entire length of the belly by inserting the knife into the neck cavity, and remove the entrails. If you're careful not to cut through the entrail membrane, you can remove the entire "pouch" easily and in one piece.

Cut off the dorsal and ventral fins with your knife. Cut about ½ inch into the flesh along both sides of the fins to make sure that you remove all the small bones around the fins.

Then remove the tail by slicing right through the body just above

To cut steaks, scale and remove head and entrails, and then cut in slices.

it. The head and tail can be used to make fish chowder or fish stock. If you wish to use the trimmings in this manner, make the soup right away—do not freeze or otherwise store them for use later; they don't keep well.

Wash the fish well to remove any scales, blood, or entrails. It is now ready to cook plain or stuffed, or to freeze, can, or dry.

Fish Steaks   This is a particularly good way to prepare large fish weighing 3 pounds or more. Salmon and swordfish, among others, are often prepared in this manner.

Clean the fish as outlined above. Then, beginning at the head end, cut cross-sectionally into about 1-inch-thick slices.

Filleting Fish   Fish about 2 pounds and under are often filleted. You need not scale, eviscerate, or trim a fish that you plan to fillet.

Begin by laying the fish on a flat surface. With a sharp knife cut above the tail just to, but not through, the bone. Then place the tip of the knife in this incision and cut all along the back ridge.

Lift up one edge of the fillet and cut along the backbone, loosening the fillet. Continue cutting the entire side of the fish away from the body so that it can be removed in one piece, then free it from the tail. Turn the fish over and remove the other fillet.

To skin the fillet, place it skin side down on a flat surface and grasp it at the tail end. Cut through the flesh about ½ inch above the tail. Then turn the knife at a slight angle on the horizontal and cut the skin away from the flesh as you work your way toward the neck.

Clams, Oysters, Mussels, and Scallops

These shellfish should be kept alive, if possible, until you are ready to eat or preserve them in some way, like canning or freezing.

Fillet fish by (1) cutting all around the back of the fish from the tail end up along the top ridge, over the neck, and then along the bottom. (2) Then lift up one end of the fillet and cut along the backbone to loosen the fillet. Continue doing the same on the other side of the backbone so the entire side comes off in one piece. Repeat this on the other side of the fish. To skin the fillet (3) place it skin side down and, (4) starting at the tail end, cut the flesh away from the skin.

One type of clam opener that is better than just a knife for shucking a quantity of clams.

Keep them in clean brine for 12 to 24 hours to allow them to clean themselves of any sand inside the shells. Make the brine by mixing ¼ cup pickling salt to each gallon of cold water. Cover them with this brine and keep them at refrigerator temperatures.

When you're ready to clean them, wash them well in either fresh or salted water (⅔ cup pickling salt to each gallon water). Then shuck them by inserting a *blunt* knife between the two shells and push down, cutting the muscles which hold the shells closed. Do not use a sharp knife, as a good edge is not necessary to sever the muscles, and you could quite easily cut yourself badly in the process.

If you plan to shell many hard-shelled clams or other shellfish, consider buying a simple device which holds the clam in place and with a little pressure on a handle, forces a blunt edge between the lips of the shells.

If you will be freezing or canning the fish or making soup, shuck them over a bowl so that you can catch all their liquid.

These shellfish can also be steamed open. The heat will kill the fish and loosen their muscles so that the shells will open. To steam, rinse off the salted water in which they were held and place in a steam basket or wire rack over boiling water. Cover the pot and reduce to a slow boil. Steam them 10 minutes and check to see if their shells have opened slightly. If not, continue to steam until they do.

Crabs, Crayfish, and Lobsters

Crabs should be alive and frisky when you get them. Keep them this way until you cook them for eating or processing.

Place crabs one at a time in rapidly boiling fresh or lightly salted water. Do this slowly so as not to reduce the water temperature too much as you add them. Make sure the crabs are entirely covered with water. Reduce the heat to simmering and cook the crabs about 25 minutes. Remove and plunge into cold water if they are to be shelled or eaten cold.

Plunge live lobsters or crayfish headfirst in a pot of fresh or lightly salted water large enough so that the fish will be completely immersed. Bring the water to a boil again, reduce heat, and simmer the fish until the entire shell is bright red: about 5 minutes for the first pound and 3 minutes for each additional pound. Remove and plunge into cold water if it is to be shelled or eaten cold.

Once cool, split the fish and clean out the viscera by running the belly under cold water. Crack the legs of crabs and the tails and claws of lobsters and crayfish and remove the meat.

Shrimp

Cut off the heads and shell by separating the underside of the shell at the neck and pulling off the entire shell in one piece, along with the tail. Then make an incision the whole length of the underside and cut or pull out the black vein (colon). To make sure all the vein is removed, hold the fish under cold running water as you take it out.

Shrimp may be cooked before or after cleaning. Most freezing and canning directions, however, suggest cleaning the shrimp first and packing them raw.

To cook shrimp, bring water to a boil. The water may be fresh or lightly salted (1 teaspoon salt to each quart water). Add shrimp slowly to the boiling water so as not to reduce its temperature too much. Reduce heat to simmering. Shrimp need only cook a few minutes, until they lose their transparency. Do not overcook them or they will become tough.

To cool them for eating cold or shelling, remove them from the hot water and plunge into very cold water.

Freezing

Because of the minimum amount of processing involved and the excellent preservation qualities of freezing, this is the recommended method for keeping fish. Although not as convenient and safe, fish can be canned, smoked, and dried (see later). Freeze fish within 24 hours after it is caught, and keep it at temperatures at or below 0°F.

For keeping more than a few weeks, it will have to be kept in a regular upright or chest freezer. A refrigerator freezer compartment will not hold it at a cold enough temperature.

Fin Fish

Fish weighing 2 pounds or less can be frozen whole, but it is best to fillet fish weighing 2 to 4 pounds and cut larger fish into steaks before freezing. (See earlier directions for cleaning and cutting fish.)

Unlike meats, fish are generally frozen in a thin layer or block of ice. This ice keeps air off the surface of the fish and prevents excessive oxidation which would otherwise deteriorate its quality. There are different ways to do this; here are three that we recommend:

Dipping   Before wrapping your fish in moisture-proof paper dip it in water. Wrap without draining.

Instead of dipping in just fresh water some people dip their fish in a weak brine solution to better preserve the fish. For the more perishable, leaner fish like:

bonito	ocean perch
carp	pompano
catfish	salmon (pink and chum)
eels	shad
flounder	smelt
herring	squid
lake trout	striped bass
mackerel	tuna
mullet	whitefish

mix:
2 teaspoons of ascorbic acid powder into 1 quart cold water.

Dip each fish for 20 seconds before wrapping.

For the fatter fish like:

bass	pike
blowfish	porgy
cod	red snapper
flounder	sole
fluke	swordfish
fresh water herring	whiting
haddock	yellow perch
halibut	yellow pike
pickerel	

mix:
¼ cup salt and 1 quart of cold water.

Dip each fish for 20 seconds.

Ice Glaze A thicker and more effective ice layer can be made by placing unwrapped fish one layer deep and not touching one another on a cookie sheet in the freezer. Allow them to freeze solid. Then dip each frozen fish or fish fillet or steak into ice water. Return the dipped fish to the cookie sheet in the freezer for a short time and then repeat the dipping process and keep repeating it until a ⅛- to ¼-inch thick ice layer has formed. Then wrap in moisture-proof paper or seal in a plastic bag and store in the freezer. This method is good if you plan to use a few pieces of fish at a time because you can freeze them all in the same package and take out just what you need—they won't stick together.

Ice Block Another way to prevent oxidation is to pack fish in an ice block. To do this, fill a loaf pan, coffee tin, or large freezer container with several small fish, fish steaks, or fillets to within a few inches of the top. Fill with water to cover the fish, making sure you leave some headspace to prevent spillage and to allow for expansion as the water freezes. When the block is solid, remove from pan and wrap in freezer paper. Then store again in the freezer.

Clams, Oysters, Mussels, and Scallops

None of these shellfish should be cooked before freezing, as the heat would toughen them. Merely wash them well in cold water to clean all the sand out and then shell them. Save their liquid. Scallops can be rinsed again after shelling, but don't rewash any of the others.

All should be packed in liquid to cover for freezing. Their liquid may be used, but if there is not enough of this liquid (and this will be the case with scallops, as they have little liquid), make up a brine of:

1 tablespoon pickling salt and 3 cups cold, fresh water.

Do not use seawater. Pack in rigid freezer containers, leaving a ½-inch headspace, and seal. Do not store for more than 4 months.

Crabs, Crayfish, and Lobster

Cook the fish as described on page 423. Cool them quickly in cold water, crack the shells, and remove the meat. Pack lightly in plastic bags or airtight containers, leaving no headspace. Seal and freeze. Use within 2 months for best quality.

Shrimp

Shrimp may be frozen shelled or unshelled; cooked or uncooked. Shrimp that have been shelled and have had their heads removed and the black vein (colon) running under the surface of their undersides cut out will keep longer. Most authorities recommend that shrimp be frozen uncooked because they tend to get tough if cooked first. For cooking and cleaning directions see page 423.

To freeze unshelled shrimp, rinse them quickly in cold water, drain, and pack tightly in freezer bags or containers, leaving no headroom. Seal and freeze. Store for no more than 2 months for best flavor.

To freeze raw, clean fish remove heads and black vein. Remove shells if you wish. Dip each shrimp briefly in a mild brine made from:

> 1 teaspoon pickling salt and 1 quart water.

Drain, pack, and seal as above.

Thawing and Cooking Frozen Fish

Frozen fish need not be thawed before cooking, unless you plan to stuff it or bread it; then thawing is necessary. Allow extra time for cooking unthawed fish, about 1¼ times as long as unfrozen fish.

If you wish to thaw fish, do it slowly so that it loses little of its own juice and delicate texture. Thaw in the fish's freezer wrapping in the refrigerator. Fish that is completely thawed should be cooked as soon as possible.

Freezer Storage Time Chart for Fish
(Freezer Temperature 0°F. or colder)

Fish	Recommended Storage Time in Months
bass	2 to 3
blowfish	2 to 3
bonito	4 to 5
carp	4 to 5
catfish	4 to 5
clams	3 to 4
cod	2 to 3
crab	2
eels	4 to 5
flounder	4 to 5
fluke	2 to 3
fresh water herring	2 to 3
haddock	2 to 3
halibut	2 to 3
herring	4 to 5
lake trout	4 to 5
lobster	2
mackerel	4 to 5

mullet	4 to 5
mussels	3 to 4
ocean perch	4 to 5
oysters	3 to 4
pickerel	2 to 3
pike	2 to 3
pompano	4 to 5
porgy	2 to 3
red snapper	2 to 3
salmon (pink and chum)	4 to 5
scallops	3 to 4
shad	4 to 5
shrimp	2
smelt	2 to 3
sole	2 to 3
squid	4 to 5
striped bass	4 to 5
swordfish	2 to 3
tuna	4 to 5
whitefish	4 to 5
whiting	2 to 3
yellow perch	2 to 3
yellow pike	2 to 3

Canning

Because of the delicate nature and high perishability of fish, canning is a tedious task; you will have to prepare several brine solutions to prevent the fish's protein from coagulating. In addition, it is necessary to first cook fish-filled jars to exhaust air and then process for a relatively long time. If you have a choice, we suggest that you freeze it instead. However, if you decide to can it, work carefully and quickly.

Please read the previous sections on preparing fin fish and/or shellfish, and the discussion of canning meats for necessary equipment and general procedures before you attempt to can fish. As in any other means of storing fish, be sure the product is as fresh as possible. Lobsters, crabs, crayfish, and clams should be alive and all fin fish kept on ice until you prepare them for canning.

Since fish are low-acid foods and vulnerable to toxic bacteria they must be canned under pressure and processed for a relatively long time. Glass half-pint and pint jars can be used. Larger canning jars should not be used, as the long processing time necessary to sterilize and safely seal them would seriously damage the quality of the fish.

Since each type of fish is canned in a slightly different fashion, we'll give directions for several kinds here.

Clams

Clams should be scrubbed and brined overnight before shucking. To make the brine, dissolve ¼ cup of pickling salt in each gallon of clean water you use. This brine should only be used once. Shuck or steam the clams open, reserving the liquid if you shuck them.

Take out the meat, and clean away the muscle that joins the clam to the shell. Remove the protruding neck if you're canning "steamers." Place the meats in another clean brine and swish them around well to clean them. If necessary, drain the clams and repeat the cleaning process. Then drain.

Mix up a solution of:

> ½ teaspoon citric acid and 1 gallon clear water.

Bring it to a boil in a good-sized pot. Place the clam meats into this boiling solution and boil for 2 minutes.

Drain again and pack into half-pint jars. Cover with boiling hot water, leaving a ½-inch headspace. Place the lids on the jars and screw the band down just until it catches the threads. *Don't screw tightly.* Leave one jar of fish open as the test jar—insert a thermometer deep into its middle. (This test jar may then be canned with the rest.)

Exhaust the air from the jars by placing them on a rack in your pressure canner. Pour hot water around the jars, halfway up. Cover the pot, but don't pressure it—leave the petcock open or don't put the weighted gauge on, depending upon your canner. Bring the water to a boil and boil for 10 minutes. Then check to see if the thermometer has reached 170°F. Continue boiling until it does, then adjust seals and *process at 10 pounds pressure for 70 minutes.*

Remove jars and allow to cool before storing.

Crabs

Cook and remove meat as described on page 423. Boil meat briskly for 15 minutes in a blanching solution made up of:

> 1 gallon clear water, 1 cup pickling salt, and
> either ¼ cup lemon juice or white vinegar.

Bring brine to a boil before adding the meat.

Then drain and cool the meat quickly under cold running water and clean in a new brine of:

> 1 gallon cold clean water and 1 cup pickling salt.

Drain, rinse, and wash in a solution of:

> 1 gallon clean water and 1 cup white vinegar.

Remove, squeeze out as much excess liquid from each piece of meat as you can, and pack into jars. Cover with boiling water, leaving ½-inch headspace. Put lids and bands on loosely and exhaust air as for clams. When the contents of the jars reach 170°F., adjust seals and *process at 10 pounds pressure for 65 minutes.*

Fresh Water Fish
(like mackerel, lake trout, whitefish, and mullet)

Clean and dress fish as described on page 419. Cut away the thin belly strip and split the fish, being careful not to remove the backbone. Then cut into pieces the length of pint jars. Prepare a brine of:

¾ cup pickling salt and 1 gallon clean cold water.

Soak fish pieces for about 1 hour.

Drain and pack fish strips lengthwise in your canning jars with the skin side of the fish facing outward. Pack as tightly as possible. Then pour over the fish a boiling brine made from:

⅓ cup pickling salt and 1 gallon of water.

Leave a 1-inch headspace. Do not cover the jars.

Exhaust the jars as for clams. Remove them from the boiling water, invert them on a rack, and let all the brine drain from them for a few minutes. You may add a bay leaf and/or 1 or 2 slices of onion to each jar for flavor. Then adjust seals. Put fresh boiling water in the canner and *process pints at 10 pounds pressure for 1 hour 40 minutes.*

Mackerel is often canned in a tomato sauce. To do this follow the directions above, but after inverting the cans and letting them drain, pour hot tomato sauce (see Index for recipes) over the fish, leaving a ½-inch headspace.

Adjust seals and *process pints at 10 pounds pressure for 1 hour 40 minutes.*

Lobster

Cook and remove meat as described on page 423. Prepare just as you would crabs.

Pack in half-pint jars as crab, but cover with boiling brine (not plain water) made from:

1 quart water and 1 teaspoon pickling salt.

Leave a ½-inch headspace.

Seal partially and exhaust as for clams. Then adjust seals and *process half-pints at 10 pounds pressure for 70 minutes.*

Shrimp

Wash the fish well, then remove the heads, shells, and black vein along the underside. Hold the cleaned shrimp for 25 minutes in a brine made of:

> 1 gallon cold clean water and 2 cups pickling salt.

Stir occasionally. Drain.

Then prepare a blanching brine by mixing together:

> 1 cup lemon juice and 1 cup pickling salt
> for each gallon of fresh water you use.

If you're canning a large quantity of shrimp, prepare several gallons of this brine, since it can be used only once. Bring a gallon of the brine to a boil at a time and when boiling, slowly pour in 1 quart of shrimp. Boil for 6 to 8 minutes (the larger the shrimp, the longer the time). Drain the shrimp and cool them on wire racks in a cool, preferably drafty place—or in the refrigerator. Add new brine and fresh shrimp and repeat as often as necessary.

When shrimp are thoroughly cool, pack them tightly in half-pint jars, leaving a ½-inch headspace. Put on lids, screw bands on slightly, and exhaust as for clams.

Then adjust lids and *process half-pints at 10 pounds pressure for 35 minutes.*

Drying Fin Fish

As with all other methods of processing fish, drying should be done with great care. Read about selecting, keeping, and cleaning fish on page 418 before you attempt to dry any fish. Only lean fin fish (see the listing on page 424) are recommended for drying. The greater amount of oil in fatty fish makes drying them difficult. Shellfish are not suitable for drying, either.

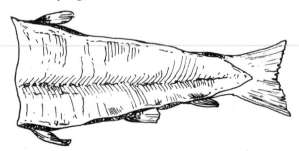

Fish split and spread open for drying and/or smoking

The directions here are for drying fish that you intend to store dry, but not smoked. Fish prepared for smoking are dried for a much shorter time first, since the smoking process further dries them. For directions for smoking fish, see page 433.

Clean, scale, and dress fish according to earlier directions. If you're drying large fish, fillet, but do not skin them. Medium and small fish should be split and spread open in one piece. To do this, insert your knife in the neck opening made by removing the head and bring the knife down along the back, cutting the fish in half right along the backbone all the way to the tail. Do not cut along the belly, but just spread the fish apart so that it lies flat in one piece. Then wash the fish in a mild brine made from:

1 cup pickling salt to each gallon of cold, clean water.

Check the fish carefully to make sure that there are no traces of blood, and no dirt, insects, or other foreign matter. Weigh the fish, and then rub pickling salt into each piece of fish, using 1 pound of pickling salt for each 4 pounds of fish. Finely grained pickling salt is best for this because it quickly penetrates into the pores of the fish and readily forms a brine. Stack the fish skin side down in a long, flat container. A wooden box with holes in the bottom for drainage of the fish juices and the brine they form is ideal. Stagger the layers of fish, sprinkling salt generously between layers, using another pound for each 4 pounds of fish.

Keep the fish indoors in a cool, dry place for 2 to 7 days. The thinner the fish, the fewer stacked layers, and the drier the weather, the shorter this first drying period.

Wait for a good drying day and then remove the fish and scrub it well so that there are no traces of pickling salt remaining. If the weather is rainy, leave the fish in the salt until the weather improves. Choose a drying area that is shaded and out of reach of animals and is as insect-free as possible. Stack the fish on clean wooden racks. Lay the pieces skin side down in a single layer.

If you have only a few fish to dry they may be hung up under an overhang where there is good air circulation. Run an S-hook through the collarbone and attach a strong string to this, or thread the string right through the bony neck.

Hopefully, the weather on the days you plan to dry fish will be warm (but not hot), dry, and breezy. Turn the fish three or four times the first day. Bring the fish in at night before dusk (and dew) comes and lay out again the next day. If the next day is rainy, leave the fish stacked under cover, but to prevent spoilage, sprinkle a thin layer of pickling salt between the fish every other day until they can be dried outdoors again. *Do not let the sun shine right on the fish, as the heat of the sun's rays can burn or cook the fish or at least encourage oxidation which will result in an inferior product.*

The fish will take anywhere from 2 to 6 days to dry, depending again on the thickness of the fish and the weather. When you suspect that it may be dry, pick up a piece and squeeze it between two fingers. If it is firm and springs back when the pressure is released, leaving no finger imprint, it's dry. Wrap individual pieces in moisture-proof paper and store in a lidded container, like a tight wooden box, in a cool, dry place. If you notice mold or rust forming on the stored dry fish, scrub the fish off in a fresh salt brine and air dry for a day or so.

Brining Fin Fish

Brining fish produces a milder salted product than does drying. The fish are kept in the brine for the entire storage period, but the brine must be changed every 3 months. As a rule, brined fish don't keep as long as those that are dried.

For this type of curing you'll need a stoneware crock (one 2-gallon crock for every 25 pounds or less of fresh fish) and a wooden tub, barrel, or box. As in dry curing, finely grained pickling salt is best because it readily penetrates the flesh.

Clean and cut up the fish as for dry curing. For best brining and storing, the fish should be of one variety and one size. Score the flesh of large pieces of fish to help the brine penetrate quicker and more evenly. Make ½-inch deep cuts at intervals of 1 to 2 inches into the flesh lengthwise.

Wash the fish well and soak it for 30 to 60 minutes, depending upon size, in a brine made from:

½ cup pickling salt to each gallon of fresh, cold water.

Then drain well for up to 10 minutes.

Weigh the fish and plan to use about 1 pound of salt for each 3 pounds of fish in all—for both the initial dredging and also the layering. Don't use much more than this.

Place a thin layer of salt in the bottom of the crock. Then dredge the fish with pickling salt. Rub more salt into the scored areas. Then arrange the salted fish over the salt in the crock, trying not to have too much overlapping. Sprinkle another layer of pickling salt over the fish and arrange another layer of fish, staggering the pieces over the layer underneath. Continue layering salt and fish until all the fish is so placed. Arrange the top layer skin side up.

Cover the crock with a lid that will sit right on top of the fish, not on the crock rim. The lid will naturally have to fit inside the crock. Then weight down the lid as you do in making sauerkraut or brining meat, with a heavy rock or large plastic bag filled with water. The fish will make its own brine. Fish should be left in this brine for anywhere from 2 to 10 days, depending upon their size.

Then remove the fish and scrub them well with a fresh brine and a stiff brush. Clean out the crock and repack as before, without dredging in salt first. Layer as before with much less salt this time. Cover the fish with fresh brine (2⅔ pounds of salt to each gallon of clean water). Cover the crock and store in a cool, dark place.

After 3 months or less if the weather is hot or fermentation begins, remove the fish and throw out the brine, relayer the fish, and cover with new brine. Repeat this overhauling every 3 months. For best quality, eat the fish within 9 months, counting from the time it was first put in the crock.

Smoking Fin Fish

Before you attempt to smoke any fish, read about cleaning, dressing, and keeping fresh fish on page 418. Since fish are smoked much like meats, read about smokehouses and smokehouse fires starting on page 395. Prepare fillets or split open the fish and spread flat as you would for drying (see page 430), but be sure to leave the collarbone on split fish that are spread open in one piece so that you have a handy place through which to insert an S-hook or thread a heavy string for hanging in the smokehouse. For best results choose one variety of fish to smoke at a time and try to cut them to approximately the same size.

Fish to be smoked must first be cured and then dried slightly before being placed in the smokehouse. The salt cure acts as a preservative, and the drying produces a protective sheen over the surface of the fish called a pellicle which will seal in the fish juices and prevent the protein in the flesh from coagulating and forming harmless but unattractive white curds on the surface.

Fish for smoking may be either just held in brine or brined for a shorter time and then dry salt cured.

Brine Cure   After the fish has been cleaned and cut properly, brine it for 8 to 12 hours, depending upon the size of the fish. Prepare the brine by mixing:

1½ cups of pickling salt and 1 gallon of fresh, cool water.

Prepare as much brine as is needed; make enough so that the fish are completely covered, and hold them at about 40°F.

Remove and drain the fish. Then rinse them off well.

Combination Brine and Dry Cure   Prepare the fish as for the brine cure, but leave them in the brine for only ½ hour. Then rinse and drain them for a few minutes and dredge in fine pickling salt. Layer them in a wooden or crockery container, placing layers of salt between each layer of fish, staggering the fish one layer on top of another. Place the top layer skin side up. Leave the fish in this dry cure for—on the average—8 hours, but longer if the day is humid and the fish are thick, and less if the day is dry and the fish are thin.

Drying   Both brine and brine- and dry-cured fish must then be dried to form the pellicle we spoke of earlier. See the directions for drying fish in a shady, outdoor place on page 431. Dry only for 3 hours or until the pellicle forms. The smoking process will further dry the fish and make it ready for storing.

Smoking   Place the fish in the smokehouse, by hanging it either from S-hooks that have been inserted or from strings that have been threaded in the collarbones and then on rods, dowels, or hooks. Fillets should be spread one layer thick on wire racks that have been lightly coated with vegetable oil to prevent sticking. Be sure that no two pieces are touching and that no fish is against a wall of the smokehouse for good smoke circulation and even drying and smoking.

As mentioned in the section on smoking meats, the fire should be kept cool and at as constant a temperature as possible throughout the smoking period. It should not get hotter than 70°F. Fish should be smoked for at least one entire day—day and night—and longer if it is to be kept more than a couple of weeks. If you plan to keep for a year, you'll have to smoke continuously for 1 week. Once the smoking has begun, it should not be interrupted until it is completed. This will mean feeding the fire if necessary in the middle of the night.

Storing   When finished smoking, cool the fish completely and store it in *porous* packaging that allows the smoked fish to "breathe," and keep in a dry, cool place. Smoked fish should not be canned or frozen. Fish smoked in this manner can be eaten without further cooking.

Meat Recipes

Baked Stuffed Apples

6 large cooking apples
1 cup sausage meat

1 teaspoon salt
honey (optional)

Clean out the center of the apples by scooping out the cores and pulp (be careful not to cut through to the bottom of the apples). Chop the pulp and combine it with the sausage meat and salt. Sweeten with a little honey, if desired. Stuff the apples with this mixture, place upright in a baking dish, and bake in a 375°F. oven until tender, about 45 minutes. Serve with brown rice or noodles.

Yield: 6 servings

Beef Stroganoff
(to freeze)

3 pounds beef fillet
2 tablespoons oil or butter
1 cup chopped onions
2 pounds sliced mushrooms
4 large green peppers, thinly
  sliced

2 garlic cloves, minced
¼ cup sifted whole-wheat flour
3 cups beef broth or stock
1 pint sour cream or yogurt

Partially freeze the beef; then slice into long, thin slices. Brown the meat in the oil or butter in a very large skillet. Remove the meat and put aside. Sauté the vegetables and garlic in the skillet. Remove vegetables and set aside.

Make a paste of the flour and drippings in the skillet. Slowly add the broth, stirring constantly. When the mixture thickens, lower the heat and blend in sour cream. Add all the vegetables and the meat.

To freeze: Put into freezer containers. Seal, label, date, and freeze.

When ready to serve: Partially thaw; pour into skillet or pan and cover. Warm gently, stirring occasionally. Simmer for 30 minutes. Serve over brown rice or noodles.

Yield: 15 servings

Savory Stuffed Cabbage Rolls
(to freeze)

16 very large or 20 smaller
   cabbage leaves
½ cup brown rice, cooked with
½ cup barley (barley will be
   slightly undercooked)
2 pounds ground beef
½ cup chopped onion
3 teaspoons salt
½ teaspoon pepper
2 eggs

For serving:
2 cups tomato sauce
2 cups stewed tomatoes
2 tablespoons oil
3 tablespoons honey
¼ cup lemon juice
¼ cup seedless raisins
   (optional)

Prepare cabbage leaves by covering entire head of cabbage with boiling water and steaming for about 5 minutes. Drain. Carefully remove the required number of leaves.

Combine grains, beef, onion, salt, pepper, and eggs. Put about 2 tablespoons of mixture on each cabbage leaf. Tuck in sides and roll up carefully. You may have to secure the roll with a toothpick.

To freeze: As soon as the cabbage rolls have been formed, pack them into a freezer container. Seal container, label, date, and freeze immediately.

When ready to serve: Place rolls in a deep casserole dish. Pour tomato sauce, stewed tomatoes, and oil over them. Then cover and bake in a preheated 350°F. oven for 1½ hours. Add the honey, lemon juice, and raisins to sauce and cook 30 minutes longer. Serve with sour cream as a garnish, and accompany the meal with black bread and butter.

Yield: 16 to 20 stuffed
cabbage rolls

Stuffed Cabbage

½ pound scrapple
1 onion, chopped
1 tablespoon arrowroot powder
1 tablespoon chopped parsley

1 cup fine whole-wheat bread
    crumbs
1 cup milk
1 egg, beaten
1 head cabbage

Put the scrapple in a bowl and mix with the other ingredients except cabbage. Dress a well-formed head of cabbage and cut the top off neatly to make a "lid." Hollow out the center to make a shell, with a hollow about 3 inches wide and 3 inches deep. Fill with the scrapple mixture and replace the "lid." Tie it in a cheesecloth and boil in salted water until tender. Such scrapple mixture as is left form into balls, dredge with flour, and fry brown. Serve them on a platter around the cabbage. If desired, make a tomato sauce to accompany it.

Yield: 4 servings

Chicken à la King
(to can or freeze)

¼ cup chicken fat or butter
½ cup flour
5 cups chicken broth
meat from 2 stewing chickens,
    cooked and diced

¼ cup chopped pimiento
1 tablespoon chopped parsley
salt and pepper to taste
mushrooms

Melt fat or butter; add flour and stir until smooth. Gradually add chicken broth and cook until thickened, stirring constantly.

To can: Add chicken and remaining ingredients, except mushrooms. Simmer 5 minutes. Pack hot into hot jars, leaving 1-inch headspace. Adjust caps. Process pints 1 hour and 5 minutes, quarts 1 hour and 15 minutes, at 10 pounds pressure.

When ready to serve: Brown in butter ¼ cup mushrooms for each pint of chicken à la king; add to meat mixture and heat through.

To freeze: Brown 1 cup mushrooms in butter. Add mushrooms, chicken, and remaining ingredients to the sauce mixture. Pour into clean freezing containers. Seal, label, date, and freeze.

When ready to serve: Reheat gently.

Yield: 4 pints

Chicken-Corn Casserole
(to can or freeze)

½ cup chicken fat or butter
1 cup sifted whole-wheat flour
1½ teaspoons salt
½ teaspoon pepper
1 small onion, minced
1½ quarts hot chicken broth

1¾ quarts (about 2 pounds)
    cooked, diced chicken meat
6 cups whole kernel cooked
    corn
¾ cup chopped pimiento
bread crumbs, butter, cheddar
    cheese (for serving)

Melt fat or butter. Stir in flour, salt, pepper, and onion. Add broth slowly, and cook until thickened, stirring constantly. Add chicken, corn, and pimientos to hot mixture.

To can: Pack hot into hot jars, leaving 1-inch headspace. Adjust caps. Process pints 1 hour and 5 minutes, quarts 1 hour and 15 minutes, at 10 pounds pressure.

To freeze: Evenly divide mixture in 4 baking pans lined with freezer paper. Freeze. Fold and seal wrapping. Label, date, and return to freezer.

When ready to serve: Preheat oven to 350°F. Pour canned casserole into baking dish, or unwrap frozen casserole and place in baking dish. Top with a mixture of ¼ cup fine dry bread crumbs mixed with 1 tablespoon melted butter and ½ cup shredded cheddar cheese. Bake until center is hot and crumbs are lightly browned (about 1¼ hours if casserole is frozen, 45 minutes if canned).

Yield: 2 quarts or 4 casseroles,
6 servings each

Chili con Carne
(to can or freeze)

3 pounds ground beef
3 tablespoons olive oil
2 cups chopped onions
2 cups chopped green peppers
3 garlic cloves, minced
3 pounds canned tomatoes,
    with liquid

⅓ cup chili powder or less
    (depending on taste)
1 tablespoon salt
1 hot red pepper
1 teaspoon cumin seed
    (depending on taste)

Brown meat in a very large skillet in oil. Add onions, green peppers, and garlic, and cook slowly until tender. Add remaining ingredients, bring to a boil. Lower heat and simmer for 25 minutes. Skim off excess fat.

To can: Pour hot into hot jars, leaving 1-inch headspace. Adjust caps. Process pints 1 hour and 15 minutes, quarts 1 hour and 30 minutes, at 10 pounds pressure.

To freeze: Divide chili into family-sized containers. Label, date, cool, and freeze.

When ready to serve: Add cooked or canned pinto or kidney beans to chili (thaw, if frozen), heat, and serve.

Yield: about 3 quarts

Curried Meatballs in Vegetable Sauce
(to freeze)

1 cup raw wheat germ
2½ pounds lean ground beef
1 pound lean ground pork
6 eggs
2 minced cloves of garlic

¼ cup curry powder
1 tablespoon salt
¼ teaspoon pepper
oil as needed

Combine ingredients, except oil. Shape into 24 balls. Heat oil and brown the meatballs, then add ½ cup water. Cover closely and simmer for 30 minutes.

Vegetable Sauce:

1½ cups finely shredded
    cabbage
4 onions, chopped
4 green peppers, chopped

3 large potatoes, diced
¼ teaspoon black pepper
1 teaspoon salt
water

Put vegetables and seasonings into 1½ cups boiling water. Cover and simmer for 20 minutes or until vegetables form a thick sauce. Stir occasionally.

To freeze: Pack 12 meatballs into each of 2 freezing containers. Pour ½ of the vegetable sauce into each container. Cool. Label, date, seal, and freeze.

When ready to serve: Reheat in a double boiler. Remove from heat and stir in 3 tablespoons yogurt for each 6 servings.

Yield: 12 servings

Eggplant Moussaka
(to freeze)

2 **large** or 4 small eggplants
salt

2 pounds lean lamb, ground (or
    ground beef)

½ cup chopped onion
2 garlic cloves, minced
6 peeled, chopped fresh
 tomatoes or 2 cups canned
 tomatoes
½ teaspoon nutmeg
½ cup dry red wine
1 teaspoon fresh chopped basil
 or ¼ teaspoon dried

½ teaspoon fresh, chopped
 thyme or ¼ teaspoon dried
2 tablespoons butter
2 tablespoons sifted
 whole-wheat flour
1 cup stock
½ cup grated Romano cheese
 (or more)

Cut off stems of eggplant and peel lengthwise, leaving ½-inch strips all around. Cut into ½-inch slices; sprinkle salt on each, and then leave them for 30 minutes or so to drain.

Brown meat, adding oil if necessary. Add onions and garlic, and sauté. Add tomatoes, seasoning, wine, and herbs. Simmer until meat is tender.

Pat eggplant slices dry. Sauté in oil until brown on both sides.

Melt butter; add the flour, and stir to a smooth paste; then add stock, cook, stirring constantly, and bring to a full boil. Lower heat; add grated cheese and simmer, stirring, until cheese melts.

To freeze: In each of 2 lined baking pans, place a layer of eggplant, then a layer of meat, a layer of eggplant, then meat, finishing with a layer of eggplant. Pour on the cheese sauce, and sprinkle on a little extra grated cheese. Freeze. When solidly frozen, wrap in freezer paper. Label, seal, date, and return to freezer immediately.

When ready to serve: Preheat oven to 375°F. Do not thaw casserole; unwrap and place in baking dish. Bake for about 1¼ hours or until casserole is heated through.

Yield: 2 casseroles each
serving 6 to 8

Fish Chowder
(to can or freeze)

5 pounds cleaned fish, filleted
 (retain the bones)
3 quarts of water
salt and pepper to taste

½ pound salt pork
1 cup chopped onions
3 cups diced potatoes
2 cups diced carrots

Cut fillets into 1-inch pieces, and refrigerate until needed. Make a stock with the fish bones by boiling them in the water until the meat falls from the bones. Season with salt and pepper; strain.

Dice the pork and brown it in a frying pan. Add the onions and sauté until they are tender and yellow. Add the pork and the onions, other vegetables, and the cubed fillets to the fish stock, and boil for 10 minutes.

To can: Season to taste with other herbs or flavorings. Pour into prepared jars. Seal and process in pressure canner at 10 pounds pressure, 90 minutes for pints, 1 hour and 40 minutes for quarts.

To freeze: Plan to season it when reheating. Allow the soup to cool, and remove excess fat. Pour into leak-proof containers leaving 1/2- to 3/4-inch headspace. Seal, label, date, and freeze.

When ready to serve: Reheat gently.

Yield: 7 pints

Goulash
(to can or freeze)

1/2 cup flour	1/2 cup cider vinegar
4 teaspoons salt	4 bay leaves
3 tablespoons paprika	20 peppercorns
3 teaspoons dry mustard	2 1/2 teaspoons caraway seeds
6 pounds boned chuck, cut in 1-inch pieces	6 medium onions, halved
	6 large carrots, halved
1/2 cup oil	10 stalks celery, halved
1 3/4 cups water	

Combine flour, salt, paprika, and mustard. Roll meat in flour mixture. Brown slowly in oil. Sprinkle remaining flour mixture over meat; add other ingredients. Cover and simmer until tender (1 1/2 to 2 hours). Remove bay leaves.

To can: Pack hot into hot jars, leaving 1-inch headspace. Adjust caps. Process pints 1 hour, quarts 1 hour and 15 minutes, at 10 pounds pressure.

To freeze: Pour into family meal-sized containers. Cool, label, date, and freeze.

When ready to serve: Reheat and serve with boiled potatoes or noodles.

Yield: about 6 pints

Stuffed Green Peppers
(to freeze)

2 dozen large green peppers	1 pound mushrooms, diced
1/2 cup butter	3 garlic cloves, minced
2 cups minced onions	1 pound pork sausage

1 pound ground beef
6 cups cooked brown rice
4 beaten eggs

salt and pepper
1 cup grated Parmesan cheese

Cut off the tops of the green peppers, and remove seeds and membranes. Blanch for five minutes.

Sauté the remaining vegetables in the butter until soft.

Brown the meat in a separate skillet, draining off excess fat. Add meat and the remaining ingredients to the sautéed vegetables. Combine. Stuff the peppers.

To freeze: Set the peppers on a tray in freezer. When completely frozen, wrap individually or in family-sized portions in plastic bags or heavy-duty foil. Seal, label, date, and return to freezer.

When ready to serve: Preheat oven to 375°F. Place peppers upright in a covered casserole dish. Bake, covered, for approximately an hour.

Yield: 24 stuffed peppers

Italian Meat Sauce
(to can or freeze)

4 pounds ground beef
1 pound ground pork or
   sausage meat
2 cups chopped onions
1 cup chopped green peppers
9 cups whole tomatoes and
   juice
2⅔ cups tomato paste (4
   6-ounce cans)

1 tablespoon honey
2 tablespoons minced parsley
1½ tablespoons salt
1 tablespoon oregano
½ teaspoon pepper
½ teaspoon ginger
½ teaspoon allspice
2 tablespoons vinegar

Brown meat; add onions and green peppers and cook slowly until tender. Add remaining ingredients and simmer several hours until thick. Skim off excess fat.

To can: Pour hot into hot jars, leaving 1-inch headspace. Adjust caps. Process pints 1 hour, quarts 1 hour and 15 minutes, at 10 pounds pressure.

To freeze: Pour into containers. Allow to cool, label, date, and freeze. Serve over spaghetti, use in casseroles or as base for lasagne.

Yield: about 3 quarts

Lamb or Beef Pie
(to freeze)

3 pounds boneless lamb or beef
   cut in 1-inch pieces
1 quart water
1 tablespoon salt
1½ cups chopped celery
3 cups cubed potatoes

3½ cups quartered onions
2 cups sliced carrots
1 cup peas, fresh or frozen
½ cup whole-wheat flour
1 recipe whole-wheat pastry
poppy seeds (if desired)

Brown the meat in its own fat. Add the water. Simmer until meat is tender. Add salt and vegetables and cook until vegetables are almost tender.

Drain the broth from the meat and vegetables, and add water, if necessary, to make 3½ cups. Add ½ cup cold water to the flour and stir until smooth. Stirring constantly, slowly add flour mixture to the rest of the broth, and cook until thickened. Combine with meat and vegetables. Cool mixture quickly.

To freeze: Line 4 8-by-8-inch baking pans with freezer paper, allowing enough extra wrap to fold over top. Pour stew for 1 meal (6 servings) into each pan and top with pastry crust. Sprinkle with poppy seeds, if desired. Freeze. When frozen, remove from pan, fold and seal wrapping, label, and return to freezer immediately.

When ready to serve: Unwrap pie, place in baking pan, and bake for 1 hour in a preheated 450°F. oven.

Yield: 24 servings

Lasagne
(to freeze)

For each lasagne serving 8 people you'll need:

6 wide lasagne noodles
2 cups ricotta cheese
½ cup grated Parmesan cheese
2 tablespoons minced parsley
2 beaten eggs
2 teaspoons salt

½ teaspoon pepper
¼ teaspoon cinnamon
1 pound thinly sliced
   mozzarella cheese
1 quart Italian meat sauce

Cook noodles 15 to 20 minutes in a large kettle of boiling salted water to which a tablespoon of oil has been added. Leave the noodles slightly undercooked. While the noodles are cooking, combine the ricotta, Parmesan, parsley, eggs, salt, pepper, and cinnamon.

To freeze: Line an 8-by-8-inch baking pan with freezer wrap, allowing enough extra wrap to fold over top. Lightly grease freezer paper. Lay ½ the noodles in the baking pan; spread with ½ of the ricotta mixture. Cover with ½ the sliced mozzarella and ½ of the meat sauce. Repeat layers. Fold and seal wrapping. Label, date, and freeze.

When ready to serve: Preheat oven to 400°F. Remove freezer wrap. Place in baking pan. Bake 1¼ hours until sauce bubbles and lasagne is heated through.

Yield: 8 servings

Meat Loaf
(to freeze)

6 pounds ground beef	½ teaspoon pepper
3 eggs, beaten	1 garlic clove, minced
3 cups soft bread crumbs	1 tablespoon Worcestershire
1 cup chopped onion	sauce
½ cup chopped green pepper	2⅔ cups milk
4 teaspoons salt	1 cup catsup

Mix all ingredients thoroughly. Shape into whatever size loaves you prefer (¼ of the recipe should serve 6).

You can freeze a meat loaf baked or unbaked. To freeze an unbaked meat loaf, mold the loaf on foil and wrap. Freeze. When ready to serve, unwrap meat loaf and put into baking dish. Bake in a preheated 350°F. oven about 1½ hours or until done.

If you prefer to bake the meat loaf before freezing, bake the mixture in loaf pans at 350°F. for about 1 hour. Let cool, wrap, seal, and freeze. When ready to serve, bake in a 350°F. oven until heated through, about one hour.

Yield: 24 servings

Meat-Vegetable Stew
(to can)

4 to 5 pounds beef, veal, or lamb, cut into 1½-inch cubes	3 cups chopped celery
	4 cups quartered onions
2 quarts sliced small carrots	3 quarts cubed, pared potatoes
	salt, seasonings, if desired

If you prefer, you can brown the meat in a small amount of fat. Then combine with vegetables and seasonings. Cover the mixture with boiling water and pack hot into hot jars, leaving 1-inch headspace. Adjust caps.

If you choose not to brown the meat, simply combine the meat, vegetables, and seasonings, and pack into clean jars, leaving no headspace. Adjust caps. Process pints 1 hour, quarts 1 hour and 15 minutes, at 10 pounds pressure.

Yield: 7 quarts or 14 pints

Old-Fashioned Mincemeat
(to can or freeze)

4 cups lean beef, chopped
1 cup beef suet
1 cup molasses
1 cup honey
2 quarts cider
1 pound currants
1 pound sultana raisins
2 pounds seedless raisins
¼ pound chopped citron
1 lemon and 1 orange rind, grated
½ lemon, juiced

4 cups chopped apple (use more if you'll be using this for pies)
½ cup chopped walnuts
1 teaspoon cinnamon
1 teaspoon mace
1 teaspoon cloves
1 teaspoon nutmeg
½ teaspoon allspice
brandy to taste (optional)
1 teaspoon salt

To can: Cook these ingredients together for 1 hour over low heat. Then pack hot into hot, sterilized canning jars, leaving 1-inch headspace. Adjust caps. Process pints and quarts at 10 pounds pressure for 20 minutes, or in a boiling-water bath for 90 minutes.

To freeze: Cook these ingredients together for two hours over low heat, stirring frequently. Pack into clean freezing containers, seal, and freeze. When using for pies, more apples may be added.

Tipsy Mincemeat
(to can)

5 cups ground, cooked beef (about 2 pounds)
1 quart ground suet (about 1 pound)
3 quarts chopped, pared tart apples (about 12 medium)

1½ cups chopped orange pulp (about 2 large)
¼ cup lemon juice
3 11-ounce packages currants
1 cup brandy (optional)
3 cups sweet cider or grape juice

⅓ cup finely chopped orange
  peel
3 pounds seeded raisins
  (mixture of light and dark)
1 8-ounce package chopped
  candied citron

honey
1 tablespoon salt
1 tablespoon cinnamon
1 tablespoon allspice
2 teaspoons nutmeg
1 teaspoon cloves
⅓ teaspoon ginger

Mix together all ingredients in a large kettle; simmer 1 hour. Stir frequently to prevent sticking. Pack hot into hot jars, leaving 1-inch headspace. Adjust caps. Process pints and quarts 20 minutes at 10 pounds pressure.

Yield: about 6 quarts

Schnitz and Knepp

This is a specialty of the Pennsylvania Dutch. Schnitz means dried apple slices, and knepp means dumplings.

2 cups dried, sweet apples
1¾ pounds hock end of ham
1 egg
2 tablespoons butter

½ cup milk
2 cups flour
3 teaspoons baking powder
½ teaspoon salt

Soak the schnitz in water overnight. Next day, scrub and dry the ham, simmer in water for about 3 hours, then add the soaked apple schnitz and the water in which they have been standing. Boil together for another hour while you make the knepp:

Beat the egg, add the butter and milk. Sift the dry ingredients and add to the first mixture. There should be just enough milk to make a fairly stiff batter. Drop by spoonfuls into the fast boiling stew, cover the kettle tightly, and steam the knepp for 12 to 15 minutes. Serve on a large platter, placing the meat in the center, the schnitz circling it, and the knepp all around the edge.

Yield: 6 to 8 servings

Stuffed Pig's Stomach

1 fresh pig's stomach
2 to 3 potatoes
1 onion

1 pound smoked sausage
salt and pepper

Clean the pig's stomach thoroughly, removing the membranes and rinsing in warm water. Dice the raw potatoes and the onion. Cut the sausage into thin slices. Mix together the potatoes, onions, sausage, salt, and pepper, and stuff the stomach with this mixture. Sew up the stomach and roast it in a 350°F. oven for 3 hours. Serve with applesauce.

Yield: 4 servings

Stuffed Pork Chops
(to freeze)

¾ cup bulgur	⅓ cup chopped parsley
12 double pork chops	1½ teaspoons salt
1½ cups chopped onions	pepper
¼ pound butter	3 eggs
1 cup chopped mushrooms	soy sauce
¾ cup finely chopped celery	
pinch thyme	

To prepare bulgur, brown the grains lightly in a little oil in a heavy skillet. Add 1½ cups water or stock, bring to a boil, lower heat, cover, and simmer for 20 to 25 minutes until grains are light and fluffy.

Have your butcher cut a pocket in each double pork chop. Sauté onions until limp in the butter. Then add the mushrooms, celery, and herbs. Cook for another 5 minutes. Remove from heat and add the bulgur, salt, pepper, and slightly beaten eggs.

Fill chop pockets with this mixture. Skewer with toothpicks to hold in stuffing.

To freeze: Wrap in a package enough stuffed pork chops for a single meal, inserting a double thickness of freezer paper between the chops. Label, seal, date, and freeze.

When ready to serve: While chops are still frozen, brush them liberally with soy sauce, and broil until lightly browned on both sides. Bake, uncovered, in a 325°F. oven for about 30 minutes, and then covered for about an hour.

Yield: 12 stuffed pork chops

Baked Sausage Meat
with Sweet Potatoes

4 large sweet potatoes	honey
1 pound loose sausage meat	milk
4 large apples	butter
salt	

Boil the sweet potatoes until tender and cut into thin slices. Cover the bottom of an oiled baking dish with half the potato slices. Make 4 flat patties from the sausage meat and brown them lightly in a skillet. Drain the patties and place them over the potato slices. Slice the apples and place them over the patties. Sprinkle salt and honey to taste over the apples, and cover with the remaining potato slices. Brush the potatoes with milk and dot with butter. Bake for 45 minutes at 350°F.

Yield: 4 servings

Beans and Sausage
(to can or freeze)

1½ quarts dry kidney beans (about 2½ pounds), soaked overnight
1 quart chopped onion
1¼ cups chopped green pepper
1 tablespoon minced garlic
4 pounds bulk pork sausage
2 tablespoons salt
2 tablespoons chili powder
2 quarts canned tomatoes

Cook beans until slightly underdone (about 1 hour). Drain, saving the liquid. If necessary, add water to make 1 quart.

Sauté onion, green pepper, garlic, and sausage. Cook until sausage is light brown, breaking it up as it cooks. Drain off excess fat. Add remaining ingredients, including beans. Simmer until thickened (about 30 minutes), stirring frequently to prevent sticking.

To can: Pour hot into hot jars, leaving 1-inch headspace. Adjust caps. Process pints 1 hour and 15 minutes, quarts 1 hour and 30 minutes, at 10 pounds pressure.

To freeze: Divide into family-sized containers. Cool, label, date, seal, and freeze.

When ready to serve: Reheat with a small amount of water in a saucepan over low heat, or bake in a 400°F. oven until heated through.

Yield: 3 quarts

Fried Sausages

2 pounds pork sausages
salt to taste
2 egg whites
½ cup flour
1 cup whole-wheat bread crumbs
1 tablespoon butter or drippings

Salt the sausages, dip in a mixture of egg whites, flour, and bread crumbs. Fry slowly in the butter or drippings to a nice, brown, crisp color.

Yield: 6 servings

Potato Sausage Pie

6 or 8 potatoes
1 onion
1 pound loose sausage meat

1 teaspoon minced parsley
1 top pie crust

Slice the potatoes into rounds and dice the onion. Put the potatoes in salted boiling water and cook for a few minutes, until tender. Sauté the meat and onion in a skillet. Drain the potatoes, saving 2 cups of water. Fill a casserole dish with alternating layers of potato slices and sausage meat. Add the water and sprinkle with parsley. Place a pie crust over the top and bake at 450°F. for 10 to 15 minutes, or until pie crust is golden.

Yield: 4 servings

Sausage and Sweet Potato Filling

2 large sweet potatoes
1 cup sausage meat
3 tablespoons chopped onion
1 cup diced celery

2 cups whole-wheat bread crumbs
3 tablespoons parsley
salt and pepper

Boil the potatoes, skin them, and mash. Brown the sausage meat in a skillet lightly and then remove from pan. Sauté the onion and celery until translucent. Mix the cooked onion, celery, and meat. Add the mashed potatoes, bread crumbs, and parsley, and season with salt and pepper, if necessary. Mix well.

Yield: enough for a 10-pound bird

Sausage with Gravy

1 pound pork sausage
1 tablespoon butter or drippings
½ cup onion

½ teaspoon flour
1 cup bouillon or soup stock
½ cup water (if necessary)

Fry the sausage over low heat in drippings or butter until brown. Remove, then add sliced onion; add the flour. When brown add bouillon or soup stock. Cook a few minutes. Add extra water if necessary. Pour gravy over sausage and serve with sauerkraut.

Yield: 2 to 3 servings

Scrapple Croquettes

1 cup scrapple
1 cup cooked brown rice or
   mashed potatoes
2 eggs, hard boiled
1 teaspoon minced parsley

salt, pepper
1 egg, beaten
½ cup whole-wheat bread
   crumbs

Mix the scrapple, the rice or potatoes, and the hard-boiled eggs, chopped fine. Season with parsley, salt, and pepper. Shape into croquettes with beaten egg and bread crumbs; fry in deep fat.

Yield: 3 to 4 servings

Peppers Stuffed
with Scrapple

½ pound scrapple
butter
1 egg
1 onion chopped
½ teaspoon paprika

1 teaspoon chopped parsley
1 cup fine whole-wheat bread
   crumbs
6 green peppers

Fry the scrapple lightly in slices, in butter, and then mix the scrapple, the drippings, the slightly beaten egg, onion, paprika, parsley, and bread crumbs into a thick paste. Remove the ends and stems of the peppers and stuff them with the mixture. Lay them on their sides in a baking pan containing a little water and bake for 30 minutes, turning occasionally.

Yield: 4 servings

Turkey Chop Suey
(to freeze)

1¼ quarts celery cut diagonally
   in ½-inch strips
½ cup sliced onion
1 tablespoon salt

¼ teaspoon pepper
1 tablespoon honey
2 quarts turkey broth
2 quarts cooked turkey, in
   pieces

¼ cup oil
1½ cups mung bean sprouts
1½ cups sliced water chestnuts
  (if desired, or you can
  substitute diced kohlrabi)

¾ cup cornstarch
½ cup cold water
½ cup soy sauce

Simmer the celery, onion, salt, pepper, and honey in the broth for 20 minutes. Heat the turkey in the oil, and add it, the bean sprouts, and the water chestnuts to the vegetable mixture. Blend cornstarch with cold water, and stir into the mixture. Simmer for 15 minutes, stirring frequently. Add soy sauce.

To freeze: Cool the food quickly. Pack in freezer containers, seal, label, date, and freeze immediately.

When ready to serve: Thaw and reheat in double boiler. Serve over brown rice.

Yield: about 3 quarts

Nuts, Seeds, and Grains

Nuts

Harvesting Tree Nuts

Nuts should be allowed to fully mature and fall from the tree naturally. Some growers, because they are impatient or want to make the harvesting easier, shake the trees to get the nuts to fall. Shake the tree if you wish, but shake gently. Harsh jolts may cause immature nuts to fall, and rough handling can damage the branches. Whatever method you use to get the nuts to the ground, be sure to gather them as soon as possible. Nuts left on the ground mold quickly, especially in damp weather. This is particularly true with the Persian walnut (also known as English or Chinese walnut) and chestnut. If you don't gather and hull the nuts of the black walnut promptly, the bitter fluid secreted in their hulls penetrates to the kernels and this discolors them and impairs their flavor. The nuts of the Persian walnut, if infested with husk maggots, will not open on the tree. Rather, the nuts will fall to the ground with the hulls still on them. In such cases, Persian walnuts should be picked from the ground soon after they fall and hulled like the black walnuts (see below) before the dark, bitter fluid of the hulls penetrates the nut meats.

Hulling

Black walnuts should be hulled before the hull turns black to secure a light-colored, high quality nut. The importance of hulling was proven to Spencer B. Chase, of the Northern Nut Growers Association, after the experiment he performed one year at harvesttime. Nuts from ten trees were collected, and the first batch was hulled within a few weeks after maturity. The remainder was hulled at weekly intervals. These hulled nuts were then stored and cured, and subsequently cracked open for inspection from the following January through March. "Without exception," says Mr. Chase, "the first batch of nuts which had been hulled within a week after maturity were light in color, mild in flavor, and could be eaten out of the hand like peanuts." This was not the case with the nuts hulled later at weekly intervals. These nuts produced "darker kernels and suffered considerably in flavor and overall quality." Most commercial walnut operations dip their nuts in a bleaching solution to get the light color which can be obtained naturally just by hulling the nuts as soon as they are mature.

There are several different ways of getting the tough but porous hulls off black walnuts. The old-fashioned corn sheller does a first-rate job when it is equipped with a flywheel and pulley and is driven by a ¼ horsepower motor. For small quantities of nuts—a few bushels or so—you can spread the nuts on a hard dirt or concrete road and drive over them with a car or truck until all the hulls are mashed. The trouble with this method is that the nuts tend to shoot out from under the wheels when the car rolls forward to crush them. Be ready to hunt for stray nuts and line them up again under the wheels. You can prevent nuts from shooting out from under the wheel by placing a small quantity of nuts in a wooden trough that is just the width of

This wooden trough, built especially for the purpose, allows you to use your car to hull black walnuts. Without the trough the nuts shoot out from under the car's tire, without cracking.

the wheel. Remove the side from one end of the trough so that the wheel can roll in and out of it. Roll the wheel only slightly when crushing the nuts so that you do not run over the other side of the trough.

Whatever method you follow to remove the hulls, it's best to wear thick rubber gloves to protect your hands from the acrid fluid secreted from the hulls that will leave persistent stains on your skin. It is this staining agent which also penetrates the hulls and stains the delicate kernels and otherwise impairs their quality and flavor. After hulling, the nuts which are still in their inner shells should be washed with a hose or placed in a tub and rinsed thoroughly.

Not all nut varieties, however, will give you as much trouble as the black walnuts. The hulls of pecans, hickory nuts, Persian walnuts (if not infested with husk maggot), filberts, and chestnuts obligingly open on the tree and let the nuts fall to the ground, so harvesting is merely a matter of picking the nuts off the ground.

Drying

Once gathered (and hulled and washed, if need be), nuts should be placed in water, and the rotten and diseased nuts which float to the top should be discarded. The nuts must then be dried or cured. Green, uncured nuts are bitter and unpalatable. To dry the nuts, spread them out rather thinly on a dry, clean surface and allow them to dry gradually by exposure to a gentle, but steady, movement of air. A clean, cool, darkened, well-ventilated attic or porch is ideal. Nuts dried this way will not be attacked by fungus or mildew. When done properly, such drying will make for light-flavored, light-colored nut kernels. The nuts may also be spread thinly on wire trays or window screens. Nuts should not be more than two nuts deep on such trays and screens. Deep wire baskets are used by some nut growers, but growers are careful to pour the nuts every 4 days from one basket into another to prevent mildew from forming.

Drying time varies with nut variety. All nuts except chestnuts contain a great amount of oil. This oil prevents nuts from drying out completely and from becoming hard and brittle. Chestnuts contain little oil, but they do contain much water and carbohydrates. Because of their high water and carbohydrate contents, chestnuts dry out easily and become hard and inedible. Dry chestnuts only for 3 to 7 days. All other nuts need several weeks to dry properly.

Walnuts, pecans, filberts, and hickory nuts are dry enough to be stored when the kernels shake freely in their shells, or when the kernels break with a sharp snap when bent between the fingers or bitten with the teeth. Avoid excessive drying which will cause the nut shells to crack.

After drying, nuts still in their shells may be stored in attics for up to a year, but cool underground cellars are preferable for longer storage. Storage containers for nuts may be made from large plastic bags with ventilation holes punched in them to allow air to circulate and excess moisture to escape. Then put these bags in tin cans lined with paper. Keep them tightly closed, but punch a small hole in the can below the lid. Store the nuts at 34° to 40°F. and check regularly for mold and mildew.

Some nut growers store their nuts in peat moss and other moisture-absorbing materials. By so doing, there is no need to allow for ventilation because the peat moss takes care of excess moisture that would cause mildew. Prepare peat moss for nut storage by moistening it with just enough water to prevent its being dusty. The peat moss should be damp, but not wet. Add about one-third as much peat moss as nuts by volume when packing nuts in lidded cans or plastic bags.

Nuts may also be packed in clean, dry sand and then stored in a cool area. Stored this way, nuts will retain their germinating powers (should you wish to plant them to start new trees), but may lose some of their flavor.

Cracking the Nuts

You can crack a nut with almost anything that does the job. You can use a hammer with a block of wood or metal. Or you can swing a heavy iron like your grandmother used to iron shirts. Every nut fan has his or her own method. Take wild foods gourmet Euell Gibbons, for example. Here are his instructions for shelling black walnuts, especially hard nuts to crack:

> If you stand the nut—pointed end up—on a solid surface and hit it with a sharp blow with a hammer, it will crack into two halves. Stand each half, again pointed end up, and strike it again, which will break it into quarters. Strike each quarter again on the pointed end, and it will break into eighths—at which point the nut meats fall out.
>
> When the nuts are well dried, I have been able to completely empty shell after shell without resorting to a nutpick with this method.

And his secret for shelling hickory nuts:

Put the unshelled nuts in the deep-freeze for a day, then remove a few handfuls at a time and shell them while they are still frozen and brittle. The nuts are narrower one way than the other, so hold them up edgewise and strike each a sharp blow with the hammer, just hard enough to crack the shell but not smash the kernel. Cracked this way, many entire halves can be removed while the others usually have to come out in quarters.

Hard-shelled nuts are more easily shelled if the shells are softened first. This may be done by pouring boiling water over the nuts and allowing them to stand for 15 to 20 minutes before shelling. Another method of softening the nuts is to place them in a container, sprinkle them with water, and cover them first with a damp cloth and then the container cover. Let them stand this way for 12 to 24 hours. The shells will be easier to crack, and the nut meats will not splinter when the shells are cracked.

Soft-shelled nuts, like chestnuts, can have their shells peeled off with a knife. To do so, first cut across the base of the nut to allow steam to escape so that the nut won't explode, and then place the nut in boiling water for 3 to 4 minutes. The shell will then pull off easily.

Harvesting and Curing Peanuts

To get the most out of your peanut crop, allow plants to remain undisturbed until heavy frost completely destroys the tops. This would be around the last of October or in November in most peanut-growing areas. Most of the pods are formed as early as the middle of September, but they are empty. The kernels will need another month or 6 weeks to develop fully. Early frost may darken the leaves of the plants, but the stems will continue to provide food for the nuts to develop.

When harvesttime arrives, lift each bush carefully out of the soil with a garden fork, and shake free of all dirt. You might want to run your finger through the dirt left in the hole to rescue any peanuts that have broken off. Pluck the pods from the roots immediately, and drop in shallow trays for drying. Later, store these filled trays in the attic of your home or garage to dry. Never store in a cool, damp basement, as the moisture still in the uncured pods will cause them to mildew and rot. Two months of curing time will make the peanuts fit for roasting (see page 461) or more permanent storage.

Storing Shelled Nuts

Shelled nuts don't store as well as those with their shells intact. Without their protective shells, nut meats are exposed to light, heat, moisture, and air—all of which cause rancidity in the nut. If you plan to keep your nuts for a while, crack them as you need them. Nut meats can be stored safely for a few months under refrigeration if they are placed in tight-closing jars or sealed plastic bags.

Freezing For longer keeping, nut meats can be frozen or canned. They will keep well at freezer temperatures for up to 2 years so long as they are stored in airtight containers or sealed plastic freezer bags. Sealed and processed in jars, nut meats will keep well for 1 year.

Canning Nut meats to be canned should be spread in a thin layer on a cookie sheet or baking pan and heated for 30 minutes in an oven preheated to 250°F. Stir the nuts every 10 minutes to make sure that they are drying evenly. Fill dry, sterilized canning jars with the hot nut meats and process them in either a boiling-water bath or at *5 pounds pressure* in a steam-pressure process.

For a boiling-water bath:

1. Pack hot nut meats in canning jars and adjust lids.
2. Place the jars on a rack in the boiling-water canner so that no jars are touching one another. If you are placing jars two layers deep, separate the layers with a metal rack and be certain that no jar is directly on top of another.
3. Have enough hot water in the canner to come up 2 inches over the top of the jars. Put the lid on the canner.
4. After the water in the canner comes to a boil, process for 30 minutes. Boil gently and steadily, adding more boiling water if needed to keep jars covered. Do not pour the water directly over the tops of the jars.
5. As soon as 30 minutes are up, remove the jars from the canner.
6. If the lids are not the common self-sealing type, complete the seals as soon as they are removed from the canner.
7. Set jars upright to cool. Place them far enough apart from one another so that air can circulate freely around them.

For steam-pressure processing:

1. Pack hot nut meats in canning jars and adjust lids.
2. Place the jars on a rack in the canner, making sure that no two cans are touching. If your jars are two layers deep, place a metal rack between the layers and be certain that no jar is directly over another.
3. Put 2 or 3 inches of hot water in the bottom of the canner.
4. Fasten the canner lid securely so that no steam escapes except at the open petcock or weighted gauge opening. Then close the petcock or weighted gauge and let the pressure rise to *five pounds.*
5. Process for 10 minutes.
6. After 10 minutes, gently remove the canner from the heat and let the canner stand until the pressure returns to zero. Wait a minute or two and slowly open the petcock or remove the weighted gauge.
7. Remove the jars and complete the seals if necessary. Allow the jars to cool.

Label the jars and store them with your other canned foods, in a cool, dry, dark place.

Roasted and Salted Nuts
(except peanuts)

Remove the nuts from their shells and skin them, if you wish. To make skinning them easier, place nuts in a bowl and pour boiling water over them. Leave them in the boiling water for about 3 minutes, until the skins begin to wrinkle. Pour the water off the nuts and rinse them under cold water to cool them. Now, rub the skins off with your fingers; they should come off easily.

Place the nuts in a bowl, and for each cup of nuts, sprinkle 1 teaspoon of oil (peanut, sesame, or safflower oil) over them. Stir the nuts so that all are coated with oil. Then spread them out on a cookie sheet, sprinkle lightly with salt, and roast in a 350°F. oven for about 10 minutes, stirring once or twice, until the nuts are just lightly browned. Watch the nuts carefully in the last few minutes of roasting so that they don't become too dark.

Dry Roasted Nuts
(except peanuts)

Skin the nuts, if you wish, as described above. Place the nuts in a bowl, and for each cup of nuts, pour 1 teaspoon of soy sauce over them. Mix the nuts so that all are coated with the soy sauce. Spread on a cookie sheet and bake in a 350°F. oven for about 10 minutes, stirring once or twice, until the nuts are lightly roasted. Watch the nuts carefully in the last few minutes of roasting so that they don't become too dark.

Roasted Peanuts

Preheat oven to 300°F. Spread peanuts in a shallow pan and bake in the middle of the oven for 30 to 35 minutes. Stir occasionally and check for brownness by removing the skin of one or two peanuts. If you have bought peanuts without skins, roasting will not take the full 30 to 35 minutes. When the peanuts are as brown as you'd like them, remove them from the oven and cool before placing in plastic bags or covered jars for storage.

Sunflower Seeds

The back of a thoroughly ripe sunflower head is brown and dry, with no trace of green left in it. But the trouble with leaving it in the field until that point is that the birds may have harvested the crop for you, or the head may have shattered and dropped many of its seeds to the ground.

Small-Scale Harvesting

In the home garden, sunflower seeds may be covered with cheesecloth to keep away birds. Or the heads may be cut off when the seeds are large enough and allowed to dry elsewhere. If cut with a foot or two of stem attached, they may be hung in a dry, well-ventilated place to finish drying. They may also be cut and spread on boards on the ground—protected with a wire screening from rodents—to dry in the sun for a couple of weeks. Heads should not be piled on top of each other, as seeds may rot or become moldy.

The heads are dry and ready to have their seeds removed when the rough stalks are brittle and the seeds separate from the head easily as you run your thumb lightly over the surface of the head. You can brush the seeds out with a stiff brush, a fish-scaler, or a currycomb; or you may remove them by rubbing the heads over a screen of 1/2-inch hardware cloth stretched over a box or barrel. If some seed is still moist it may be spread out to complete drying after being removed from the head.

Large-Scale Harvesting

For large-scale harvesting the larger-sized varieties of sunflowers, which are the types grown for seeds, present a problem. They are too high and too heavy to be handled by a combine, and the toughness of the heads makes threshing difficult. The smaller sunflower varieties can be combined, but their yields are not big and their seeds are small.

It is best to allow the sunflowers that must be harvested by hand to dry in the fields on the stalk. The birds and animals will primarily pick at those heads that fall to the ground. If the heads are taken indoors to dry, rats may get to them.

The dry heads are harvested by driving around the edge of the field with a wagon or truck and clipping the heads off with pruning shears. The heads are then run through a corn fodder shredder which knocks the heads apart and loosens the seeds. The seeds should then be screened to separate them from pieces of the stalk and the head.

Hulling the Seeds

For hulling small amounts of sunflower seeds you can make the hand hulling job easier by putting the seeds in boiling water for a short time. You can also crack the shells with pliers or a clothespin. Lay the seed between the prongs as you would between your teeth, so that it will be cracked lengthwise.

To hull large quantities, use any small farm hammermill. A 10-inch size is adequate. Remove screens and set the mill to run at 350 revolutions per minute. You can use a tachometer (which measures the speed of rotation) to judge so no hulled seeds go up the dust

A home-improvised arrangement, such as this wire brush attached to a small electric motor, can simplify removing the seeds from the pods—often a difficult and tiring task for gardeners with a large quantity of sunflowers to harvest.

collector. Slowly pour in about 2 gallons of seeds, then speed up the mill to 1 200 to 1,500 revolutions per minute. This clears the mill so that it won't clog at the bottom. Such a mill will hull out about 80 percent or more of the seeds.

Local feed mills may be willing to shell sunflower seeds in amounts too large for hand hulling if you haven't a hammermill or quantity large enough to make purchasing one practical.

Storing the Seeds

Dried seed should be stored in a cool, dry place in small containers and should be stirred once or twice a week to prevent mustiness. Seeds that are stored in large bins may heat up and lose some of their vitamin content.

Hulled seeds may also be canned or frozen for long-time keeping. If packed in freezer containers or freezer bags, sunflower seeds will keep for a year or more at freezer temperatures. Can seeds as you would nut meats; see page 459.

Grains

Knowing that the breads and cereals you eat have been made from grains you have grown yourself is truly satisfying. Compared to growing fruits and vegetables, growing grain is almost effortless. What little hard work there is comes at the end, when you must harvest and prepare the grains for storage.

If you were a big operator, you would wait until your grain was dead ripe and had dried down to a moisture content of about 14 to 14½ percent, and then get out the combine that would do the whole job of harvesting: cut the stems, beat the grain loose from the hulls, and separate the straw and chaff from the grains. Then you would run the grain through a mechanical fan seed cleaner to remove the wild plants and seeds that were harvested with the grain. After the grain was cleaned you might machine dry it to prevent it from heating up and spoiling and to retard mold and fungus growth during storage. Then you would pour the cleaned, dry grain into 100-pound sacks and store them in an atmospherically controlled warehouse.

Harvesting by Hand

A combine costs several thousand dollars, and unless a neighbor has one that can be borrowed, a grower with only a small amount of grain to harvest will have to do it the old-fashioned way and harvest the grains by hand. (The cutting can be done with a sickle-bar on a garden tractor, but the machine running over the heads may cause the grain to shatter onto the ground.)

Harvesting by hand involves cutting the grain before it is completely ripe (contrary to combine harvesting), then binding the stalks into sheaves, arranging the bundles into shocks, and allowing the grain to remain drying in the field until it is ripe. The best way to tell if the wheat is ready to cut is to pull a few heads and shell out the grain in the palms of your hands with a rubbing motion. The wheat should come free from its husks fairly easily, but not too easily. Blow the chaff out of your cupped hand and chew a few grains. They should feel hard when you bite into them. If the grain is still milky, wait a few more days. If the grains are very hard and shatter out of the husks very easily, you've waited too long. The stalks should be nearly all yellow, with only a few green streaks remaining. When wheat is dead ripe, there is no more green in the stalk.

Rye should be harvested in the almost-ripe stage, which it usually reaches about a week before the wheat does. Oats are hand harvested when the kernels can be dented by the thumbnail—not too hard and not too soft—when the heads are yellow and some leaves are still green. Buckwheat, which takes longer to mature, should be harvested after the first seeds mature. Ideally, this is after one or two frosts have made the grains easier to separate from the plants.

Cradling

To harvest by hand, cut the stems near the ground after the dew is off, on a dry day. To do this properly, according to homesteader Gene Logsdon, you need a grain cradle. A grain cradle is a scythe equipped with three or four long wooden tines arranged about 6 inches apart above the scythe blade. When you swing the cradle, the cut stalks of grain gather against the wooden tines as you make your stroke through the standing grain. The cut stalks then fall in a neat little pile to the left of the swath you are cutting as you complete your stroke. These little piles are then easily tied into bundles.

It takes practice to develop the proper rhythm for cradling. But if you've ever done scything, you can catch on to it in a hurry. The trick of scything is to cut a rather narrow strip, letting the scythe blade slice through the standing stalks at a sidewise, 45° angle. Don't try to whack off the stalks with the blade at right angles to the stalks. The blade should be very sharp and always held parallel to the ground. Don't let the point dip down to catch the ground.

If a grain cradle isn't available, an ordinary scythe will do. This cuts the stalks, but instead of letting the grain fall in convenient piles, it lets the stalks fall where they are cut. This means that you have to gather the stalks in bundles, which could become a laborious job if you've got a big area to harvest. A hedge trimmer could conceivably be used to cut down the grain, too. Like a scythe, a hedge trimmer would let the stalks fall haphazardly, making gathering difficult. It wouldn't give you the aching arm and back a manually operated cradle or scythe would, but you would have to consider the practicality of using a power tool in the field. You would have to have an extra long extension cord for an electrically powered trimmer or get one that runs on gasoline.

After the stalks are cut, tie them into bundles with ordinary binder twine, available from farm stores. A bundle should measure about 8 inches in diameter at the tie. The bundles are then set up in shocks.

Building the Shocks

To build a shock, grab a bundle in each hand, sock the butt ends firmly down on the ground, and then lean the tops against each other. Two more bundles are set the same way on either side of the first ones. Sometimes your beginning shocks will fall over until you get the knack of it. With the first four bundles in place, you can then stand about six or eight more evenly around them. When you have a fairly sturdy shock of twelve bundles or so, you can tie a piece of twine around the whole shock to make it stand more solidly. After the shocks dry in the sun for about 10 days (a few days longer for oats) put the bundles in an airy shelter to finish ripening where no rain will fall on them.

A grain cradle, like the one here, makes hand harvesting grains easier.

Threshing

Before the advent of the combine, a steam-powered threshing machine would make the rounds and farmers would take their sheaves down out of the mow and feed it to the roaring, hungry monster. Before that—for thousands of years, and it's still done this way in many parts of the world—people beat the grain out of the hulls with flails.

You can make a simple flail by taking an old broom handle and drilling a ¼-inch hole near one end. Then similarly drill a ¾-inch stick about 1 foot long and attach loosely to the broom handle with a leather or wire loop through the drilled hole. Throw a sheaf of wheat on the threshing floor—a clean garage or cellar floor or a hard-packed piece of ground free of growth or debris. Then beat the heck out of the sheaf with the loose end of the flail; the grain will easily fall out.

A flail isn't the only instrument you can use for this job, though. Gene Logsdon uses a homemade flail, but recommends trying a plastic baseball bat instead. His son used it and found that it worked better than the flail because it was firm enough to knock the kernels

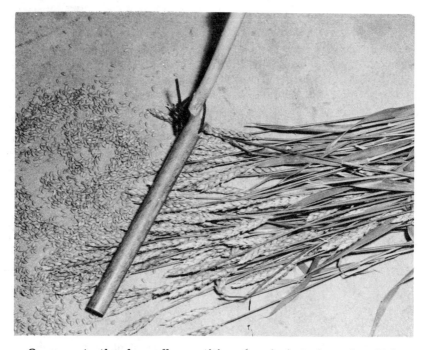

One way to thresh small quantities of grain is to lay a handful on a hard surface and beat the grains with a homemade flail, like the one pictured here.

out, but flexible enough not to crack the grains. David Criner in Arkansas uses a compost shredder to thresh his grain. He places a tarpaulin under the shredder to catch the grain and feeds the bundles of wheat—one at a time, head first—into the shredder chute. Another simple way to thresh is to wham a handful of stems against the inside of a barrel. The grain falls neatly into the barrel instead of flying around and landing in a wide area on the floor.

Removing the Chaff

Once the straw bundles are relieved of their grain and removed, there still remains lots of chaff with the grain. Traditionally, grain was winnowed simply by pouring it slowly into a bucket in a brisk wind. The breeze blew away the light husks, and the heavier grain fell directly to the bucket. If the day you pick is windless, use an electric fan to generate a winnowing wind; it works like a charm. Repeat this process until the grain is clean, or nearly so. A few grains may not hull out completely, but this will not affect the flour.

Storing the Grain

Curing

If you harvested your grain when it was almost ripe and allowed it to dry in the field for at least a week or so, the moisture content of your grain is probably somewhere between 12 and 15 percent. If this is the case, your grain is dry enough to store as is, although it should be cured for at least 1 month before you try to mill it for flour. Green grain heats up and gums up a flour mill, and the flour is difficult to work with. Grain makes much better bread if it is allowed to cure longer than a month.

If the weather should be damp when you harvest the grain, and it rains while the grain is curing in the field, there is a good chance that its moisture content is somewhere above 15 percent. If you store high-moisture grain you are inviting all kinds of trouble. Mold, fungi, beetles, and mites thrive in moist grain. Wet grain should be dried before it is stored. The drier the grain, the less susceptible it is to damage by bacteria, fungi, and insects.

A small quantity of wet grain may be dried by putting it into a box that is enclosed with screen wire on all six sides. Ordinary house screen wire should be reinforced with ½-inch mesh wire to support the weight of the grain. The wet grain may also be poured into bags and placed on end on slats in a dry place, preferably not on a concrete or earth floor. After a few days the boxes or bags should be inverted and the grain disturbed to permit air to get to all the kernels. Invert again in a week or so and allow it to cure at least 1 month before attempting to store or mill it.

Keeping Rats Out of Your Grain

Dry grain keeps best when stored in a cool, dry place. It may be stored in bags on slats, so long as precautions are taken against rat infestation. Rats are capable of doing severe damage to stored foods. Not only can they consume many pounds of food in 1 year, they can also contaminate foods by carrying into the storage area insects and bacteria. You'll know you have a rat problem when you see droppings, tracks, and grease marks. In dusty locations you may notice tracks left by the trail of a tail and the four-toed front paws and the five-toed rear paws. Storing bags of grains in a cellar or outbuilding that has double walls and space between floors and ceilings invites rat problems because such structures provide ideal homes for these rodents. Choose a storage area that is rat-tight and provides no hidden areas in which rats can nest.

Many commercial warehouses take care of rat problems by using rodenticides. These poisons are not only toxic to rats, but to domestic animals and humans as well. For the homesteader, rat infestation can be prevented by making sure the storage area is clean and rodent-proof. Having a couple of good-working felines around to patrol the area is also a good idea. Perhaps the simplest and most logical way to protect your grain is to store it in metal drums in a dry place, not in the cellar, if possible, although this is all right, too, if the cellar is dry and the drums are kept a couple of feet above the floor in case of flooding. Small quantities of cured grains may be kept in glass jars; light helps to inhibit mold.

Preventing Insect Infestation

Weevils, moths, beetles, and the other insects that feed on grain are unlikely to cause you any trouble if your grains and storage containers are clean and dry and kept relatively cool. If insects should invade your grain, remove and destroy that grain which is infested. If, after examining the grain, you find that little damage has been done, you may kill the insects by putting the grain on trays and heating it in a 140° to 160°F. oven for 30 minutes. Or you may put your grain in a freezer that maintains a 0°F. temperature for 3 or 4 days. Such temperature extremes will destroy insects. Carefully clean out your storage containers, making sure that you have not overlooked any place where insects may be hiding. Pour the cleaned grain back into the containers and check frequently for future infestations.

Refrigerated Storage

At Walnut Acres, organic grower Paul Keene keeps all his grains under refrigeration. This not only assures him of fresh grains, but also makes insect and rodent damage nonexistent. Once, when he spotted some grain moths flying about in the refrigerator as grain was coming in from the weeding out process, he became worried. But the moths settled on the refrigerator wall, only to fall dead on the floor. The 40°F. temperature did it.

Refrigeration of grains is a good idea, especially if you want to hold grain over the summer, when the cool weather is gone. Insects that attack grains multiply at temperatures greater than 70°F. Insect infestations of this type are controlled at the commercial level with fumigants like methyl bromide and methyl parathion, both deadly poisons that are pumped into airtight sealed bins. Experiments are being done with carbon dioxide fumigation, which, while deadly to insects and rodents, will leave no toxic residues. But, like all fumigants, it requires sealed storage bins, and these are not very practical for the grower with small grain acreage because they are usually too

You can keep unmilled grain in glass jars for up to one year. Make sure that the grain is dry and free from insects before storing it this way.

big and too expensive to construct. The same insects that multiply at 70°F. lie dormant at 35°F. and will eventually be killed by extended periods of low temperatures. Refrigeration also aids in controlling fungus attack: fungus grows very slowly at temperatures of 45°F. and below.

Organically Grown Grains Store Better

Organic growers may be happy to learn that Paul Hawkin, president of Erewhon Trading Company, which is one of the country's largest distributors of organic grains, claims that organic grains have a higher resistance to heat and humidity than chemically grown grains:

> For example, if you fill a pint jar half full with organically grown and produced rice, fill it with water and cap it tightly and then do the same thing with the chemically grown rice, you'll discover a strange thing. After a period of several months the rice that was not grown organically will have dissolved and turned into a white powder, whereas the organically grown rice has not. Now, I have no scientific explanation for this, but it is pretty graphic. I spent a year in Japan studying natural foods and farming methods, and one of the most stunning demonstrations I saw was this rice experiment done by the Messian people in Shikoku. They performed dozens of similar experiments, using other grains as well.

Grinding Grains

Grinding for flour is best done just before baking, for grinding will expose the germ which will turn rancid in the presence of oxygen and warm temperatures. Rancidity destroys vitamins E, A, and K in the human body and has been found to destroy several of the B vitamins. Ground flour should be stored in the refrigerator or freezer if kept for any extended length of time. However, do not use cold flour when you're baking with yeast. Leave the cold flour at room temperature for a few hours before you use it for baking so that the yeast will not be chilled and become inactivated.

If flour stored in a pantry becomes infested, it can still be used, provided the infestation is not severe. Follow the directions on page 470 for killing grain-loving insects. Then clean the flour by straining it through a sieve.

Storing Bread

Before the time of factory-made bread, all bread was made at home and 1 day a week was set aside as bread-baking day. On that day, the baker of the household would rise before dawn and build a roaring fire in the brick oven. By midmorning, when the oven's inner walls were thoroughly heated, the glowing embers would be shoveled out. Then the bread, which had been mixed, kneaded, left to rise, shaped into loaves, and left to rise again, would be placed in the oven. The heavy iron oven door would be closed and sealed tightly with clay so that no heat could escape, and the bread would be left to bake for the rest of the morning and afternoon. The farmhouse soon would be filled with the sweet yeasty smell of baking bread, and by early evening there would be enough loaves cooling in the kitchen to last a family of ten or twelve for a week, until the next bread-baking day came around again.

It made good sense then—and still does today—to bake a number of loaves at one time. Although we're spared the chore of building a fire in the brick oven, we still must mix and knead the dough, let it rise twice, and bake it. It takes just about as long to prepare and bake a double or triple bread batter as it does a single one. Once you have the kneading board out, the bread pans lined up, and your hands covered with flour, you might as well go ahead and bake enough bread to last your family at least a week.

What do you do with the eight or twelve loaves after they're baked? Farmwives used to put them in a wooden box especially made for keeping bread, cover them with a clean towel, and store them in a cool place. The bread tasted just great for the first few days, but as the week wore on the bread started to get a little stale—no preservatives to keep bread "tasting" fresh forever, then.

Freezing Baked Bread

Today we're more fortunate. We have a way of keeping bread fresh for months, not by adding sodium or calcium propionate or another bread freshener to our breads, but by using the freezer.

Baked yeast breads retain their just-baked quality at freezer temperatures for 6 months and more if they are stored properly. Prepare and bake your loaves according to the recipe. Allow them to cool in a draft-free place (drafts tend to shrink baked goods) thoroughly before wrapping for the freezer. Warm or hot breads which are wrapped tightly will emit water vapor while they cool, and this water vapor will condense on the inside of their wrappings. This moisture can lead to the growth of mold even under the most sanitary of conditions. Wrap the breads intended for the freezer after they are cool in aluminum foil or heavy-duty plastic bags. Then freeze them.

Thawing the Bread

Frozen breads should be thawed while still in their wrappers. If you like to thaw your breads in a low oven, your freezer wrap should be aluminum foil. Otherwise, let your bread thaw at room temperature. It should take at least an hour to thaw normal-sized loaves. Allow extra thawing time for large, heavy loaves. If you're toasting your bread, there's no reason to thaw it first. Slice it (you'll need a good sharp knife to get through a loaf of bread which is frozen solid) and place it still frozen in the toaster. The slices will thaw and toast at the same time and taste just as good as fresh toasted bread.

Start with Moist Breads

Bread should be eaten soon after it has thawed, for thawed bread dries out quickly. To keep your bread from becoming stale too rapidly after thawing, begin with a recipe that makes a moist bread. One that calls for honey will make a moister loaf than one that uses sugar, corn syrup, cane syrup, maple sugar, or molasses as a sweetener. The more honey you use, the moister your bread will be and the longer it will keep. Honey is hygroscopic. It absorbs moisture from the air and holds this moisture in bread. Recipes that call for a good amount of oil or butter will also make a moist bread that will keep for a longer time because shortening doesn't dry out. Don't freeze French and Italian breads unless you know that you're going to finish off the entire loaf soon after it is thawed, or unless you have a use for stale, dried-out bread. French and Italian breads do not contain oil or butter, so they dry out very quickly.

Foods with a high oil content, like nuts, sunflower and sesame seeds, and soybeans, can be ground in a grinder or blender and used to replace some of the flour in your bread. In addition to making your bread moister, these foods will make your bread more nutritious because they are good sources of protein and the B vitamins. You can replace about one-eighth of the flour in any dark bread recipe with ground nuts, seeds, or soybeans without making an appreciable change in the taste or texture of the finished product. In place of the ground nuts or seeds, you can add a few tablespoons of a nut or seed butter (like peanut butter or tahini) to your bread ingredients. Another way to make your bread moister is to replace some of the flour in the recipe with cooked cereal, like oatmeal or farina.

Softening Dried-Out Bread

If your thawed bread starts to dry out and get stale, do what the farm wives used to do with their 6-day-old bread: rejuvenate it by sprinkling a little water on it and putting it in the oven for a few minutes. You can also lay a flat strainer above a pot of boiling water or cooking vegetables or soup or anything you've got on the stove, and place slices of bread in the strainer. Put the lid over the bread. The steam rising from the pot below will moisten and warm your dry pieces of bread.

Freezing Unbaked Bread Dough

If you're the kind of person who goes crazy over the tastes and smells of bread right from the oven and would rather take the time to bake every day just so that you can enjoy just-baked bread all the time, don't despair; there is a shortcut for you, too. Just freeze your bread *before* it's baked, and then when you're ready for a loaf, let it thaw and rise, and bake it as you would freshly risen dough. Your kitchen can smell like it's bread-baking day every day even though you only get your hands and the kneading board floury once a week.

Unbaked yeast dough can only be stored in the freezer for 2 to 3 weeks. Mix and knead it the way you usually do, then allow it to rise once. When it's risen to double its bulk, punch it down, shape it into loaves, and freeze it. Your loaves should be thinner than usual. They should be no more than 2 inches deep so that they will thaw quickly when taken from the freezer. When you're ready for a fresh loaf, thaw the dough in a low oven (250°F.) for 45 minutes, then reset the oven for the normal baking temperature and bake as usual. Once cooked, this bread will also dry out quickly, so eat it soon after baking.

Refrigerating Unbaked Bread Dough

Floss and Stan Dworkin, in their book BAKE YOUR OWN BREAD (Holt, Rinehart and Winston, 1972), recommend storing unbaked bread dough in the refrigerator instead of the freezer. It keeps just as long at 40°F. as it does at 0°F.—2 to 3 weeks. In the refrigerator you can get the dough to rise an extra time because the yeast is still working at refrigerator temperatures, even though it does work slower when it is chilled. This extra rise can only improve the texture of the bread.

To prepare the dough for keeping in the refrigerator, let it rise once until double in bulk, and then punch it down in the bowl. Cover the bowl loosely with plastic wrap or with a clean towel and put it in the refrigerator. If the dough rises above the rim of the bowl, punch it down and cover it again. When you're ready to bake the bread, take the dough from the refrigerator and punch it down. Shape it into loaves and put it into bread pans. Place the pans in a warm place and let the dough rise until it is double in bulk. Then bake the bread as normally.

Sprouts

Although sprouting is not actually a means of preserving any food, we're including a discussion of it here because we think it's a wonderful way to turn grains, seeds, and even beans that you grow yourself (or buy) into vitamin-rich vegetables that you can have fresh whenever you want—any day of the year. Sprouting is fast, simple, and requires no special equipment.

Most all vegetable and herb seeds, grains, and beans can be sprouted. The exceptions are potato and tomato seeds; they are members of the nightshade family and grow sprouts that are toxic. Seeds from tree fruits, sorghum, and Sudan grass are also poisonous when sprouted. The best seeds to use are those that you've grown yourself or have gotten from a natural foods store that sells special seeds for sprouting. Packaged beans sold in regular food stores are fine, too. Some mail-order seed companies sell seeds for sprouting and will advertise them as such in their catalogs. Be sure that seeds you use have not been treated with chemicals or dyes. *Such seeds should not be sprouted.*

If you want to use some of the grains, seeds, or beans that you grow for sprouting, harvest and prepare them as you do for drying. Follow directions for drying that can be found elsewhere in this book (check Index), but do not use an oven or dryer that gets above 175°F. Do not pasteurize dried food that you plan to sprout. Store in airtight containers in a cool, dry place.

Many long-time sprouters stick to just alfalfa and mung beans, but all the following can be sprouted fairly easily:

alfalfa	kidney beans	radish
barley	lentils	rye
black-eyed peas	lima beans	soybeans
buckwheat	marrow beans	sunflower
chia	mung beans	triticale
chick-peas	navy beans	watercress
flax	pintos	wheat

The How-To's of Sprouting

Begin sprouting by soaking your beans, seeds, or grains overnight to soften the seed case and make sprouting easier. You'll only need a small quantity of seeds, as they increase their volume many times over as they sprout. Two tablespoons of the smallest seeds (like alfalfa, radish, flax, and watercress), or about ¼ cup of the medium seeds (the grains), or ½ cup of the larger ones (the beans) should be sufficient.

Place them in a small bowl, cover with water, and let them set overnight. Flax, chia, and watercress seeds tend to stick together when moistened, so they should not be soaked like the others. Rather, sprinkle the seeds over a saucer of water, being careful not to clump them together. After soaking, pour the seeds through a strainer or your sprouter. Remove any that stick to the saucer with a good stream of water. Then follow the regular sprouting procedure, but rinse them at least 4 times daily and 6 times if you can.

After soaking, drain this water off the seeds and put them into something that will allow them to continue to drain now and every time that you rinse them. You can place them in a strainer and prop the strainer over a bowl. Or you can put the seeds in a glass jar and cover the jar with cheesecloth or nylon netting which is held in place with a string or rubber band. Tilt the jar on an angle upside down so moisture can run out into a bowl. There are commercially made sprouting containers that are variations on this strainer and the covered jar that work quite well, too.

Once in their container, rinse the seeds again and let them drain. Put the container in a nearby cabinet, closet, or other place that is out of bright light and that stays at about 70°F. Bright light or direct sunlight will make the seeds dry out too quickly, too high a temperature will promote mold, and too low a temperature will discourage sprouting. Make sure the place is convenient, because you have to rinse the sprouts at least 2 times and preferably 3 times a day for the entire sprouting time. This rinsing prevents fungus from growing on the sprouting seeds. Never eat any sprouts that have even a trace of mold on them, as it could prove to be toxic.

When the sprouts are a good size you can rinse them once more and leave them in a sunny window for a day so they grow tiny leaves. These leaves make the sprouts look pretty and also add chlorophyll to your diet. Sprouts that are particularly nice with little leaves include alfalfa, flax, and rye.

Storing and Using Sprouts

All sprouts but bean sprouts (not including lentils and mung beans) can be and are best eaten fresh—on salads, sandwiches, or any other food. The other bean sprouts are more palatable and digestible if they're cooked before they're eaten. To cook the sprouted beans, drop them into boiling water slowly (so as to keep the water boiling), then turn down the heat and simmer, covered, for 30 to 45 minutes, or until they are as tender as you like them. Kidney and marrow beans take almost an hour to cook. The others take about 30 minutes.

There are so many ways you can use or cook sprouted beans. Add stock, vegetables, and seasonings for soup; cool and marinate them for a 3-bean salad; freeze them and put them in the Boston baked bean pot, in a cassoulet, or a lima bean bake. Or you may want to stick to your old chili con carne recipe and try your sprouted beans in it. If you want to be more adventuresome, try sprout-burgers, using the uncooked sprouted beans, or stir fry some vegetables Chinese-style, adding uncooked sprouts, steaming them at the end until tender.

All bean sprouts should be refrigerated and will keep for at least a week. You can freeze them, but plan to use thawed sprouts in soup, bread, or other cooked food, as freezing destroys their crispness.

Sprout Powder or Flour   To make a sprout powder or flour for baking, thickening soups and gravies, and adding to casserole toppings, dry the sprouts first and then grind them in a blender. Stay away from bean sprouts, though, because their high oil content doesn't make them good drying and grinding candidates. Store the sprout powder in a cool, dark place in an airtight container.

To dry sprouts for grinding, spread them on a cookie sheet and either put them in a very low oven (no higher than 175°F.), in a food dryer, or in another warm place, like on the top of a radiator, yogurt maker, plate warmer, or on a wood stove.

Sprouting Timetable

Food	How Long to Sprout (in days)	Recommended Length of Sprout
Alfalfa	4 to 5	1 inch (place in sun last day and let tiny green leaves appear)
Barley	3 to 5	length of the seed
Beans (navy, kidney, pinto, fava, lima, marrow—for others see separate listings)	2 to 5	1 inch (the bigger the bean, the smaller the sprout should be so that it remains tender—most beans, especially the larger and tougher ones, are better cooked before eating)
Black-eyed pea	3 to 4	length of the pea
Buckwheat	2 to 4	¾ inch
Chia	1 to 2	¼ inch (tend to get sticky when they are wet: see special instructions in the following section)
Chick-pea	6 to 7	¾ to 1 inch (these are best cooked before eating)
Flax	3 to 4	¾ to 1 inch (nice to place in sun last day and let tiny green leaves appear; tend to get sticky when they are wet: see special instructions in the following section)
Lentil	3 to 4	¼ to ½ inch
Mung bean	3 to 5	1 to 2 inches or longer
Radish	3 to 5	⅛ inch or less
Rye	3 to 5	1 to 1½ inches (nice to place in sun last day and let tiny leaves appear—these are remarkably sweet when they are long)
Soybean	4 to 6	1 inch (these are best cooked before eating)
Sunflower (unhulled)	5 to 8 (perhaps more)	⅛ inch or less
Triticale	1 to 3	length of the seed
Watercress (or garden cress)	2 to 4	¾ to 1 inch
Wheat	4 to 5	length of the seed (these are quite sweet)

Nut, Seed, Bread, and Sprout Recipes

Nuts

Almond Loaf

Preheat oven to 350°F. Pour 1 cup of hot stock or potato water over 1 cup of oatmeal, blend together, cover, and let stand until cool. Add the following:

2 eggs
1½ cups ground almonds
½ cup ground sunflower seeds
1 medium onion, chopped
½ cup diced celery and tops
½ chopped bell pepper or
    pimiento

1 teaspoon marjoram (fresh or
    dried)
1 tablespoon kelp or salt
1 tablespoon oil

Mix together well and add as much more of the stock or potato water as is needed for loaf consistency. Pour into an oiled bread tin and bake from 30 to 45 minutes. Turn out on a meat platter and garnish with red pepper and parsley. Serve hot or cold.

Yield: 4 servings

Nutty-Rice Loaf

Preheat oven to 350°F.

2 onions, chopped
½ green pepper, chopped

¼ cup celery and tops, chopped
2 tablespoons oil or butter

1 cup chopped nuts
1 cup cooked brown rice
1 cup wheat germ
2 eggs, beaten
1 cup water
¼ cup soy or wheat flour

½ teaspoon salt
½ teaspoon sage
freshly grated pepper

Sauté the onion, pepper, and celery in butter or oil until transparent. Blend all ingredients, and turn into a greased loaf pan. Bake for 45 minutes. Serve with tomato sauce or your favorite sauce.

Yield: 6 servings

Walnut Loaf

Preheat oven to 325°F. Cover 1½ cups of wheat germ with milk and let the wheat germ absorb all the liquid it will take up. Drain slightly to remove excess milk.

Sauté 1 chopped onion and 1 chopped pepper in 2 tablespoons oil until transparent. Place in a mixing bowl with the wheat germ and mix in:

1 cup chopped walnuts
1 chopped tomato

juice of one lemon
1 egg, well beaten

Form into a loaf, place in oiled pan, and bake about 30 minutes. Serve with tomato or mushroom sauce.

Yield: 4 servings

Chestnut Stuffing

3 cups chestnuts
3 cups diced celery
2 tablespoons butter
3 cups whole-wheat bread
  crumbs

⅓ cup light cream
1 teaspoon salt
¼ teaspoon mace

Boil the chestnuts, peel, and blanch by steaming over water for 5 minutes. Break into small pieces. Brown the celery in the butter, add the chestnuts and the crumbs, and mix. Moisten with the cream, sprinkle with salt and mace, and stuff loosely into turkey or chicken.

Yield: 6 cups—enough to stuff a 12-pound bird

Eggplant-Walnut Delight

Preheat oven to 350°F.

1 medium-sized eggplant	salt and oregano to taste
1 medium-sized onion	½ cup tomato sauce
½ cup walnut meats	

Peel eggplant and cut into chunks. Put eggplant, onion, and nut meats through a food grinder or chop in a blender. Add seasonings and tomato sauce and mix thoroughly. Form into a loaf and bake in an oiled casserole for 20 minutes. Serve hot.

Yield: 4 servings

Nutty Burgers

1 cup nuts, ground fine	1 stalk celery, ground
½ cup sunflower seeds, ground fine	1 small onion, ground
½ cup grated carrots	½ green pepper, ground
	1 sprig parsley, ground

Put the above ingredients through the food grinder or in the blender with enough liquid (a tomato or ½ cup broth) and then add the following:

2 eggs	
1 teaspoon kelp or salt	whole-wheat bread crumbs (if
1 pinch of sage or marjoram	necessary)

Blend the ingredients, and if they are too dry, add more broth or tomato pulp; if too moist, add whole-wheat bread crumbs. Shape into patties and broil until browned on each side.

Yield: 4 servings

Nutty Waffles

1¾ cups whole-wheat pastry
    flour
2 teaspoons baking powder
½ teaspoon salt
3 egg yolks
1 tablespoon honey

4 tablespoons melted butter or
    oil
1½ cups milk
½ cup chopped nuts
3 egg whites

Mix together the flour, baking powder, and salt. In another bowl, beat together the egg yolks, honey, butter or oil, and milk.

Make a well in the dry ingredients and pour in the milk mixture. Combine them with a few swift strokes. Mix in the nuts.

Beat the egg whites until stiff and fold them into the batter briefly. Pour the batter by the tablespoonful into a hot waffle iron and cook for about 4 minutes or until brown and crispy.

Yield: 6 waffles

Cucumber, Nut, and Olive Spread

2 cucumbers, chopped
2 cooked eggs
1 cup stuffed olives, chopped

½ cup chopped pecans or
    almonds
½ cup chopped celery
yogurt

Toss all chopped ingredients with yogurt or salad dressing. This is good on rice wafers, rye crackers, or turnip slices.

Yield: 4 servings

Peanut Butter

2 cups roasted peanuts
¼ cup oil
¼ teaspoon salt

Pour the roasted peanuts into an electric blender and blend about 1 minute. Gradually add the oil, 1 tablespoon at a time, blending after each addition. To be sure all the peanuts are getting blended, turn off the blender from time to time and with a knife or a narrow rubber spatula push the peanuts away from the sides and blades of the blender.

Then add the salt and blend quickly once more.

Store peanut butter in a covered jar in the refrigerator.

Yield: 1½ cups

Chewy Squares

Preheat oven to 350°F.

1 egg, separated
⅓ cup honey
3 tablespoons whole-wheat pastry flour

⅛ teaspoon salt
½ cup nuts, chopped
¼ cup ground sunflower seeds

Beat egg yolk until thick. Blend in honey. Combine with flour, salt, nuts, and seeds. Fold in stiffly beaten egg white. Turn into oiled square pan. Bake for 20 to 25 minutes, until light brown. Cool. Cut into squares.

Yield: 9 squares

Coconut-Nut Balls

Preheat oven to 350°F.

2 cups whole-wheat pastry flour
3 tablespoons oat flour
¼ teaspoon salt
4 tablespoons honey

1 cup oil
2 cups ground nuts
3 tablespoons coconut
coconut

Sift flours and salt. Blend in honey and oil. Add nuts and coconut and mix well. Shape into ½-inch balls. Place on lightly oiled cookie sheet. Bake about 30 minutes. When cool, roll in coconut.

Yield: 3 dozen ½-inch balls

No-Bake Nut-Coconut Pie Shell

½ cup nuts, ground
½ cup shredded coconut
⅓ cup oil

Blend all ingredients. Press into 9-inch pie pan. Chill and fill with a filling that needs no cooking.

Walnut Torte

Preheat oven to 325°F.

6 eggs, separated
¾ cup honey
½ cup powdered skim milk

½ cup wheat germ
2 cups ground walnuts*
1 teaspoon vanilla extract

Prepare two 9-inch layer cake pans by oiling bottoms of pans with pastry brush. Cut two circles from heavy brown paper. Place a circle of heavy brown paper on the bottom of each pan. Brush the paper thoroughly with oil.

Separate eggs, putting whites into a large bowl and yolks into smaller bowl, reserving two of the yolks for filling.

Beat whites until stiff peaks form when beater is slowly raised. Set to one side. With same beater, beat yolks until thick and lemon colored. Gradually blend the honey into the yolks. Stir into the beaten yolks the powdered milk, wheat germ, and ground walnuts. Blend together. Add vanilla extract.

With wire whisk or rubber scraper, gently fold yolk mixture into beaten egg whites, using an under-and-over motion, until well combined. Pour the batter into the prepared pans, spreading evenly to edge.

Bake in a slow oven for 30 minutes.

Remove from oven and loosen sides with a spatula to ease the cake out of the pan. Invert pans on a wire rack and remove paper immediately. Cool cakes completely before frosting.

Make custard filling (recipe follows). Cool. Put layers together with filling. Frost top and sides with frosting (recipe follows).

Yield: 10 to 12 servings

* Ground pecans or ground almonds may be substituted for the walnuts. When using ground almonds, use ½ teaspoon almond extract in torte and frosting in place of the vanilla extract.

Custard Filling:

⅓ cup cornstarch
¼ cup instant nonfat dry milk
¼ teaspoon salt
1¾ cups water

⅓ cup honey
2 egg yolks, slightly beaten
1 teaspoon vanilla extract

In medium saucepan, combine cornstarch, powdered milk, and salt. Add ¼ cup of water gradually, stirring with wooden spoon until mixture is smooth and free of lumps. Add remaining 1½ cups of water, mixing thoroughly.

Add honey to mixture and place over medium heat, stirring constantly until custard thickens; this should take from 10 to 12 minutes.

Remove custard from heat. Add 3 tablespoons of hot mixture to beaten egg yolks. Mix well. Gradually pour yolk mixture into custard, blending well. Return to medium heat and cook 3 minutes, stirring constantly. Remove from heat. Add vanilla. Cool custard completely before using to fill cake.

Yield: approximately 2 cups of custard

Frosting:

> 2 egg whites
> ½ cup honey
> ½ teaspoon vanilla extract

In top of double boiler, combine egg whites and honey. Beat one minute with rotary beater (hand or electric type), to combine ingredients.

Cook over rapidly boiling water (water in bottom should not touch top of double boiler), beating constantly until soft peaks form when beater is slowly raised. Allow about 10 minutes for beating to get proper consistency.

Remove from boiling water. Add vanilla. Continue beating until frosting is thick enough to spread—about 4 minutes.

Yield: enough frosting for a 9-inch 2-layer cake

Seeds

Roasted Sunflower Seeds

3 tablespoons oil   1 pound sunflower seeds
3 tablespoons soy sauce

In a shallow pan combine the oil and soy sauce and mix well. Then add the seeds and toss so that they are all coated with the oil and soy sauce.

Bake in a 350°F. oven for 20 minutes, stirring occasionally. When done, the seeds should be dry and rich brown in color.

Cool and store in a covered jar or sealed plastic bag. Store in the refrigerator for best keeping.

Yield: about 2 cups

Sunburgers

Preheat oven to 350°F.

1 cup ground sunflower seeds
½ cup finely chopped celery
2 tablespoons chopped onions
  or chives
1 egg (not absolutely necessary)
1 tablespoon chopped raw
  green pepper

1 tablespoon chopped parsley
½ cup grated raw carrots
1 tablespoon oil
½ teaspoon salt
¼ cup tomato juice
1 pinch basil

Mix all ingredients together well and add more tomato juice if necessary so that the patties hold a good formed shape. Arrange in an oiled shallow baking dish and bake until browned, about 5 minutes, then turn and brown on the other side.

They can also be broiled if coated with oil on both sides before cooking. They can be served with mushrooms, grated cheese, or fixed up in various ways, but they make a perfect protein dish.

Yield: 2 servings

Sunflower Baking
Powder Biscuits

½ cup sunflower seed flour
½ cup whole-wheat pastry
  flour
¾ teaspoon salt

1¾ teaspoons baking powder
2 or 3 tablespoons lard
⅓ cup milk (about)

Sift the dry ingredients, then cut in the lard with a pastry knife. Three tablespoons of lard make a very rich biscuit, 2 tablespoons make a puffier and a little lighter one. Mix just enough milk in the dough to make it soft but firm. Drop from a spoon onto a greased, floured pan and bake at 350°F. for 10 to 12 minutes.

Sunflower Drop Biscuits

Preheat oven to 375°F.

½ cup whole-wheat flour
½ cup sunflower seed flour
2 teaspoons baking powder

¾ teaspoon salt
¼ cup sour cream
3 tablespoons milk (about)

Mix dry ingredients together and cut in the cream with a fork as you would fat. Then moisten with just enough milk to mix the dough.

Drop from a spoon on a greased, floured pan and bake 8 to 10 minutes at 375°F.

Yield: 12 small biscuits

Sunflower and Whole-Wheat Bread

1 package yeast
⅓ cup lukewarm water
¼ cup honey
1¼ cups boiling water
2 teaspoons salt

2 tablespoons butter
1¾ cups whole-wheat flour
1¾ cups sunflower seed flour
⅓ cup raisins

Soak yeast in the lukewarm water and honey and let stand a few minutes until foamy. Boil water, salt, and butter and then let them cool to lukewarm. Mix the flours and raisins together. Combine the yeast mixture with the water and add it to the flour mixture to form a sponge. This sponge should be stirred well and be soft, but with a body.

Set the bowl in a pan of quite warm water and cover with a tea towel to rise. This takes about 1¼ hours.

Beat down a couple of minutes with a wooden spoon and then put in a greased, floured bread pan to rise again in a warm place, covered, about 45 minutes.

Light the oven when you put the bread in and let it come up to 375°F., then turn down to 325°F. to finish the baking which takes about 45 minutes altogether.

Yield: 1 loaf

Sunflower Corn Bread

1 cup sunflower seed flour
1 cup cornmeal
2 teaspoons baking powder
1 teaspoon salt
½ teaspoon soda

2 egg yolks, beaten
⅓ cup sour cream or yogurt
2 tablespoons honey
⅓ cup buttermilk
2 egg whites, beaten

Mix all dry ingredients together and combine with the mixed liquids. Fold in the whites last. Bake in a greased, floured pan 5 minutes at 375°F., then turn down to 325°F. and bake 10 minutes more.

Serve hot with butter. It keeps well and is good cold.

Yield: 1 loaf

Sunflower Seed Bread

Preheat oven to 350°F.

1 package yeast
⅓ cup warm water
⅓ cup oil
⅓ cup honey

1½ cups hot water
½ cup raisins
1½ cups ground sunflower
   seeds

Dissolve yeast in ⅓ cup warm water. In another bowl, dissolve the oil and honey in 1½ cups hot water; add the raisins and let the mixture cool. When cool, add the sunflower seeds and then the dissolved yeast, plus the following:

2 teaspoons salt
1 teaspoon nutmeg
½ cup wheat germ

1 teaspoon cinnamon
3 cups whole-grain flour

Stir into moderately stiff dough, grease the top and let rise. Stir down and spoon into two small greased bread tins. Let rise again and bake about 50 minutes.

Yield: 2 small loaves

Sunflower Seed Cookies

Preheat oven to 275°F.

2 egg whites
1 cup honey (thinned down
over hot water)

1 cup chopped sunflower seeds
1 teaspoon vanilla

Beat the egg whites until they will hold a peak, then gradually beat in the honey, then the seeds, which may be ground to your particular taste, and finally add the vanilla.

Drop cookies on a well-oiled baking sheet and bake for 30 to 40 minutes in the middle of the oven. Test and take out when toothpick comes out clean. Loosen at once as they like to stick to cookie sheet.

Keep in a container with a tight lid.

Sunflower Goulash

Preheat oven to 350°F.

2 onions, sliced
2 tablespoons oil
1 cup chopped sunflower seeds
2 cups mushrooms, halved
1 cup fresh sprouts
1 cup cooked and cooled millet

2 cups fresh or home-canned
tomatoes
1 cup lima or butter beans
1 teaspoon chili powder

Sauté the onions in the oil until tender. Add all other ingredients, pour in casserole dish, bake about 30 minutes, and serve.

Yield: 4 servings

Sunflower Seed Loaf

Preheat oven to 350°F.

1 cup finely ground sunflower
   seeds
½ cup ground or chopped
   walnuts
1 cup cooked and seasoned
   lentils
2 tablespoons minced onion or
   chives
2 large or 3 small eggs, beaten
   slightly

1 tablespoon oil
½ cup whole-wheat bread
   crumbs
½ cup grated raw carrots
½ cup diced green pepper
1 teaspoon salt
½ teaspoon paprika
¼ teaspoon thyme
2 teaspoons lemon juice

Mix together ingredients and press into an oiled baking dish. Bake 45 minutes or until done in the middle, and turn out when cooled a bit. Serve with mushroom or tomato sauce.

Yield: 1 loaf

Sunflower Seed Muffins

Preheat oven to 400°F.

1 egg
2 tablespoons honey
2 tablespoons oil or melted
   butter
1 cup milk

1½ cups whole-wheat pastry
   flour
½ teaspoon salt
½ cup ground sunflower seeds
2 teaspoons baking powder

Blend egg, honey, and oil. Add milk and the rest of the ingredients and pour in heated muffin tins. Bake 20 minutes.

If you object to using baking powder, use gradually less powder and depend on egg white to leaven the batter each time you make these muffins. This is how it is done: Use 2 eggs, separated. Mix the yolks in with the batter, then beat the whites stiff and fold in just before baking. Use less baking powder, even using 3 egg whites if needed, until you need no baking powder at all.

Sunflower Sour
Cream Muffins

Preheat oven to 375°F.

¾ cup sunflower seed flour
1 cup rye, whole-wheat, or
    oatmeal flour
2½ teaspoons baking powder
1 teaspoon salt
½ teaspoon baking soda

¼ cup currants
1 egg beaten
2 tablespoons honey
¾ cup sour cream
⅓ cup buttermilk (about)

Mix all dry ingredients with the currants, then combine with the egg, honey, and sour cream, stirring as little as possible. Add buttermilk last, as needed, to make a soft dough. Rye takes about ⅓ of a cup of buttermilk, but there is a difference in flours; whole-wheat or oatmeal may take more or less. Bake in a greased muffin tin for 5 minutes at 375°F., then 10 to 13 minutes more at 325°F.

Yield: 9 muffins

Sunflower Seed Omelet

4 eggs beaten very light
1 cup sunflower seed meal

½ teaspoon kelp or salt
½ teaspoon caraway seeds

Heat a heavy skillet. Blend ingredients. Oil the hot skillet, pour in the mixture, let brown on bottom, cut in quarters, turn, and brown on the other side.

Yield: 2 to 3 servings

Sunflower Pancakes

¾ cup sunflower seed flour
¾ cup whole-wheat flour
1 teaspoon baking powder
¾ teaspoon salt

2 eggs, beaten
1 tablespoon honey
2 tablespoons melted butter
1½ cups buttermilk

Sift all dry ingredients together. Mix the liquids together and combine with the flour, stirring just until mixed. Do not overmix. Fry a delicate brown on both sides.

They are very tender and good with just butter, but they're also fine for filling with creamed chicken or seafood and then rolled and kept warm a minute in the oven. They have an unusual flavor which can be enjoyed without syrup.

Yield: 20 to 22 pancakes

Breads

Banana Corn Bread

1¼ cups cornmeal
¼ cup wheat germ
¼ cup potato flour
½ cup soy flour
1 tablespoon or package of yeast
    softened in 2 tablespoons
    of warm water

2 eggs
½ cup mashed banana
7 tablespoons honey
2 tablespoons corn oil

Place all dry ingredients in a bowl. In another bowl, beat the eggs and all other ingredients. Add the dry ingredients to the moist mixture and mix well.

Place in an oiled bread pan and let rise about 1 hour. Bake in a 325°F. oven for 50 minutes.

Yield: 1 loaf

Carrot Corn Bread

1 cup yellow cornmeal
1 cup grated carrots
2 tablespoon oil
1 tablespoon honey

¾ cup boiling water
2 eggs, separated
2 tablespoons cold water

Mix thoroughly cornmeal, carrots, oil, and honey. Stir in the boiling water. Add 2 tablespoons water to the 2 egg yolks beaten together. Add this to the above mixture. Fold in stiffly beaten egg whites.

Pour into a warm, oiled 8-by-8-inch pan, and bake at 400°F. for 25 minutes, or until knife comes out clean.

Yield: 6 servings

Corn Bread

¾ cup cornmeal
1 cup unbleached flour
5 teaspoons baking powder
¾ teaspoon salt
1 egg, well beaten

1 cup milk
⅓ cup honey
2 tablespoons melted bacon fat
or oil

Mix and sift dry ingredients. Add beaten egg, milk, honey, and oil. Stir just enough to mix. Bake in 2 greased 8-inch pie pans at 425°F. for 20 minutes.

Yield: 2 flat breads

Cream, Honey, and Wheat Bread

⅓ cup honey (more if desired)
1 cup very hot water
1 cup heavy sweet cream
2 packages yeast

2 cups whole-wheat flour
4 large eggs
3 cups whole-wheat flour

Dissolve honey in water and add cream. Let this mixture cool to warm, then add the yeast. Add 2 cups whole-wheat flour and mix well. Add the eggs and mix. Add rest of flour, mix, then turn out on floured board and knead until smooth.

Let rise until double in bulk, shape into loaves, buns, or sweet rolls, let rise, and bake about 20 to 30 minutes at 375°F. Do *not* let dough rise too much in pan, as it will rise in the oven more than ordinary bread.

Yield: 2 loaves

This quick bread and the previous ones are very moist and make good freezers. They stay fresh and moist many days after they are thawed. Freeze them after they have been baked and cooled.

Crusty Loaves

12½ to 13½ cups flour
(unbleached, whole-wheat,
or a combination of your
favorites)
½ cup honey
2 tablespoons salt

⅔ cup instant nonfat dry milk
solids
4 packages active dry yeast
¼ cup oil
4 cups very warm tap water
(120° to 130°F.)

In a large bowl thoroughly mix 4 cups flour, honey, salt, dry milk solids, and undissolved active dry yeast. Add oil.

Gradually add very warm tap water to dry ingredients and mix well. Stir in enough additional flour to make a stiff dough. Turn out onto lightly floured board; knead until smooth and elastic, about 15 minutes. Cover with a towel; let rest 15 minutes.

Divide dough into four equal pieces. Form each piece into a smooth ball. Flatten each into a mound 6 inches in diameter. Place on oiled baking sheets. Cover sheets tightly with plastic wrap. Freeze until firm. Transfer to plastic bags. Keep frozen up to 4 weeks.

Remove from freezer; unwrap and place on ungreased baking sheets. Cover; let stand at room temperature until fully thawed, about 4 hours. Roll each mound to a 12-by-8-inch rectangle. Beginning at an 8-inch end, roll dough as for jelly roll. Pinch seam to seal. With seam side down, press down ends with heel of hand. Fold underneath. Place each, seam side down, in an oiled 8½-by-4½-by-2½-inch loaf pan. Cover; let rise in warm place, free from draft, until doubled in bulk, about 1½ hours.

Bake at 350° F. 30 to 35 minutes, or until done. Remove from pans and cool on wire racks.

To make round loaves: Let thawed mounds rise in warm place, free from draft, until doubled in bulk, about 1 hour. Bake as for loaves.

Yield: 4 loaves

Dill Bread

1 tablespoon yeast	2 teaspoons dill seed
¼ cup warm water	1 teaspoon salt
1 cup cottage cheese	¼ teaspoon baking soda
2 tablespoons honey	1 unbeaten egg
1 tablespoon minced onion	2¼ to 2½ cups whole-wheat
1 tablespoon butter	flour

Soften the yeast in the warm water. Warm the cottage cheese to lukewarm, and add the honey, onion, butter, dill seed, salt, soda, and egg. Add flour until a stiff dough is formed. Beat well.

Let mixture rise in a warm place until it has doubled—about an hour. Punch down, and turn into a well-greased round casserole or 2 small, greased bread pans. Let dough rise in a warm place for 40 minutes. Bake for about 45 minutes in a 350°F. oven. Spread butter and salt on top of loaf after removing from oven.

Yield: 1 loaf

Honey Applesauce
Oatmeal Bread

½ cup honey
½ cup oil
1 cup applesauce
6 tablespoons lukewarm milk
1 tablespoon honey
2 tablespoons yeast (2 yeast cakes)

2 eggs
3 cups sifted unbleached flour
1 cup rolled oats
1½ teaspoons salt
applesauce topping (see below)
cinnamon, nutmeg, nuts (for topping)

Combine ½ cup honey, oil, and applesauce in a small saucepan. Heat to lukewarm. Combine milk, 1 tablespoon honey, and yeast, stirring until yeast dissolves. Let stand 5 to 10 minutes.

Beat eggs in a large bowl. Add lukewarm applesauce mixture, yeast mixture, and flour. Mix to smooth batter. Add oats and salt; mix well.

Cover and let rise until double in bulk. Beat batter again, and then spread in a greased 8-inch round spring-form pan. Spread applesauce topping on dough, and sprinkle with nuts, cinnamon, and nutmeg. Cover and let rise until double. Bake in preheated 375°F. oven 50 to 60 minutes.

Yield: 1 loaf

Applesauce Topping
(for Honey Applesauce Oatmeal Bread)

1 cup applesauce
2 tablespoons butter

¼ cup honey
½ cup coconut

Slowly cook applesauce down to ½ cup; combine with remaining ingredients. Then spread on bread, above.

Honey-Glazed Nut Bread

3 cups whole-wheat flour
3 teaspoons baking soda
1 teaspoon salt
¼ teaspoon nutmeg
1 cup chopped nuts

⅔ cup honey
2 cups buttermilk or yogurt
additional honey, butter, chopped nuts for topping

Combine flour, baking soda, salt, nutmeg, and 1 cup chopped nuts. Blend honey with buttermilk, and mix into dry ingredients. Pour into well-greased loaf pan. Bake in a preheated 350°F. oven for 45 minutes.

Just before taking out, glaze top of loaf with 1 tablespoon honey mixed with 1 tablespoon melted butter. Sprinkle with more chopped nuts. Return to oven and let glaze for additional 5 minutes.

Yield: 1 loaf

Millet Bread

3 eggs
¾ cup boiling water
1½ cups millet flour
1 cup grated carrots

1 tablespoon honey
1 teaspoon kelp or salt
3 tablespoons oil
3 tablespoons cold water

Separate the eggs and beat the whites very stiff. Set aside. Preheat the oven to 350°F. and place an oiled bread pan in the oven at the same time.

Pour the boiling water over the millet flour, then add the carrots, honey, kelp or salt, and oil. Beat the egg yolks and add 3 tablespoons cold water. Add to the millet mixture. Fold in the egg whites last. Pour the batter in the by now very hot oiled bread pan and bake about 45 minutes.

This bread tastes and looks like corn bread but is a richer yellow and is moister because of the carrots.

Yield: 1 loaf

Oatmeal Bread

1 package yeast
2 cups warm milk
½ cup honey
½ cup oil
1 beaten egg

1 teaspoon salt
4 cups whole-wheat flour
    (or favorite combination
    of flours)
2 cups rolled oats or oat flour

Sprinkle the yeast into the warm milk; add the honey. Mix these ingredients together well, then add the oil, egg, and salt. Stir in the whole-wheat flour and add the oat flour so that the dough is dry enough to leave the sides of the bowl.

Turn out onto a floured board and knead until dough is smooth and elastic, about 10 minutes. Place the dough in a large, oiled bowl; cover and allow to rise until double in bulk.

Punch down and cut into 3 pieces. Shape into loaves and place in bread pans. Let rise until double in bulk, then bake at 375° F. for about 45 minutes. Yield: 3 loaves

Peanut Bread

2 packages yeast
1¾ cups warm water
3 tablespoons honey
1½ teaspoons salt
¼ cup instant nonfat dry
   milk powder

1½ cups peanut flour
4½ cups whole-wheat flour
   (or favorite combination
   of flours)

Sprinkle yeast over the water. When the yeast is dissolved, add the honey, salt, milk powder, and peanut flour. Add enough whole-wheat flour to form a moderately stiff dough.

Turn the dough out onto a wooden board and knead until smooth and elastic, about 7 minutes. Place dough in a large, oiled bowl, cover, and allow to rise until double in bulk.

Punch down dough, shape into one loaf, and place in a large oiled bread pan. Cover and allow to rise until double in bulk.

Bake at 350°F. for about 50 minutes.

Yield: 1 large loaf

Plymouth Bread

½ cup yellow cornmeal
2 cups boiling water
2 tablespoons butter
½ cup molasses

1 teaspoon salt
1 cake quick-acting yeast
½ cup lukewarm water
4¾ cups unbleached flour

Stir cornmeal very slowly into boiling water, stirring constantly to prevent lumps. Boil 5 minutes, and add butter, molasses, and salt. Cool. When lukewarm, add yeast which has been softened in ½ cup water. Add flour until dough is stiff. Knead well, and let rise until double.

Shape into 2 loaves, place in well-greased pans, and let rise again until double. Bake 1 hour at 350°F.

Yield: 2 loaves

Pumpkin Bread

4 beaten eggs
2 cups honey

1 cup oil
2 teaspoons baking soda

2 cups cooked pumpkin, fresh
   or canned
⅔ cup water
3½ cups unbleached flour

1½ teaspoons salt
1 teaspoon cinnamon
1 teaspoon nutmeg

Beat all the wet ingredients together well. Mix the dry ingredients separately in another bowl. Then pour the dry mixture into the wet one and mix thoroughly. Pour the mixture into two large, oiled bread pans. Bake 1 hour and 10 minutes at 350°F.

Yield: 2 loaves

Raisin Bread

1 medium-sized potato
1 quart water
2 tablespoons butter
2 teaspoons salt
2 tablespoons yeast
1 cup warm water

½ cup honey
11 to 12 cups sifted unbleached
   flour
1 pound raisins
½ teaspoon ground cloves
2 teaspoons cinnamon

Peel potato and cut into pieces. Cook until tender in quart of water. Mash potato, return it to the water in which it was cooked, and add butter and salt. Cool until lukewarm.

Dissolve yeast in 1 cup warm water. Let stand 5 to 10 minutes. Add honey.

Add 6 cups flour to the potato mixture, beating until smooth. Mix in the yeast. Beat thoroughly. Cover and let rise for about 2 hours.

Work in remaining flour to make a soft dough. Stir in raisins and spices. Knead until smooth on floured board. Put dough in a greased bowl and grease top. Cover and let rise until double.

Punch down, and divide in 3 portions. Place each in an oiled loaf pan. Cover, and let rise again until double. Bake in a preheated 375°F. oven for 40 minutes.

Yield: 3 loaves

Shallot Rolls

1 tablespoon yeast (1 cake)
3 cups whole-wheat flour
2 tablespoons wheat germ
2 tablespoons yeast flakes
4 tablespoons sour dough
   starter (if desired)

4 tablespoons sesame oil
4 stalks green shallot shoots,
   chopped
salt
4 tablespoons sesame seeds

Dissolve yeast with a little warm water, add 2 cups flour, wheat germ, and yeast flakes. Add sour dough starter if available. Cover and let rise for 3 or 4 hours in a warm place. Knead until smooth.

Shape the dough into a long roll, and cut into 2-inch sections. Cover sections with a cloth to prevent their drying out. Roll each section thin with a rolling pin. Brush surface with sesame oil, sprinkle lightly with chopped shallot greens, and salt to taste. Roll section into a log, and fold the ends toward the center so the shallots are hidden inside.

Pressing lightly with a rolling pin or the palm of your hand, flatten dough into a round shape. Brush surface with water, sprinkle sesame seeds on top, and press gently with rolling pin to make the seeds stick better. Bake in a preheated oven at 420°F. for 20 minutes.

Yield: 12 shallot rolls

Sprouted Wheat Bread

3 tablespoons dried yeast
1 cup lukewarm potato water or
    other liquid
1 tablespoon maple syrup or
    honey
1 tablespoon salt
4 tablespoons molasses
5 tablespoons oil
5 cups warm water

1 cup sprouted wheat
1 cup raisins, chopped black
    figs, or currants
1 cup chopped nuts
12 cups flour, including 1 cup
    each of 2 grain products
    such as rice polishing or
    oatmeal

Dissolve yeast in lukewarm water with honey or maple syrup. Let set until foamy. Then add salt, molasses, oil, water, sprouted wheat, fruit, and nuts. Starting with 1 cup each of 2 grain products, mix in about 12 cups of flour products. The other 10 cups can be a combination of other flours or all whole-wheat, as long as at least 6 cups are whole-wheat. Knead until dough is no longer sticky.

Allow dough to rise in a warm place until double in volume; punch down. Divide dough, and place in greased pans. Allow to rise until double again. Bake for 1 hour at 400°F.

Yield: 7 small or 6 large loaves

Sunflower Seed Bread

1 tablespoon yeast (1 cake)
⅓ cup water
⅓ cup oil

⅓ cup honey
1½ cups very warm water
½ cup raisins

1½ cups ground sunflower
   seeds
2 teaspoons salt
1 teaspoon nutmeg

½ cup wheat germ
1 teaspoon cinnamon
3 cups whole-wheat flour

Dissolve yeast in ⅓ cup warm water. Dissolve oil and honey in 1½ cups very warm water; then add the raisins, and allow to cool. Add ground sunflower seeds to cooled mixture; then add the dissolved yeast and the remaining ingredients.

Stir until a moderately stiff dough forms, oil the top lightly, and allow to rise in a warm place. When risen, stir it down, and spoon into 2 greased bread pans. Let rise once more, and bake 50 minutes in a preheated 350°F. oven.

Yield: 2 loaves

Sprouts

Wheat Sprout Balls

2 cups sprouted wheat (or rye)
1 cup walnuts
1 large onion
1 cup skim milk

1¼ cups fine, dry whole-wheat
   bread crumbs
½ teaspoon salt
½ teaspoon chopped parsley
2 teaspoons vegetable oil

Force first three ingredients through coarse blade of food chopper or blend slightly in a blender.

Stir in milk, then add remaining ingredients and mix well. Let stand 10 minutes to allow crumbs to absorb liquid, then shape in 1¼-inch balls.

Put on greased cookie sheet and bake at 400°F. about 15 minutes.

These are good with gravy or tomato sauce.

Yield: 3½ dozen

Bean Sprouts, Peppers,
and Tomatoes

2 medium-sized tomatoes
boiling water
1 green pepper
2 tablespoons soy sauce
¼ teaspoon honey

2 tablespoons oil
½ teaspoon salt
1 pound mung bean sprouts (or
   other bean sprouts)

Drop tomatoes into pot of boiling water. Turn off heat. After 3 minutes, remove tomatoes, cool slightly, and peel. Cut into ½-inch cubes.

Discard seeds of green pepper and cut pepper into thin slivers. Thoroughly blend soy sauce and honey in a cup.

Heat oil until hot in a wok or large, heavy frying pan. Add salt, then bean sprouts. Stir fry for 1 minute. Add green peppers and stir fry for 2 minutes. Pour in soy sauce-honey mixture. Mix well. Add tomato cubes, and stir until heated through. Serve immediately.

Yield: 4 to 6 servings

Wheat or rye sprouts can be added to your favorite whole-wheat bread recipe. In order to add one cup of sprouts, eliminate ½ cup of flour and ½ cup of liquid. Small amounts of sprouts can be added without changing the recipe.

Sprouted Rye No-Knead Bread
(wheatless)

1½ cups lukewarm water	½ cup soy flour
4 teaspoons dry yeast	½ cup rye flour
1 tablespoon molasses or	1 teaspoon salt
sorghum syrup	1 cup rye sprouts
3 cups rye flour	2 tablespoons oil

Put lukewarm water into electric mixer bowl. Sprinkle yeast over surface of water. Add molasses or sorghum syrup. Let soak a few minutes. When yeast is dissolved, stir in 3 cups rye flour. Beat in electric mixer for 10 minutes on low speed.

Mix soy flour, remaining rye flour, and salt. Add sprouts and oil. Add all ingredients to yeast mixture. Stir until just combined. Cover bowl with damp cloth and set in warm place. Let rise until double in bulk (about 45 minutes).

Stir dough down and turn out into very well oiled 9-by-5-by-3-inch bread pan. Shape into loaf with wet spatula.

Let rise again for 30 minutes or until almost to top of pan. Meanwhile preheat oven to 350°F.

Bake for 45 to 50 minutes, or until toothpick comes out clean. Remove from pan and cool completely before cutting. Store in refrigerator.

Yield: 1 loaf

Egg Foo Yung with
Mung Bean Sprouts

2 tablespoons safflower oil
2 finely chopped onions
2 cups mung bean sprouts (or
  other bean sprouts)

6 slightly beaten eggs
½ teaspoon salt

Pour oil into skillet and heat at medium temperature.

Mix remaining ingredients. Fry about 2 tablespoons of egg mixture at a time; do not stir, but cook until the pancake is lightly browned on both sides. Continue frying pancakes one at a time.

The pancakes can be kept warm in the oven in a flat pan at low temperature.

Yield: 4 servings

Meat Loaf with Sprouts

Preheat oven to 350°F.

1½ cups mung bean sprouts,
  coarsely ground
1½ pounds ground chuck
  (beef)
1 finely chopped onion
¼ cup bread crumbs

¼ cup wheat germ
1 well-beaten egg
2 teaspoons tamari soy sauce
1 teaspoon garlic salt
pinch pepper

In a large bowl combine all ingredients. Form into a loaf in shallow baking pan, or pack mixture into lightly greased loaf pan.

Bake for about 1 hour, until meat is browned.

Yield: 6 servings

Meat Patties and Sprouts

Mung bean sprouts give a moist quality to ground meat when added to make meat patties.

¾ pound ground chuck
½ cup fresh sprouts (soy or
  mung), chopped

1 tablespoon chopped onion
2 tablespoons grated carrot
salt and pepper to taste

Combine all ingredients to make patties. Broil or fry according to preference.

Yield: 4 large patties

Sprout Omelet with Tomatoes

6 eggs
2 chopped green onions
2 tablespoons fresh dill or 1
  teaspoon dried dill
salt and pepper to taste
¼ cup milk

1 tomato, chopped
½ cup sunflower seed sprouts
  (or mung, lentil, or alfalfa
  sprouts)
alfalfa sprouts (for garnish)

Beat eggs, onions, dill, salt, pepper, and milk with wire whip in bowl.

Heat mixture in pan with oil until mixture begins to set. Add 1 chopped tomato and ½ cup or more sunflower seed sprouts. Cover with a lid to get top to set. Fold omelet over and top with alfalfa sprouts.

Yield: 3 to 4 servings

Sprouted Lentils, Bean, and Rice Salad

½ pound pinto beans or kidney
  beans
1 pound fresh green beans,
  cooked
2 cups cooked brown rice
1 cup celery, diced
½ green pepper, diced
¼ cup pimiento, chopped
¼ cup lentils, sprouted (or
  mung beans or alfalfa
  sprouts)

½ cup oil
½ cup wine vinegar
1 tablespoon honey
1 teaspoon salt
1 teaspoon pepper
1 medium-sized red onion
  sliced thin, for garnish

Soak pinto beans or kidney beans overnight in water to cover. Drain. Cook the beans just until tender. Don't overcook. Drain. Save broth for soup.

Combine green beans and pinto beans or kidney beans, rice, celery, pepper, pimiento, and lentil sprouts.

Combine oil, vinegar, honey, and seasonings. Toss salad in dressing with the onion rings.

Yield: 10 servings

Sprout Salad

1 cup cottage cheese
2 tomatoes, cubed
1 cucumber, sliced or diced
¼ cup mung bean sprouts

½ cup alfalfa sprouts
¼ cup sesame seeds
herb dressing

Put cottage cheese in center of serving bowl. Arrange tomatoes, cucumbers, and mung bean sprouts around it. Top with alfalfa sprouts and sesame seeds. Serve with your favorite herb dressing.

Yield: 2 servings

Bean Sprout Soup

2 eggs
1½ tablespoons cornstarch
¼ cup water
1 scallion
6 cups chicken stock

2 cups mung bean (or other bean) sprouts
½ teaspoon honey
salt to taste

Beat the eggs. Then mix cornstarch and water. Mince the scallion. In a large saucepan heat chicken stock to boiling and add bean sprouts. Reduce heat and simmer for 3 minutes. Flavor with honey, and add salt to taste.

Stir the cornstarch mixture well and pour it into hot soup. Stir until thickened. Slowly pour in beaten eggs, stirring with a fork. Remove soup from heat immediately and serve garnished with scallions.

Yield: 5 to 6 servings

Potato and Alfalfa
Sprout Soup

5 cups potatoes, peeled and diced
½ cup celery, diced
⅓ cup onion, diced
salt to taste
2 quarts water

white pepper to taste
1 cup instant nonfat dry milk powder
1 cup cold water
1 cup alfalfa sprouts
chopped parsley for garnish

In a heavy 6-quart saucepan, place diced potatoes, celery, onion, salt, and water. Place over medium heat and bring to a boil (keep saucepan partially covered); turn down heat to low and simmer for 30 minutes, or until vegetables are tender. Season with a few grindings of fresh pepper, and add additional salt if necessary.

Combine dry milk powder with 1 cup water. Stir into soup mixture and simmer over low heat for about 5 minutes, stirring constantly. Do not boil.

Add the alfalfa sprouts just before serving. Ladle the soup into a tureen or individual soup bowls. Garnish with finely chopped fresh parsley.

Yield: 8 cups

Index

Canning Foods in High Altitude Areas

In high altitude areas adjustments must be made in the *length of time* food is processed in a *boiling-water bath* and in the *amount of pressure* foods are subjected to in *pressure canning*. The higher the altitude, the longer the processing time in boiling-water canning, and the greater the pounds of pressure in pressure canning.

		you *boiling-water* *can* this amount of time longer than directed:		or	you *pressure can* at this number of pounds:		
if you are at this altitude:							
Feet	Meters	If 20 min. or less	If over 20 min.		If 5 lb.	If 10 lb.	If 15 lb.
1,000	305	1 min.	2 min.		5½ lb.	10½ lb.	15½ lb.
2,000	610	2 min.	4 min.		6 lb.	11 lb.	16 lb.
3,000	914	3 min.	6 min.		6½ lb.	11½ lb.	16½ lb.
4,000	1219	4 min.	8 min.		7 lb.	12 lb.	17 lb.
5,000	1524	5 min.	10 min.		7½ lb.	12½ lb.	17½ lb.
6,000	1829	6 min.	12 min.		8 lb.	13 lb.	18 lb.
7,000	2134	7 min.	14 min.		8½ lb.	13½ lb.	18½ lb.
8,000	2348	8 min.	16 min.		9 lb.	14 lb.	19 lb.
9,000	2743	9 min.	18 min.		9½ lb.	14½ lb.	19½ lb.
10,000	3048	10 min.	20 min.		10 lb.	15 lb.	20 lb.